Eugene Stefanovich

Elementary Particle Theory

De Gruyter Studies in Mathematical Physics

Volume 47

Eugene Stefanovich

Elementary Particle Theory

Volume 3: Relativistic Quantum Dynamics

DE GRUYTER

Mathematics Subject Classification 2010
Primary: 81-02, 81V25, 83A99; Secondary: 81V10, 81Q99

Author
Dr Eugene Stefanovich
San Jose, California
USA
eugene_stefanovich@usa.net

ISBN 978-3-11-049090-9
e-ISBN (PDF) 978-3-11-049322-1
e-ISBN (EPUB) 978-3-11-049139-5
ISSN 2194-3532

Library of Congress Control Number: 2018016481

Bibliographic information published by the Deutsche Nationalbibliothek
The Deutsche Nationalbibliothek lists this publication in the Deutsche Nationalbibliografie; detailed bibliographic data are available on the Internet at http://dnb.dnb.de.

Typesetting: VTeX UAB, Lithuania
Printing and binding: CPI books GmbH, Leck

www.degruyter.com

Contents

List of figures

https://doi.org/10.1515/9783110493221-201

List of tables

https://doi.org/10.1515/9783110493221-202

Postulates, statements, theorems, assertions

https://doi.org/10.1515/9783110493221-203

Conventional notation

See also Conventional notation of Volumes 1 and 2.

Υ	nondecay probability of unstable system (page 47)
τ_0	lifetime of unstable system at rest (page 66)
$\mathscr{H}_\alpha$	subspace of states of unstable particle (page 48)
Ξ	generator of the dressing transformation (page 29)
$\lvert\boldsymbol{p}\gg$	normalized state with sharp momentum (page 52)
$(\mathbb{n}, \mathbb{l}, \mathbb{m})$	quantum numbers of atomic states (page 40)
H^d	Hamiltonian of the dressed theory (RQD) (page 13)
$r = \lvert\boldsymbol{r}_1 - \boldsymbol{r}_2\rvert$	distance between particles (page 103)
$\boldsymbol{E}(\boldsymbol{r}, t)$	"electric field" (page 99)
$\boldsymbol{B}(\boldsymbol{r}, t)$	"magnetic field" (page 99)
$\boldsymbol{\mu}$	magnetic moment (pages 92, 115)
γ	relativistic factor $\equiv 1/\sqrt{1 - v_1^2/c^2}$ (page 140)
$\mathfrak{g}$	generic one-particle observable (page 156)
$\lvert 1\rangle$	one-particle state (page 3)

https://doi.org/10.1515/9783110493221-204

Preface

In the first [241] and second [242] volumes of this book, we presented a fairly traditional view on relativistic quantum mechanics and quantum field theory (QFT, QED). This approach proved itself well and achieved remarkable success in many important areas of high-energy physics, in particular in the description of scattering processes. At the same time, both QFT and QED have a number of serious problems. First of all, there is the problem of ultraviolet divergences. The idea of self-acting bare particles governed by a divergent Hamiltonian seems rather dubious. Such a Hamiltonian is not suitable for describing the time evolution of wave functions and observables of interacting particles.

Calling itself a fundamental theory of physics, QFT must describe phenomena in the entire spectrum of energies, distances and time intervals, rather than confine itself to a limited set of questions related to energies of stationary states or to the scattering matrix.

In this third volume, we will offer solutions to these problems. Our research will lead us to a new theory of electromagnetic phenomena, which we call *relativistic quantum dynamics* or RQD and which differs from traditional approaches in two important aspects: (i) the *primary role of particles* (rather than fields) and (ii) the *dynamical character of boosts*.

Modern quantum field theories (such as renormalized QED) encounter serious difficulties when trying to describe the temporal evolution of even the simplest physical systems, such as the vacuum or free elementary particles. A formal application of the QED time evolution operator to such states leads to the spontaneous production of spurious (virtual) particles, which have not been observed in experiments. The problem lies in the fact that bare particles of QED have no relationship to the physically observed electrons, protons, etc. We will solve this problem by using the "dressing" formalism, which is the cornerstone of our RQD. The "dressed" RQD Hamiltonian is obtained by means of a unitary transformation from the traditional QED Hamiltonian. This transformation does not change the S-operator of QED; therefore, its excellent agreement with the experimental data is preserved in our theory.

The RQD Hamiltonian describes electromagnetic phenomena in terms of physical particles (electrons, photons, etc.) interacting with each other directly, i. e., without the mediation of fictitious virtual particles or fields. In this formulation, quantum fields only play an auxiliary, technical role. Our theory will allow us to move beyond the S-matrix and study the time evolution of systems of interacting particles. In addition, this theory can calculate not just energies, but also wave functions of bound states. All RQD calculations are performed according to the rules of standard quantum mechanics without ultraviolet divergences and without resorting to artificial cutoffs, regularization and renormalization.

https://doi.org/10.1515/9783110493221-205

Of course, the idea of particles interacting instantaneously at a distance is not new. The Newtonian theory of gravity had this form, and many approaches in modern physics use the idea of (quasi)particles. However, the established consensus is that such approaches can be only approximate, since instantaneous interactions violate the principles of relativistic invariance and causality. Textbooks are trying to convince us that these principles can be reconciled with quantum mechanics only within a theory based on local (quantum) fields and retarded interactions. In our book, we challenge this established misconception and intend to demonstrate that point particles interacting at a distance do not contradict the sacred principles of relativity and causality.

Instantaneous forces acting between dressed particles imply a real possibility of sending signals with superluminal speeds. Hence, our (obviously, relativistic!) theory finds itself in conflict with the special theory of relativity, where the propagation of anything faster than the speed of light is absolutely forbidden. This paradox prompts us to take a new look at the existing derivations of Lorentz transformations for space–time coordinates of events, which (transformations) are the most fundamental relations of Einstein's special theory. These formulas are usually derived for elementary events, associated with either light signals or with free (noninteracting) particles. Nevertheless, special relativity tacitly assumes that these formulas can be extended to events with interacting particles, regardless of the forces acting between them. We will show that this assumption is in fact wrong. In our studies we will be using Wigner's theory of unitary representations of the Poincaré group [269] and Dirac's approach to relativistic interactions [56]. We will show that boost transformations of particle observables should depend on interactions between the particles. Thus, the conventional universal formulas of special relativity are only approximations, and the concept of the four-dimensional Minkowski space–time is neither rigorous nor exact.

The essence of our theory can be summarized in few phrases:

> The physical world consists of point particles that obey the laws of quantum mechanics and interact with each other by instantaneous potentials, depending on distances between the particles and on their momenta. Generally, these potentials can also lead to the creation and destruction of particles. This picture does not at all contradict the principles of relativity and causality. To understand this point one should take into account that boost transformations of particle observables depend on forces acting between the particles and on the state of the system. Therefore, the usual Lorentz transformation formulas are not accurate, and the standard "proof" of the impossibility of superluminal propagation of interactions is not convincing.

Unlike the two previous "traditional" volumes of this book, in this third volume we will often contradict generally accepted views. This volume is opened by Chapter 1, *Three ways to look at QFT*, where we discuss three approaches to QFT: the traditional "bare-particles" approach, the "clothed-particles" approach of Greenberg and Schweber and the "dressed-Hamiltonian" approach, which is the foundation of our RQD theory.

In Chapter 2, *Dressing*, we critically evaluate the ideas of bare particles and renormalization in QFT. The main provisions of our corpuscular approach will be formulated here. The Hamiltonian of the renormalized QED will be unitarily transformed into the so-called "dressed" Hamiltonian, which is free from divergences and has a transparent physical meaning. In other words, from the picture of all-pervasive, but unobservable quantum fields, we turn to the picture of point particles interacting with each other by instantaneous potentials.

In Chapter 3, *Coulomb potential and beyond*, we derive the simplest dressed interaction potential between charged particles in the second order of perturbation theory. This is the famous Coulomb–Darwin–Breit potential leading to the classical spectrum of the hydrogen atom with relativistic corrections. Our derivation will show that in the dressed representation QFT is virtually indistinguishable from relativistic quantum mechanics. The only fundamental difference is that quantum mechanics is not suitable for describing processes with a variable number of particles. This gap will be filled in the next two chapters.

Chapter 4, *Decays*, will be devoted to the relativistic description of unstable quantum systems. Special attention will be paid to decays of rapidly moving particles. We will see that the usual Einstein "time dilation" formula cannot describe this process accurately. In principle, deviations from the predictions of special relativity are experimentally observable. However, the required accuracy of measurements is beyond reach for modern experimental equipment.

The mathematical formalism developed to describe decaying systems will be applied to radiative transitions in the hydrogen atom in Chapter 5, *RQD in higher orders*. In this chapter, we will also discuss infrared divergences and their cancellation in higher orders of perturbation theory. In particular, we will calculate the electron anomalous magnetic moment and the Lamb shift in the hydrogen atom. This will demonstrate that the corpuscular approach of RQD is equivalent to the standard renormalized QED in everything related to scattering cross sections and energy spectra.

We also want to apply our theory to the dynamics of charged particles and their force fields. To demonstrate these possibilities, we turn to the macroscopic (classical) limit. In Chapter 6, *Classical electrodynamics*, we will show how one can use RQD results to reformulate classical electrodynamics in the form of a Hamiltonian theory with instantaneous interaction forces. These forces depend not only on the distances between charges, but also on their velocities and spin orientations. In our formulation, electromagnetic fields, potentials and Maxwell's equations are not present at all. This will allow us to solve many theoretical paradoxes, inherent in classical electromagnetic theory, and at the same time remain in agreement with experimental data.

We will continue our discussion of electromagnetic phenomena in Chapter 7, *Experimental support of RQD*, where we briefly describe several important experiments. The famous Aharonov–Bohm effect will find its explanation as an effect of interparticle interactions on the phases of quantum wave packets. We will be especially interested in experiments demonstrating superluminal propagation of the near-field electromagnetic radiation. Finally, we shall analyze experiments of Professor Pizzella with relativistic electron beams [51, 53, 52]. We claim that these results give the most clear confirmation of our conclusion about the instantaneous nature of electromagnetic forces.

In Chapter 8, *Particles and relativity*, we will discuss real and imaginary paradoxes associated with our corpuscular interpretation. In particular, we will need to resolve the contradiction between the instantaneous interaction potentials and the special-relativistic prohibition of superluminal signals. In this discussion, we will rely on Dirac's approach to interactions from Volume 1. In particular, we will claim that the presence of interaction operators in boost generators must be interpreted as the inapplicability of Lorentz transformations to space–time coordinates of events in interacting systems. By the same argument, Einstein's ban of superluminal interactions will be rejected as well. We will show that superluminal action-at-a-distance can coexist with causality, if we correctly understand the nature of relativistic invariance in the presence of interactions. This is the main novelty of the theory presented in this book.

The small Chapter 9, *Conclusions*, will sum up our main results and briefly outline possible directions for future research.

Some useful mathematical facts and cumbersome calculations are collected in *Appendix*.

In this volume we use notation and terminology from the two previous volumes. References to these books will be prefixed by "**1**-" and "**2**-". For example, (**1**-4.32) denotes formula (4.32) from Volume 1, and Section **2**-2.4 is from the second volume.

Ironically, the development of our corpuscular theory did not require the introduction of radically new physical ideas. All key components of this approach have been known for a long time, but for one reason or another they did not attract the well-deserved attention. For example, the fact that in addition to time translations, some other inertial transformations (space translations, rotations or boosts) should have a dynamical dependence on interactions was first established in Dirac's work [56]. These ideas were developed further in "direct interaction" theories of Bakamjian, Thomas [12], Foldy [84], Sokolov [227, 229], Coester, Polyzou [35] and many others.

The clothed-particles approach was advanced by Greenberg and Schweber [101] (see also [88, 89]). The first indications that this approach can solve the problem of ultraviolet divergences in QFT are contained in articles by Ruijgrok [202], Shirokov and Vişinesku [216, 259]. The formulation of RQD presented in this book simply combined all these good ideas into a single theory, which is intended to be a next step towards a natural and consistent unification of quantum mechanics with the principle of relativity.

The new ideas presented in this book were partially published in eight articles [233–240].

1 Three ways to look at QFT

The first principle is that you must not fool yourself – and you are the easiest person to fool.
Richard Feynman

In this chapter, we continue our discussion of quantum electrodynamics (QED) – the theory of charged particles and photons. At the end of the second volume, we found out that the traditional QED has considerable difficulties in constructing the Hamiltonian and in describing the time evolution of interacting systems. In this volume we would like to convince the reader that the QED formalism can be substantially improved by discarding ideas of unobservable bare and virtual particles and radically changing the interaction Hamiltonian. We will call it the *dressed Hamiltonian* approach, and in Chapter 2, we will see how it can fix the divergence problems in QED. In this chapter, we are going to prepare ourselves for this effort by giving a short summary of three points of view on QFT: the renormalized bare theory, the clothed-particle theory and the dressed-Hamiltonian theory.

We call the approach developed in Volume 2 the *bare-particle* theory, because it was based on the same (bare) definitions of the vacuum and one-particle states as the noninteracting theory.[1] These bare states were eigenvectors of the noninteracting mass operator $M_0 = c^{-2}\sqrt{H_0^2 - P_0^2 c^2}$ with experimentally measured masses, i. e.,

$$M_0|\text{vac}\rangle = 0|\text{vac}\rangle, \tag{1.1}$$

$$M_0 a_{\boldsymbol{p}}^\dagger|\text{vac}\rangle = m_e a_{\boldsymbol{p}}^\dagger|\text{vac}\rangle, \tag{1.2}$$

$$M_0 d_{\boldsymbol{p}}^\dagger|\text{vac}\rangle = m_p d_{\boldsymbol{p}}^\dagger|\text{vac}\rangle. \tag{1.3}$$

We immediately face a paradox, because these states are not eigenvectors of the full QED Hamiltonian H^c and the corresponding mass operator $M^c = c^{-2}\sqrt{(H^c)^2 - P_0^2 c^2}$. This means that the bare electrons and protons experience self-interaction,[2] and their time evolutions involve troublesome "fluctuations" (see Subsection 1.1.2), which have not been observed in experiments.

Would it not be more consistent to demand that the vacuum and single-particle states are eigenstates of the full interacting Hamiltonian? Yes, this can be done. We call this the *clothed-particle* formalism.[3] The idea is to find zero-particle and one-particle

1 Recall that in Volume 2 we used the bare states $a_{\boldsymbol{p}}^\dagger|\text{vac}\rangle, d_{\boldsymbol{p}}^\dagger|\text{vac}\rangle$ to calculate matrix elements of the S-operator, thus assuming that these are the correct asymptotic states of the colliding particles in the remote past and the distant future.

2 Which is often described as emission and reabsorption of virtual photons.

3 One comment on terminology. In the literature, the names "clothed particles" and "dressed particles" were used interchangeably to describe physical particles in the Greenberg-Schweber approach to QFT. Only the former expression will be used in our book. This will help us to distinguish the

https://doi.org/10.1515/9783110493221-001

eigenvectors of M^c as linear combinations of bare states. For example, the clothed (or physical) one-electron state is a complex linear combination, i. e.,[4]

$$\begin{aligned}|\boldsymbol{p}\rangle^{\text{cl}} &= C^{(1)}(\boldsymbol{p})a^\dagger_{\boldsymbol{p}}|\text{vac}\rangle + \int d\boldsymbol{k}C^{(2)}(\boldsymbol{p},\boldsymbol{k})a^\dagger_{\boldsymbol{p}-\boldsymbol{k}}c^\dagger_{\boldsymbol{k}}|\text{vac}\rangle \\ &+ \int d\boldsymbol{k}d\boldsymbol{h}C^{(3)}(\boldsymbol{p},\boldsymbol{k},\boldsymbol{h})a^\dagger_{\boldsymbol{p}-\boldsymbol{k}-\boldsymbol{h}}a^\dagger_{\boldsymbol{k}}b^\dagger_{\boldsymbol{h}}|\text{vac}\rangle + \cdots. \end{aligned} \tag{1.4}$$

In the theory of clothed particles, this change of basis vectors is achieved by applying a unitary transformation $e^{-i\Xi}\dots e^{i\Xi}$ to the vacuum state and particle operators of the bare theory (see Table 1.1). So, in terms of the new (clothed) definitions[5] the lowest eigenvalues of the QED mass operator M^c are as follows[6]:

$$M^c|\text{vac}\rangle^{\text{cl}} = 0|\text{vac}\rangle^{\text{cl}}, \tag{1.5}$$
$$M^c a^{\text{cl}\dagger}_{\boldsymbol{p}}|\text{vac}\rangle^{\text{cl}} = m_e a^{\text{cl}\dagger}_{\boldsymbol{p}}|\text{vac}\rangle^{\text{cl}}, \tag{1.6}$$
$$M^c d^{\text{cl}\dagger}_{\boldsymbol{p}}|\text{vac}\rangle^{\text{cl}} = m_p d^{\text{cl}\dagger}_{\boldsymbol{p}}|\text{vac}\rangle^{\text{cl}}. \tag{1.7}$$

The huge benefit of this approach is that when the QED Hamiltonian H^c is expressed through the new particle operators

$$H^c = \mathcal{F}(a^{\text{cl}\dagger}, a^{\text{cl}}, \dots), \tag{1.8}$$

it becomes finite and does not exhibit self-interaction.

The clothed-particle approach still requires us to imagine the vacuum as a "boiling soup" and physical particles as surrounded by clouds of virtual photons and electron–positron pairs. We believe that these "vacuum polarization" effects are just theoretical artifacts, and not real phenomena.[7] The distinction between bare and physical particles looks totally artificial, and, clearly, there must be only one (physical) kind of particles in nature.

The distinction between bare and physical particles disappears in the *dressed-Hamiltonian* theory, which will be the basis of our discussions in the rest of this book. This theory uses the original (bare) definitions of the vacuum $|\text{vac}\rangle$ and particle operators $\{a^\dagger, a, \dots\}$. The main difference with respect to the bare-particle theory is the redefinition of the Hamiltonian. In particular, the dressed Hamiltonian H^d is expressed by the same function of particle operators $H^d = \mathcal{F}(a^\dagger, a, \dots)$ as the function (1.8) de-

---

Greenberg-Schweber "clothing" of particles from the "dressing" of the Hamiltonian, which is in the center of our own modification of QFT. These two approaches are compared in Subsection (1.3.4).

4 Here the first term is the bare electron, the second term is electron+photon, the third term is electron+$(e^- + e^+)$ pair.

5 They are labeled by the superscript "cl."

6 These relations will be proved at the end of this chapter.

7 It is sometimes claimed that these effects are real and they are responsible, in particular, for the *Casimir forces*. However, we prefer a different interpretation [120] of these experiments.

Table 1.1: Comparison of three approaches to QFT.

	Bare-particle theory	Clothed-particle theory	Dressed-Hamiltonian theory
vacuum	$\|\text{vac}\rangle$	$\|\text{vac}\rangle^{cl} = e^{-i\Xi}\|\text{vac}\rangle$	$\|\text{vac}\rangle$
e^- creation op.	$a^\dagger$	$a^{cl\dagger} = e^{-i\Xi}a^\dagger e^{i\Xi}$	$a^\dagger$
e^- annih. op.	a	$a^{cl} = e^{-i\Xi}ae^{i\Xi}$	a
1-electron state	$\|1\rangle = a^\dagger\|\text{vac}\rangle$	$\|1\rangle^{cl} = a^{cl\dagger}\|\text{vac}\rangle^{cl}$	$\|1\rangle = a^\dagger\|\text{vac}\rangle$
1-electron position	$\boldsymbol{r}$	$\boldsymbol{r}^{cl} = e^{-i\Xi}\boldsymbol{r}e^{i\Xi}$	$\boldsymbol{r}$
Hamiltonian	$H^c = H_0 + V^c$	$H^c = H_0 + V^c$	$H^d = e^{i\Xi}H^c e^{-i\Xi} = H_0 + V^d$
Interaction	$V^c = a^\dagger ac + \cdots$	$V^c = a^{cl\dagger}a^{cl\dagger}a^{cl}a^{cl} + \cdots$	$V^d = a^\dagger a^\dagger aa + \cdots$
Mass operator	$M^c = c^{-2}\sqrt{(H^c)^2 - P_0^2c^2}$	$M^c = c^{-2}\sqrt{(H^c)^2 - P_0^2c^2}$	$M^d = c^{-2}\sqrt{(H^d)^2 - P_0^2c^2}$
vacuum mass	$M^c\|\text{vac}\rangle \neq \text{const}\|\text{vac}\rangle$	$M^c\|\text{vac}\rangle^{cl} = 0$	$M^d\|\text{vac}\rangle = 0$
1-electron mass	$M^c\|1\rangle \neq \text{const}\|1\rangle$	$M^c\|1\rangle^{cl} = m_e\|1\rangle^{cl}$	$M^d\|1\rangle = m_e\|1\rangle$
S-operator	S^c	S^c	S^c

scribing the dependence of H^c on $a^{cl\dagger}, a^{cl}, \ldots$. So, the dressed theory also has a well-defined finite Hamiltonian without self-interaction effects. In particular, bare ($\equiv$ physical) zero-particle and one-particle states are eigenstates of the dressed mass operator $M^d = c^{-2}\sqrt{(H^d)^2 - P_0^2c^2}$ with experimentally measured eigenvalues

$$M^d|\text{vac}\rangle = 0|\text{vac}\rangle, \tag{1.9}$$

$$M^d a_{\boldsymbol{p}}^\dagger|\text{vac}\rangle = m_e a_{\boldsymbol{p}}^\dagger|\text{vac}\rangle, \tag{1.10}$$

$$M^d d_{\boldsymbol{p}}^\dagger|\text{vac}\rangle = m_p d_{\boldsymbol{p}}^\dagger|\text{vac}\rangle. \tag{1.11}$$

The physical equivalence of the clothed-particle theory and the dressed-Hamiltonian theory will be demonstrated in Subsection 1.3.4.

Finally, we would like to stress that all three approaches to QFT produce exactly the same scattering operator S^c, which agrees very well with experimental observations. However, we expect quite different predictions regarding the time evolution of states and observables. In the rest of this chapter we will consider the relations between these three approaches in more detail; see Table 1.1.

1.1 Bare-particle theory

Let us briefly review the material of Chapter **2**-4 and repeat the logic of renormalization in QED. Recall that the renormalization was introduced as a transition from the naïve interaction Hamiltonian V^n to the interaction with counterterms V^c.[8]

8 Our arguments in this chapter apply equally to QED formulations in the Coulomb and Feynman gauges. In the latter case, the mentioned interaction operators are $V^n \equiv V_1$ (**2**-3.30) and $V^c \equiv V_1 + Q$ (**2**-4.44).

1.1.1 Once again about renormalization

The naïve Hamiltonian $H^n = H_0 + V^n$ was not acceptable, because it lead to incorrect (even divergent) scattering amplitudes in high perturbation orders. The most damaging was the presence of *renorm* terms in the scattering operators Φ^n and Σ^n.[9] We can trace the appearance of *renorm* terms in Φ^n to the presence of *unphys* terms in the original interaction V^n. Indeed, according to Table **2**-1.2, products or commutators[10] of *unphys* potentials usually contain *renorm* terms.

The presence of *unphys* interactions is the most significant deviation of QFT from all previous physical theories. These terms are responsible for self-interactions, which are the reason why definitions of particles cannot be fixed in advance, but should be corrected (renormalized) after introduction of the interaction. The self-interactions are the source of divergence difficulties. We did not have these difficulties in ordinary quantum mechanics, where interactions were purely *phys* and did not affect one-particle states.

The idea of renormalization offers a partial solution to this paradox: change the interaction operator from V^n to $V^c = V^n + Q$ by adding divergent *unphys* and *renorm* counterterms Q. Two physical requirements (the no-self-scattering and charge renormalization conditions) are sufficient to uniquely determine all counterterms in V^c. Then all *renorm* terms in the scattering operator cancel out, and (quite mysteriously), the S-operator obtained with the new Hamiltonian,

$$H^c = H^n + Q = H_0 + V^n + Q = H_0 + V^c,$$

is in good agreement with experiments.

There are two important groups of experimental data associated with the S-operator and, therefore, directly comparable to QFT calculations: scattering cross sections and energies of bound states.[11] These experiments cover most of what can be observed in the microworld, so the renormalized S-operator formalism in QFT is very successful.

1.1.2 What is not calculable in QFT?

So, the renormalization did not change the *unphys* character of QFT interactions. The renormalization simply suppressed the self-scattering of bare particles,[12] but their

9 This violated the no-self-scattering condition (Statement **2**-4.1).

10 Found in definitions of Φ^n and Σ^n.

11 As we know from Subsection **1**-7.1.6, energies of bound states coincide with positions of poles of the S-operator $S(E)$ on the complex energy plane.

12 This means that renormalization suppressed the cumulative effect of self-interaction during the infinite time interval $(-\infty, +\infty)$.

self-interaction at finite time intervals remained. As long as *unphys* self-interactions are present, we will have to redefine (renormalize) the notions of the vacuum and particles.

Therefore, the renormalization program did not solve the problem of ultraviolet divergences in QFT. The introduction of counterterms simply moved the problem of divergences from one place to another. The divergences were removed from the S-operator, but they surfaced in the Hamiltonian H^c in the form of infinite counterterms. The divergent Hamiltonian $H^c = H_0 + V^c$ can be used in S-matrix calculations, since all divergences cancel out there. However, the use of this Hamiltonian for studying bound states[13] or the time evolution[14] is very problematic.

In Section **2**-1.1, we introduced vectors

$$|\mathrm{vac}\rangle,\ a^\dagger|\mathrm{vac}\rangle,\ d^\dagger|\mathrm{vac}\rangle \ldots, \tag{1.12}$$

which were (bare) zero-particle and one-particle eigenstates of the noninteracting Hamiltonian H_0 and the mass operator M_0 (see (1.1)–(1.3)). However, these states are not eigenvectors of the full interacting Hamiltonian $H^c = H_0 + V^n + Q$, which makes their properties quite peculiar. Let us forget for a moment that the counterterms in H^c are infinite and apply the time evolution operator $U^c(t \leftarrow 0) = \exp(-\frac{i}{\hbar}H^c t)$ to the vacuum state $|\mathrm{vac}\rangle$. Expressing V^c in terms of creation and annihilation operators, we get[15]

$$\begin{aligned}|\mathrm{vac}(t)\rangle &= e^{-\frac{i}{\hbar}H^c t}|\mathrm{vac}\rangle = \left(1 - \frac{it}{\hbar}(H_0 + V^c) + \cdots\right)|\mathrm{vac}\rangle \\ &\propto |\mathrm{vac}\rangle + ta^\dagger b^\dagger c^\dagger|\mathrm{vac}\rangle + td^\dagger f^\dagger c^\dagger|\mathrm{vac}\rangle + \cdots \\ &\propto |\mathrm{vac}\rangle + t|abc\rangle + t|dfc\rangle + \cdots.\end{aligned} \tag{1.13}$$

We see that various multiparticle states ($|abc\rangle$, $|dfc\rangle$, etc.) appear from the vacuum during the time evolution. So, in this interpretation the QED vacuum is not an empty state without particles, as we might naïvely assume. It is full of "virtual" particles, antiparticles and photons.

Likewise, the time evolution of the bare one-electron state $a^\dagger|\mathrm{vac}\rangle$ is accompanied by the appearance and disappearance of multitudes of virtual particles [32]. Of course, no one has ever seen such strange vacuum polarizations and virtual clouds in experiments. If the theory is incapable of describing the time evolution of even such simple states, then it has no chance with more complex (and important) multiparticle systems.

---

13 By diagonalization of H^c.

14 By forming the time evolution operator $\exp(-\frac{i}{\hbar}H^c t)$.

15 Here we focus on the presence of *unphys* interaction operators $a^\dagger b^\dagger c^\dagger$ and $d^\dagger f^\dagger c^\dagger$ in V^c (**2**-4.44). We omitted all factors and coefficient functions that are not relevant to our arguments.

It is often argued that in scattering experiments, interactions occur almost instantaneously, so that a detailed experimental study of the interacting dynamics is practically impossible. Thus, the absence of a well-defined Hamiltonian and the inability of the renormalized QED to describe the interacting regime are completely tolerable.

> The more one thinks about this situation, the more one is led to the conclusion that one should not insist on a detailed description of the system in time. From the physical point of view, this is not so surprising, because in contrast to nonrelativistic quantum mechanics, the time behavior of a relativistic system with creation and annihilation of particles is unobservable. Essentially only scattering experiments are possible, therefore we retreat to scattering theory. One learns modesty in field theory – G. Scharf [209].

But there is no doubt that time-dependent processes will sooner or later become available for study by advanced experimental methods.[16] Moreover, the time evolution of interacting systems is easily observed in our daily "macroscopic" life. Therefore, a complete and consistent theory of subatomic phenomena is obliged to describe time-dependent processes. The renormalized QED cannot do this, hence it cannot be regarded as a comprehensive theory of electromagnetism.

The fundamental problems of renormalization worried many outstanding physicists, including Dirac and Landau. For example, Rohrlich wrote:

> Thus, present quantum electrodynamics is one of the strangest achievements of the human mind. No theory has been confirmed by experiment to higher precision; and no theory has been plagued by greater mathematical difficulties which have withstood repeated attempts at their elimination. There can be no doubt that the present agreement with experiments is not fortuitous. Nevertheless, the renormalization procedure can only be regarded as a temporary crutch which holds up the present framework. It should be noted that, even if the renormalization constants were not infinite, the theory would still be unsatisfactory, as long as the unphysical concept of "bare particle" plays a dominant role – F. Rohrlich [198].

1.1.3 Effective quantum field theory

Often one can hear arguments that the divergence of the renormalized Hamiltonian H^c in the limit $\Lambda \to \infty$ is not such a big problem. The idea is that, in fact, the large momentum (or, equivalently, small distance) limit in loop integrals is unattainable, because in this limit the fundamental premises of the local QFT[17] lose their meanings. There is hope that some yet unknown effects at the Planck scale will establish a natural cutoff in momentum integrals and hence ensure finite values of the renormalization constants. This set of ideas is called the *effective field theory* [157]. In particular, this

---

16 See, for example, recent measurements of dynamics of nonstationary atomic wave functions with attosecond time resolution [57, 153, 45].

17 Such as the locality of interactions or the continuity of space.

approach claims that QFT is just a low-energy approximation to some yet unknown microscopic truly fundamental theory that will be free from divergences and contradictions.[18]

There are various assumptions about the possible content of this fundamental and self-consistent theory that will one day replace the internally contradictory QFT. Some think that the future theory will reveal a nontrivial – possibly discrete or noncommutative – structure of space at distances comparable to the Planck length $\approx 10^{-33}$ cm. Others hope that paradoxes will be solved if extended objects, such as strings, are used instead of point particles. However, so far these ideas have not raised above the level of guesswork.

Speculations of this kind cannot be refuted, because nothing is known about the physics at the Planck scale. However, the good news is that such speculations are completely unnecessary in the dressed-Hamiltonian approach, which we are going to develop in this volume. The dressed Hamiltonian H^d remains finite even in the limit of the infinite cutoff parameter Λ. Therefore, our RQD is a completely autonomous and self-consistent theory, which is formally applicable throughout the entire range of distances, energies and momenta, even beyond the hypothetical Planck boundary.

1.2 Clothed-particle theory

As we already know, the time evolution of bare vacuum and one-particle states involves unrealistic creation of virtual photons and electron–positron pairs, because these bare states are not eigenstates of the full interacting Hamiltonian H^c. One fruitful idea was that the virtual clouds are inherent parts of the physical vacuum and particle states, so that we should focus on the dynamics of these linear combinations as separate entities. In other words, we can redefine the notions of the zero-particle and one-particle states, so that they coincide with eigenvectors of H^c. Then the entire theory can be expressed in the language of the new "clothed" or "physical" particles.

We already know that in QFT this redefinition is necessary, because *unphys* terms are present inevitably in QFT interaction operators.[19] Thus *any* QFT is doomed to suffer

---

18 It should be noted that a similar situation takes place in nonrelativistic theories, for example, in solid-state physics, where the continuous quantum field description is really approximate and only works within the limits of low energies and large distances. For example, the quantum field description of crystal lattice vibrations in terms of *phonons* is applicable only when deviations of atoms from their equilibrium positions are much smaller than interatomic distances. The concept of renormalization is also physically justified in these theories. For example, the *polaron* (= the electron moving through the crystal while interacting with lattice vibrations) has a renormalized mass that differs from the effective mass of the electron in the conduction band of the "fixed" lattice. In this book we discuss only fundamental relativistic quantum fields where such considerations and analogies are inapplicable.

19 Indeed, any quantum field is a sum of creation and annihilation parts $\phi \propto (\alpha^\dagger + \alpha)$. In field theories, interaction is built as a polynomial in quantum fields (see Section **2**-3.1), and any field monomial would

from renormalization problems. These problems are much deeper than just the divergence of loop integrals. The redefinition of the vacuum and one-particle states would be necessary even if all loop integrals were convergent.

1.2.1 A bit of history

The idea of *clothed particles* has a long history. Van Hove was the first to express his thoughts about "persistent perturbations" in QFT [257, 258]. The first clear formulation of the clothed-particle concept and its application to model QFTs is contained in the excellent article by Greenberg and Schweber [101]. This formalism was extended to various quantum field models, including the scalar-field model [265], the Lee model [73, 83, 54, 55, 6] and the Ruijgrok–Van Hove model [202, 159]. Faddeev [74] proposed a method for constructing the Hamiltonian of clothed particles in the form of a perturbation series for a general QFT (see also [246, 208, 76]). This method was called the "Faddeev construct" in [226]. Shirokov and coauthors [216, 259, 219, 220, 214] developed these approaches further and, in particular, discussed the removal of ultraviolet divergences from the Hamiltonians of clothed particles. Such Hamiltonians have found many fruitful applications in the theory of nuclear forces [149–151, 58].

1.2.2 Clothed (physical) particles

In the clothed-particle formalism, the definitions of zero-particle and one-particle states are rather straightforward. For example, the clothed vacuum could be defined as an eigenvector of the mass operator $M^c \equiv c^{-2}\sqrt{(H^c)^2 - P_0^2c^2}$ with the lowest possible eigenvalue. The set of states of the clothed electron (1.4) could be defined as an eigensubspace of M^c with the lowest eigenvalue under the conditions that the lepton number is one ($L = 1$), the baryon number is zero ($B = 0$) and the electric charge is $Q = -e$.[20] However, there is much less certainty when we want to define two-particle, three-particle, etc. states, i. e., the full Fock structure of the clothed world. To solve this problem, the clothed-particle theory makes three crucial assumptions:

necessarily contain *unphys* terms after normal ordering. For example, $\phi^4 \propto (\alpha^\dagger + \alpha)^4 = \alpha^\dagger\alpha^\dagger\alpha^\dagger\alpha^\dagger + \cdots + \alpha\alpha\alpha\alpha$ [223].

20 In the literature one can find claims that due to Haag's theorem [105, 61, 59], the physical vacuum, physical particles and the entire Fock space of the interacting theory are orthogonal to the original Fock space of bare particles. Indeed, all coefficients in expansions like (1.4) tend to zero. However, this does not mean that the entire sum tends to zero as well. This is because the number of components in the sum is infinite, which results in the uncertainty of the type $0 \times \infty$. We believe that both bare and clothed states can be accommodated within one Fock space. Perhaps, a rigorous mathematical formulation of this idea would require a more careful handling of infinitely small and infinitely large numbers, as in nonstandard analysis [87]. See also [223].

(1) There is a unitary operator $e^{-i\Xi}$ that connects between bare, $\{|\mathrm{vac}\rangle, a^\dagger, a, \ldots\}$, and clothed,

$$|\mathrm{vac}\rangle^{\mathrm{cl}} = e^{-i\Xi}|\mathrm{vac}\rangle, \tag{1.14}$$

$$a^{\mathrm{cl}\dagger} = e^{-i\Xi}a^\dagger e^{i\Xi}, \tag{1.15}$$

$$a^{\mathrm{cl}} = e^{-i\Xi}a e^{i\Xi}, \tag{1.16}$$

ingredients of the two theories.

(2) After clothing, the interaction V^c expressed as a function (1.8) of the new (clothed) particle operators becomes purely *phys*.

(3) The scattering operator of the clothed particles is exactly the same (S^c) as in the bare renormalized theory.

1.3 Dressed-Hamiltonian theory

The current wisdom is that electromagnetic interactions can be described only within a QFT, such as QED. Three arguments are usually put forward to support this statement:

(1) QED enjoys a phenomenal agreement with experiments.

(2) The structure of the QED (field-based) Hamiltonian is dictated by physical features of electromagnetic interactions and follows unambiguously from the application of quantum mechanics to Maxwell's equations.

(3) QFT is the only possibility to combine quantum mechanics with relativity.

Here we would like to challenge these arguments. First, as we mentioned above, all successes of QED are associated with the S-matrix, but from Subsection **1**-7.2.1 we know that one can always find an infinite number of Hamiltonians (thus, an infinite number of physically nonequivalent theories) that yield exactly the same S-matrix. Some of these theories may have a nonfield character.

The validity of item (2) is questionable as well. Renormalization and clothing are good examples of how a naïve quantum field theory must be drastically modified in order to comply with observations. The renormalization adds counterterms to the original Hamiltonian. The clothing redefines the vacuum and single-particle states. Are these steps really necessary? Perhaps, our problem was that we started from a wrong place[21] and then had to spend enormous efforts to correct this bad theory to make it consistent with observations? Can we do better and formulate a good (no infinities, no internal contradictions) theory from the beginning?

21 That is, a local quantum field theory with the Hamiltonian H^n. In Subsection 1.3.1 we will briefly analyze the ideas that historically lead to the choice of V^n as the interaction operator in QED.

Finally, argument (3) is not convincing either. For example, the clothed-particle theory is not a field theory.[22] Nevertheless, it is perfectly relativistic [213]. We will provide other examples of relativistic nonfield theories in Subsection 1.3.2.

In this section we will present a third approach, which we call the *dressed Hamiltonian* formalism. It is based on three ideas:

(1) Stop making the distinction between bare and physical states. We claim that there is only one sort of particles in nature,[23] and we will use the original bare operators $\{\alpha^\dagger, \alpha, \ldots\}$ to create and annihilate them.
(2) Drop the requirements that the interaction operator should be built as a polynomial in field-like combinations $(\alpha^\dagger + \alpha)$. Instead, we will demand that (like in the clothed-particle theory) this operator is composed only of *phys* potentials. This will guarantee the absence of vacuum and single-particle self-interaction effects and, therefore, remove the divergences and the need for renormalization.
(3) Demand that the S-operator of the new theory is exactly the same as the scattering operator (S^c) of the renormalized QED, which has been thoroughly verified by precision experiments.

1.3.1 On origins of QED interaction

We have already established that the main cause of problems in QED is the presence of *unphys* terms in the interaction operator V^c. But where did this operator come from? How strong is our confidence that the QED Hamiltonian was chosen correctly?

In Volume 2, we simply postulated the QED interaction operator (**2**-3.13)–(**2**-3.15) and its renormalized version (**2**-4.44). What is the physical basis for these expressions? Are there any physical principles that require this specific form of electromagnetic interaction? In textbooks, one can find the standard answer that the deep reason for the interaction between charged particles and photons is the principle of *local gauge invariance*. It is usually postulated that the field Lagrangian of electromagnetic theory must be invariant with respect to certain simultaneous "gauge" transformations of the fermion (charged) fields $\psi_i(t, \boldsymbol{x})$ and the photon field $\mathcal{A}_\mu(t, \boldsymbol{x})$. As it turns out, the Lagrangian of the free theory does not satisfy this requirement, and the local gauge invariance can be ensured only after some "minimal" interaction is added to the theory. Many textbooks on QFT explain this idea,[24] so we will not dwell on it. Suffice it to say that the same principle of local gauge invariance was used to derive Lagrangians of non-Abelian theories of electro-weak interactions and quantum chromodynamics.

---

22 The interaction operator V^c cannot be expressed as a polynomial in field-like combinations $(\alpha^{\mathrm{cl}\dagger} + \alpha^{\mathrm{cl}})$ of clothed operators.

23 That is, the physical particles with experimentally measured masses and charges.

24 See, for example, p. 343 in [266] and Section 15.1 in [183].

Despite its wide acceptance in theoretical physics, the idea of gauge invariance is rather vague. In its original form, this idea arose in Maxwell's classical electrodynamics. However, in Chapter 6 we will see that Maxwell's theory can be replaced by an alternative approach in which charges interact directly, while electromagnetic fields, potentials and gauges play absolutely no role. Moreover, the physical meaning of quantum fields is not clear, and the very necessity of using fields and electromagnetic potentials for the formulation of physical theories raises questions.[25] For these reasons, in our book we reject the usual postulate that assigns fundamental roles to fields, potentials and gauges. We maintain that the local gauge invariance should be considered only as a heuristic device, whose remarkable efficiency still awaits its proper explanation.

An alternative (or, rather, complementary) justification for the QED Hamiltonian was put forward by Weinberg.[26] He suggested that the interaction operator should be a polynomial function of field components. Further, he empirically substantiated the claim that this polynomial should be linear in the photon field $\mathcal{A}_\mu(t, \boldsymbol{x})$. Another important requirement is that the action integral should be invariant with respect to inertial transformations from the Lorentz subgroup. If $\mathcal{A}_\mu$ were transformed as a 4-vector, then the latter condition would be easily satisfied by choosing the interaction Hamiltonian density in the form $\propto \mathcal{A}_\mu J^\mu$, where J^μ is any 4-vector composed of fermion fields. However, the Lorentz transformations of $\mathcal{A}_\mu$ are different from the 4-vector law by the presence of an additional 4-gradient term (**2**-C.34). This difficulty can be overcome if we choose as J^μ a *conserved* fermionic 4-vector. Then the action integral turns out to be a Lorentz scalar, even though $\mathcal{A}_\mu$ is not a 4-vector. The simplest choice for J^μ is the density of the fermionic current (**2**-D.1) $\mathrm{j}^\mu = -e\overline{\psi}\gamma^\mu\psi + e\overline{\Psi}\gamma^\mu\Psi$. This line of reasoning leads to the interaction (**2**-3.13)–(**2**-3.15), consistent with gauge invariance arguments.

Unfortunately, neither the local gauge invariance arguments nor Weinberg's reasoning seems convincing, especially if we take into account the fact that the (*unphys*) trilinear interaction density operator

$$V_1(\tilde{x}) = \mathcal{A}_\mu \mathrm{j}^\mu = -e\mathcal{A}_\mu\overline{\psi}\gamma^\mu\psi + e\mathcal{A}_\mu\overline{\Psi}\gamma^\mu\Psi \tag{1.17}$$

is not final. First, it has to be modified by adding renormalization counterterms, including those of the *renorm* type. Then it should be unitarily transformed into the clothed-particle (or dressed) representation.

So, we have to admit that the modern formulation of QED does not have a rigorous theoretical justification. Quantum fields and gauges are only heuristic tools, and the whole construction is supported rather by a successful agreement of results with experiments than by reliance on well-understood physical principles.

---

25 We will discuss these questions in more detail in Section 8.6.

26 See, for example, Section 8.1 in [266].

1.3.2 Earlier attempts at construction of nonfield theories

We also do not agree with the claim that QFT is the only way to combine quantum theory with the principle of relativity. Usually this statement is formulated as a self-evident postulate that does not require a proof. In this respect Weinberg was not so categorical, when he wrote the following:

> There have been many attempts to formulate a relativistically invariant theory that would not be a local field theory, and it is indeed possible to construct theories that are not field theories and yet yield a Lorentz-invariant *S*-matrix for two-particle scattering [12], but such efforts have always run into trouble in sectors with more than two particles: either the three particle *S*-matrix is not Lorentz-invariant, or else it violates the cluster decomposition principle – S. Weinberg [266].

But what were these attempts at constructing nonfield theories that Weinberg mentioned? Could these theories fulfill the three conditions (relativistic invariance, cluster separability and a variable number of particles), formulated in the preface to Volume 2? What problems did they encounter? Perhaps these problems are solvable, and it would make sense to take a second look at these nonfield models?

An important class of nonfield theories, mentioned by Weinberg, constitute "direct-interaction" models.[27] Two-particle models of this kind were first constructed by Bakamjian and Thomas [12]. Later, Sokolov and Shatnii [227, 229], Coester and Polyzou [35] and Mutze [172] showed how this approach can be extended to multiparticle systems, while satisfying the principle of cluster separability. We also note more recent attempts [189] to apply this formalism to systems with a variable number of particles. Despite these achievements, the "direct-interaction" theory is currently applicable only to model systems. One of the reasons is that the important condition of cluster separability has a very complex formulation in this approach. This mathematical difficulty was obvious when we discussed the simplest nontrivial example of a three-particle system in Subsection **1**-6.4.6.

In direct interaction theories, interparticle potentials were formulated as functions of dynamical particle variables, i. e., particle positions and momenta. However, this approach seems to be not optimal. It turns out that it is more convenient to write interactions in the form of polynomials (**2**-1.57)–(**2**-1.58) in creation and annihilation operators in the Fock space. In this case, cluster separability is satisfied as soon as coefficient functions of interaction potentials depend smoothly on particles' momenta (Theorem **2**-2.5). Moreover, in this approach there are no difficulties in constructing potentials that change the number of particles in the system.

For interacting theories in the Fock space, the most difficult part is to ensure their relativistic invariance, i. e., the commutation relations of the Poincaré Lie algebra.

---

27 We discussed them in Sections 6.3 and 6.4 of Volume 1.

One solution to this problem was developed in the works of Kita[28] [143–145, 147, 146, 148] and Kazes [128] (see also [212]). These models were not based on the rigid framework of quantum fields, so it was possible to avoid difficulties associated with divergences and renormalization. Coefficient functions of interaction potentials were determined by solving a system of differential equations resulting from Poincaré commutators (**1**-6.26)–(**1**-6.30). Kita and Kazes demonstrated that such systems have an infinite number of solutions and considered few nontrivial examples. Unfortunately, this approach is not yet applicable to realistic physical systems.

Another fruitful idea was to fit nonfield "effective" particle interactions to reproduce QFT scattering amplitudes and/or energy spectra. Such approaches can be found in a variety of works, starting from classic articles of Tamm [245], Dancoff [46] and Okubo [176] and continuing to more recent studies [112, 184, 185, 103, 104, 77]. In theoretical nuclear physics, an entire industry has been developed for producing internuclear potentials from quantum field models [67, 58]. The difference of our dressed approach from these works is philosophical rather than fundamental. We do not consider quantum fields as fundamental physical objects and do not regard effective potentials simply as approximations to the "rigorous" field-based description. For us, the idea of particles interacting directly through dressed potentials V^d is the most adequate way of describing nature. We think that QFT is just one intermediate step on the way to achieve this description. We will try to go beyond this step.

1.3.3 Dressed interactions

In the dressed-Hamiltonian approach we simply pretend that the bare states $|vac\rangle$, $|1\rangle$, etc. are not different from the physical states that we want to study. So, we will use the same set of creation and annihilation operators $\{a^\dagger, a, \ldots\}$ as in the bare-particle theory (see Table 1.1). The only difference between the two theories lies in their Hamiltonians.[29] We will obtain the dressed Hamiltonian H^d by applying to H^c the (reverse) unitary transformation that was used in the clothed-particle method (1.14)–(1.16), i. e.,

$$H^d \equiv H_0 + V^d = e^{i\Xi} H^c e^{-i\Xi}. \tag{1.18}$$

By doing that we will ensure that the dressed interaction is a *phys*-type polynomial in the particle operators $\{a^\dagger, a, \ldots\}$. This will also guarantee the desirable property that the vacuum and single particles do not experience self-interactions, so there will be no need for state redefinitions and renormalization. Other desirable features are that the interaction V^d is finite[30] and the S-operator calculated with V^d is exactly the same as the well-tested S^c.

28 I am thankful to Dr. M. I. Shirokov for drawing my attention to Kita's works.

29 And interacting boost operators.

30 That is, independent of the ultraviolet cutoff Λ.

1.3.4 Comparison with clothed-particle theory

In the dressed-Hamiltonian approach, the unitary transformation $e^{i\Xi}$ is applied to the field-theoretic Hamiltonian H^c of QED, while creation and annihilation operators $\{a^\dagger, a, \ldots\}$ of (bare) particles do not change. This means that the same function $\mathcal{F}$ expresses the dependence of the dressed Hamiltonian H^d on the bare-particle operators $H^d = \mathcal{F}(a^\dagger, a, \ldots)$ as the function expressing the dependence of H^c on the clothed-particle operators $H^c = \mathcal{F}(a^{\text{cl}\dagger}, a^{\text{cl}}, \ldots)$. Indeed, from (1.15), (1.16) and (1.18) we obtain (1.8)

$$
\begin{aligned}
H^c &= e^{-i\Xi} H^d e^{i\Xi} = e^{-i\Xi} \mathcal{F}(a^\dagger, a, \ldots) e^{i\Xi} = \mathcal{F}(e^{-i\Xi} a^\dagger e^{-i\Xi}, e^{-i\Xi} a e^{i\Xi}, \ldots) \\
&= \mathcal{F}(a^{\text{cl}\dagger}, a^{\text{cl}}, \ldots).
\end{aligned} \tag{1.19}
$$

Therefore, these two approaches are mathematically equivalent.

Let us confirm this equivalence by a simple example. Suppose that we are interested in the trajectory (= the time dependence of the expectation value of the position operator) of a physical electron in a two-particle (electron + proton) system. In the dressed-Hamiltonian approach, the initial state of the system has two particles,

$$
|\Psi\rangle = a^\dagger d^\dagger |\text{vac}\rangle,
$$

and the expectation value of the electron position $\boldsymbol{r}$ is given by the Heisenberg picture formula

$$
\langle \boldsymbol{r}(t) \rangle = \langle \Psi | \boldsymbol{r}(t) | \Psi \rangle = \langle \text{vac} | d a (e^{\frac{i}{\hbar} H^d t} \boldsymbol{r} e^{-\frac{i}{\hbar} H^d t}) a^\dagger d^\dagger | \text{vac} \rangle, \tag{1.20}
$$

where the time evolution is governed by the dressed Hamiltonian H^d.

However, we can also rewrite (1.20) in the following form, which is characteristic of the clothed-particle formalism:

$$
\begin{aligned}
\langle \boldsymbol{r}(t) \rangle &= \langle \text{vac} | d a (e^{i\Xi} e^{\frac{i}{\hbar} H^c t} e^{-i\Xi}) \boldsymbol{r} (e^{i\Xi} e^{-\frac{i}{\hbar} H^c t} e^{-i\Xi}) a^\dagger d^\dagger | \text{vac} \rangle \\
&= \langle \text{vac} | e^{i\Xi} (e^{-i\Xi} d a e^{i\Xi}) e^{\frac{i}{\hbar} H^c t} (e^{-i\Xi} \boldsymbol{r} e^{i\Xi}) \\
&\quad \times e^{-\frac{i}{\hbar} H^c t} (e^{-i\Xi} a^\dagger d^\dagger e^{i\Xi}) e^{-i\Xi} | \text{vac} \rangle \\
&=^{\text{cl}} \langle \text{vac} | d^{\text{cl}} a^{\text{cl}} (e^{\frac{i}{\hbar} H^c t} \boldsymbol{r}^{\text{cl}} e^{-\frac{i}{\hbar} H^c t}) a^{\text{cl}\dagger} d^{\text{cl}\dagger} | \text{vac} \rangle^{\text{cl}}.
\end{aligned} \tag{1.21}
$$

In this case, the time evolution is generated by the original renormalized Hamiltonian H^c of QED, while "clothed" definitions are used for the vacuum vector, particle operators and the position observable (see Table 1.1).

Despite differing formalisms, physical predictions of the two theories coincide. So, the choice between them is a matter of convenience. From the author's point of view, it seems more economical and physically transparent to perceive dressing simply as a change of the Hamiltonian (1.18) and use a single (bare ≡ physical) type of particles throughout the formalism. Hence, we choose the dressed-Hamiltonian approach.

1.4 Effective potential in toy model

In Chapter 2 we will apply the dressed-Hamiltonian theory to QED. Before moving to these nontrivial calculations, we would like to discuss this approach with a simpler example of the toy model from Section **2**-2.1. This will give us the opportunity to clarify important principles without being distracted by purely computational issues.

Recall that our toy model describes systems of interacting electrons and photons with the bare-particle Hamiltonian (**2**-2.35)

$$H^c = H_0 + V_1 + Q, \tag{1.22}$$

where

$$V_1 = \frac{e\hbar c}{(2\pi\hbar)^{3/2}} \int \frac{d\boldsymbol{p}d\boldsymbol{k}}{\sqrt{2\varepsilon_{\boldsymbol{k}}}} a^\dagger_{\boldsymbol{p}} c^\dagger_{\boldsymbol{k}} a_{\boldsymbol{p}+\boldsymbol{k}} + \frac{e\hbar c}{(2\pi\hbar)^{3/2}} \int \frac{d\boldsymbol{p}d\boldsymbol{k}}{\sqrt{2\varepsilon_{\boldsymbol{k}}}} a^\dagger_{\boldsymbol{p}} a_{\boldsymbol{p}-\boldsymbol{k}} c_{\boldsymbol{k}} \tag{1.23}$$

is the naïve interaction operator and Q combines all divergent renormalization counterterms. In this model, we calculated the S-operator for nonrelativistic electron–electron scattering in the second order (**2**-2.23)

$$S^c_2[a^\dagger a^\dagger aa] \approx \frac{ie^2}{8\pi^2\hbar} \int d\boldsymbol{p}d\boldsymbol{q}d\boldsymbol{k} \frac{\delta(\omega_{\boldsymbol{p}-\boldsymbol{k}} + \omega_{\boldsymbol{q}+\boldsymbol{k}} - \omega_{\boldsymbol{q}} - \omega_{\boldsymbol{p}})}{\lambda^2 c^2 + k^2} a^\dagger_{\boldsymbol{p}-\boldsymbol{k}} a^\dagger_{\boldsymbol{q}+\boldsymbol{k}} a_{\boldsymbol{q}} a_{\boldsymbol{p}}. \tag{1.24}$$

Now recall that in Section **1**-7.2 we found out that one and the same scattering matrix can be obtained from many different Hamiltonians with different interaction operators. Here we want to illustrate this idea by defining a second-order *effective interaction* between our toy electrons,

$$V^{ee}_{2\text{eff}} = \int d\boldsymbol{p}d\boldsymbol{k}d\boldsymbol{p}' D(k) a^\dagger_{\boldsymbol{p}'-\boldsymbol{k}} a^\dagger_{\boldsymbol{p}+\boldsymbol{k}} a_{\boldsymbol{p}} a_{\boldsymbol{p}'}, \tag{1.25}$$

with the coefficient function

$$D(k) = -\frac{e^2}{2(2\pi)^3\hbar(\lambda^2 c^2 + k^2)}. \tag{1.26}$$

Indeed, if we calculate the second-order S-operator by substituting interaction (1.25) instead of V_1 in (**2**-2.8), we get[31]

$$S_{2\text{eff}}[a^\dagger a^\dagger aa] = \underbrace{V^{ee}_{2\text{eff}}},$$

i. e., exactly the same result as in (1.24). The two S-operators S^c_2 and $S_{2\text{eff}}$ coincide despite the fact that the new *effective* Hamiltonian $H_0 + V^{ee}_{2\text{eff}}$ is completely different from

31 Since operator (1.25) is already of the second order, we should be interested only in the second term of the expansion (**2**-2.8).

the original Hamiltonian (1.22). In particular, the interaction (1.25) is *phys*, while (1.23) is *unphys* + *renorm*. Replacing the renormalized (divergent) *unphys* + *renorm* interaction of field theories by a scattering-equivalent finite *phys* potential is the central idea of our dressed approach to QFT.[32]

We have already met the potential (1.25) in equation (**2**-1.88). We also found that in the position representation this potential multiplies two-electron wave functions by the *Yukawa potential* (**2**-1.89), which depends on the interparticle distance ($r \equiv |\boldsymbol{r}_1 - \boldsymbol{r}_2|$), i. e.,

$$V^{ee}_{2\text{eff}}(r) = \frac{e^2}{4\pi r} e^{-\lambda c r/\hbar}. \tag{1.27}$$

In the limit of zero photon mass $\lambda \to 0$ it transforms into pure Coulomb repulsion, i. e.,

$$V^{ee}_{2\text{eff}}(r)|_{\lambda \to 0} = \frac{e^2}{4\pi r}. \tag{1.28}$$

Similarly, we can derive an effective electron–photon potential $a^\dagger c^\dagger ac$, by fitting it to the Compton scattering ($S_2[a^\dagger c^\dagger ac]$) part of the field-theoretical S-operator. Thus, the total second-order effective potential is a sum (Yukawa + Compton), i. e.,

$$V_{2\text{eff}} \propto a^\dagger a^\dagger aa + a^\dagger c^\dagger ac, \tag{1.29}$$

and our toy quantum field model can be replaced in a reasonable approximation by the elementary quantum-mechanical Hamiltonian $H_0 + V_{2\text{eff}}$.

1.4.1 Higher orders of perturbation theory

It is not difficult to extend this approach to higher orders of perturbation theory. In particular, in the third order of the field-based S^c-operator we expect to obtain contributions

$$S^c_3 \propto a^\dagger a^\dagger c^\dagger aa + a^\dagger a^\dagger aac + a^\dagger c^\dagger c^\dagger ac + a^\dagger c^\dagger acc \tag{1.30}$$

that describe absorption and emission of a photon in electron–electron and electron–photon collisions, i. e.,

$$e^-e^- \leftrightarrow e^-e^-\gamma, \tag{1.31}$$

$$e^-\gamma \leftrightarrow e^-\gamma\gamma. \tag{1.32}$$

32 Note that the coefficient function (1.26) of the effective interaction can be obtained from the coefficient function of the S-operator (1.24) in two steps: (i) discarding the energy delta function and (ii) dividing by $(-2\pi i)$. In the rest of this book, we will often use this rule to fit an effective (=dressed) interaction to the known S-operator.

In the effective theory, these processes can be described by third-order *phys* potentials having the same operator structures,

$$V_{3\text{eff}} \propto a^{\dagger}a^{\dagger}c^{\dagger}aa + a^{\dagger}a^{\dagger}aac + a^{\dagger}c^{\dagger}c^{\dagger}ac + a^{\dagger}c^{\dagger}acc. \tag{1.33}$$

Their coefficient functions should be adjusted so that S-operator contributions (1.30) are consistent with the formula $S_3^c = \underbrace{V_{3\text{eff}}}$. The third-order interactions are much weaker than the Yukawa + Compton potential, so they could be regarded as small corrections to (1.29).

Later we shall see that this fitting procedure can be continued in higher perturbation orders. Ideally, this should lead to the construction of an effective *phys* Hamiltonian that accurately reproduces the field theory scattering matrix in all orders.

1.4.2 Interactions preserving numbers of particles

Let us now analyze what we have achieved in the two preceding subsections. We are going to compare the action of three Hamiltonians, i. e.,

$$H_2^d = H_0 + V_{2\text{eff}}, \tag{1.34}$$

$$H_{2+3}^d = H_0 + V_{2\text{eff}} + V_{3\text{eff}}, \tag{1.35}$$

$$H^c = H_0 + V_1 + Q, \tag{1.36}$$

in the bare Fock space, which is shown schematically in Figure 1.1. Sectors with fixed particle numbers are indicated by squares in the plane. Each square/sector is labeled by two numbers: N_{ph} is the number of photons and N_{el} is the number of electrons.

We already know that from the point of view of scattering, the Hamiltonian H_2^d (and especially H_{2+3}^d) is very close to the quantum field Hamiltonian H^c. However, there are striking differences in how these theories describe the time evolution of states. First we turn to the time evolution $\exp(-\frac{i}{\hbar}H_2^d t)$ generated by the second-order effective Hamiltonian (1.34). As shown by the arrows in Figure 1.1 (a), this evolution

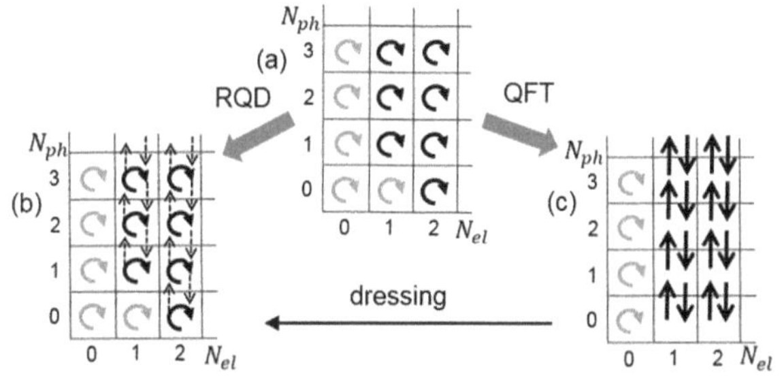

Figure 1.1: Time evolution in our toy Fock space. (a) Ordinary quantum mechanics. (b) Reasonable expectations from a theory with a variable number of particles. (c) A theory with the field Hamiltonian (1.22).

does not affect the number of particles: it leaves all sectors invariant.[33] There is no direct or indirect photon–photon interaction in our model, so the time evolution is noninteracting in all sectors without electrons ($N_{el} = 0, N_{ph} \geq 0$). The single-electron sector ($N_{el} = 1, N_{ph} = 0$) is also noninteracting due to the *phys* character of the dressed potential $V_{2\text{eff}}$. Time evolutions in these noninteracting sectors are shown by gray arrows in Figure 1.1. The nontrivial dynamics in other sectors is generated by the interaction (1.29). This description is a quite reasonable approximation for the behavior of real charges when they move with low accelerations, so that radiation effects are suppressed. Thus, our simplest effective potential (1.29) copes quite well with its duties and provides a fairly realistic model of interacting charged particles.

1.4.3 Interactions changing numbers of particles

Of course, the above description is approximate, primarily because it does not take into account the possibility of emission and absorption of photons. In an effective interaction model, this problem is solved by adding the third-order potential $V_{3\text{eff}}$ as in (1.35).

With this more accurate Hamiltonian H^d_{2+3}, the time evolution in the Fock space is depicted in Figure 1.1 (b), where the additional processes (1.31)–(1.32) that change the numbers of photons and connect different Fock sectors are shown by thin dashed arrows. As before, the renormalization condition is satisfied. There is no self-scattering in the vacuum and in one-particle states. Obviously, this desirable property follows from the fact that interaction potentials (1.29) and (1.33) were chosen to be *phys*.

A physically transparent and realistic approach with the Hamiltonian H^d_{2+3} is in stark contrast with the description of dynamics by the field-based Hamiltonian (1.36). The quantum field interaction V_1 (1.23) consists of trilinear (*unphys*) terms,

$$e^- \leftrightarrow e^- \gamma, \tag{1.37}$$

which do not correspond to any real physical process. In addition, the presence of divergent counterterms Q does not add realism to this picture. The dynamics generated by the interaction $V_1 + Q$ is shown schematically in Figure 1.1 (c). It is completely different from the physically justified dynamics like in Figures 1.1 (a) and (b). One-electron states $a^\dagger|\text{vac}\rangle$ are not eigenstates of the Hamiltonian H^c. This can be interpreted as a self-interaction of the free electron.

This simple example explains why we decided to abandon the quantum field description of electrodynamics (H^c) in favor of the dressed-interaction theory (H^d).

33 This is equivalent to ordinary particle number-preserving quantum mechanics being applied in each Fock sector separately.

1.4.4 Energy–momentum spectra

The dressed Hamiltonian H^d can be diagonalized to obtain energies and wave functions of bound states. Let us discuss the results of such a diagonalization at a qualitative level. For simplicity, here we restrict ourselves only to multielectron states, ignoring the presence of photons.

In Figure 1.2 (a) we showed the energy–momentum spectrum of our theory in the absence of interaction (compare with Figure **1**-6.2 (a)). The no-electron state (vacuum) has zero energy and momentum. One-electron states have energies $E = \sqrt{(m_e c^2)^2 + P^2 c^2}$, shown in the figure by the hyperbola "1". The energies of two-electron states depend not only on the total momentum of the system, but also on the relative momentum of its parts. This relative momentum can be arbitrarily large, so the energy is not bounded from above for each value of $\boldsymbol{P}$, and the two-electron states occupy an infinite (hatched) region in Figure 1.2 (a), bounded from below by the hyperbola "2" ($E = \sqrt{(2m_e c^2)^2 + P^2 c^2}$). Similarly, the energy momenta of three-electron states form a region (with double hatching) bounded from below by the hyperbola "3" ($E = \sqrt{(3m_e c^2)^2 + P^2 c^2}$), etc.

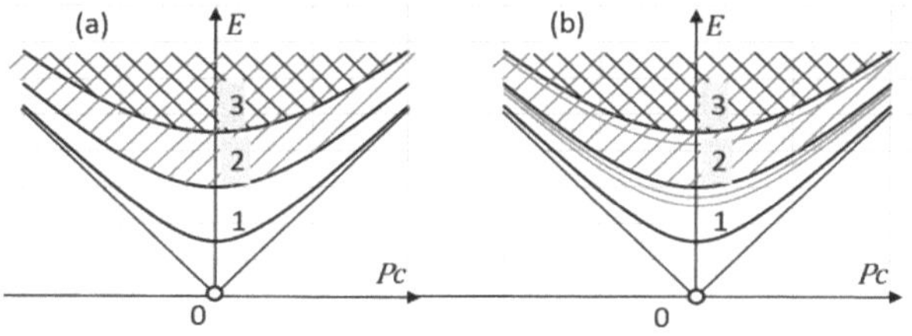

Figure 1.2: Typical energy–momentum spectrum of (a) noninteracting theory and (b) interacting dressed theory.

In Figure 1.2 (b) we showed how the energy–momentum spectrum changes in the presence of an effective *phys* interaction $V_{2\text{eff}}$. We know that this interaction has no effect on zero-electron and one-electron states, so that the corresponding energies "0" and "1" do not change. However, this interaction affects states with two or more electrons. In particular, if the attractive potential in $V_{2\text{eff}}$ is not too strong, then we can expect the splitting of bound-state hyperbolas under the main regions of n-electron spectra with $n = 2, 3, \ldots$.[34] In Section 3.2 we will illustrate the formation of such bound states in the case of the hydrogen atom.

Interactions in higher perturbation orders can be regarded as small corrections to the strong potential $V_{2\text{eff}}$. They can be taken into account by means of perturbation theory from Subsection **1**-6.5.2 and decay theory from Chapter 4. In particular, the

34 Compare with Figure **1**-6.2 (b). For the purposes of this discussion we assumed that electrons attract each other and multielectron bound states can be formed.

third-order potentials (1.33) are responsible for the instability of excited bound states. These levels have finite lifetimes: they experience spontaneous radiative transitions with the emission of photons, and their energies acquire some uncertainty (spread). We will have detailed discussions of these effects in Chapters 4 and 5.

From the above discussion we conclude that the dressed Hamiltonian H^d provides a physically satisfactory description of both scattering properties and energy–momentum spectra in multiparticle systems.

Let us now turn to the field Hamiltonian H^c. We already know that (by definition) the zero-particle and one-electron eigenvectors of H^c are $|\text{vac}\rangle^{\text{cl}}$ and $a^{\text{cl}\dagger}|\text{vac}\rangle^{\text{cl}}$. But what can we say about the corresponding eigenvalues? The answer is surprisingly simple if we recall the unitary relationship (1.18) between H^c and H^d. This relationship guarantees that both operators have the same spectrum.[35] In particular, for zero-particle, one-electron and one-photon states this means

$$\begin{aligned} H^c|\text{vac}\rangle^{\text{cl}} &= 0|\text{vac}\rangle^{\text{cl}}, \\ H^c(a_{\boldsymbol{p}}^{\text{cl}\dagger}|\text{vac}\rangle^{\text{cl}}) &= \sqrt{m_e^2c^4 + p^2c^2}(a_{\boldsymbol{p}}^{\text{cl}\dagger}|\text{vac}\rangle^{\text{cl}}), \\ H^c(c_{\boldsymbol{p}}^{\dagger}|\text{vac}\rangle^{\text{cl}}) &= \sqrt{\lambda^2c^4 + p^2c^2}(c_{\boldsymbol{p}}^{\dagger}|\text{vac}\rangle^{\text{cl}}). \end{aligned}$$

The spectra of clothed multiparticle states should also coincide with those shown in Figure 1.2 (b), but this conclusion is not at all obvious from the form of the field Hamiltonian, because direct diagonalization of the divergent H^c seems almost impossible [160, 161].

35 See Lemma **1**-G.6.

2 Dressing

This work contains many things which are new and interesting. Unfortunately, everything that is new is not interesting, and everything which is interesting, is not new.
Lev D. Landau

In this chapter we will construct a finite "dressed" QED Hamiltonian H^d, which, in addition to exact scattering amplitudes, provides a good description of both the time evolution and bound states. We call this approach *relativistic quantum dynamics* (RQD). We use here the word "dynamics," because, unlike traditional QFT, concerned with calculations of the time-independent S-matrix and energy spectra, RQD emphasizes the dynamical, i. e., time-dependent, nature of physical processes.

2.1 Dressing approach to QED

In Section 1.4, we saw that our toy field model can be reformulated in terms of effective *phys* interactions. As a result, the Hamiltonian of the theory acquires a more clear physical meaning, while all scattering characteristics do not change.

This approach is applicable also to the QED interaction, which basically has the same trilinear structure $V_1 = a^\dagger c^\dagger a + a^\dagger ac + \cdots$ as our toy potential (1.23). Our goal in this chapter is to apply the effective interaction idea to QED. We will try to construct an effective (dressed) Hamiltonian H^d that replaces the renormalized quantum field Hamiltonian H^c.

As in Section 1.4, here we will use the S-operator as a link between the Hamiltonians H^c and H^d. We know that the S-operator S^c, calculated in the framework of the renormalized QED, reproduces experimental results very accurately. We do not want to destroy this perfect agreement in any way. In other words, we require that our new Hamiltonian H^d leads exactly to the same S-matrix as H^c. As we noted in Chapter 1, QFT deals exclusively with S-matrix properties. So, by replacing the Hamiltonian we are not going to lose anything in terms of the correspondence theory $\leftrightarrow$ experiment. But if our new Hamiltonian H^d succeeds in avoiding the paradoxes and divergences characteristic of QED, then we will acquire a lot. This is the idea of our approach.

We assume that all involved operators can be expanded in perturbation series, and that all relevant series converge, i. e.,[1]

$$H^c = H_0 + V_1^c + V_2^c + \cdots, \tag{2.1}$$

$$S^c = 1 + S_2^c + S_3^c + S_4^c + \cdots, \tag{2.2}$$

$$H^d = H_0 + V^d = H_0 + V_2^d + V_3^d + \cdots. \tag{2.3}$$

1 In (2.3) we already take into account that $V_1^d = 0$; see Subsection 2.3.2.

https://doi.org/10.1515/9783110493221-002

As usual, the subscript denotes the degree of the coupling constant e (= the perturbation order).

2.1.1 No-self-interaction condition

In Chapter 1 we saw that the presence of *unphys* and *renorm* terms in the interaction operator V^c is the direct cause of many troubles in QFT, such as the need for renormalization and the lack of a well-defined time evolution. One simple possibility to get rid of these problems is to demand that the true interaction operator V^d contains only *phys* potentials, i. e.,[2]

$$V^d = \alpha^\dagger \alpha^\dagger \alpha\alpha + \alpha^\dagger \alpha^\dagger \alpha\alpha\alpha + \alpha^\dagger \alpha^\dagger \alpha^\dagger \alpha\alpha + \cdots. \tag{2.4}$$

In particular, this means that operator V^d yields zero when applied to zero-particle and one-particle states (see Theorem **2**-1.4), i. e.,

$$V^d|\text{vac}\rangle = 0,$$
$$V^d\alpha^\dagger|\text{vac}\rangle = 0.$$

According to Table **2**-1.2, commutators of *phys* potentials can be only *phys*. Thus, when we calculate the scattering phase operator Φ^d, by substituting *phys* V^d in the Magnus formula (**1**-7.19), only *phys* terms will arise in each perturbation order. Therefore, the operator Φ^d is also *phys* and we have

$$\Phi^d|\text{vac}\rangle = 0,$$
$$\Phi^d\alpha^\dagger|\text{vac}\rangle = 0.$$

This agrees with the renormalization condition **2**-4.1, which demands the absence of self-scattering. So, in our dressed theory, this condition would be satisfied automatically, and there would be no need for the mass renormalization. Moreover, interacting time evolutions of the vacuum and one-particle state vectors would coincide with their noninteracting counterparts, i. e.,

$$\begin{aligned}|\text{vac}(t)\rangle &= e^{-\frac{i}{\hbar}H^d t}|\text{vac}\rangle = \left(1 - \frac{it}{\hbar}(H_0 + V^d) + \cdots\right)|\text{vac}\rangle \\ &= \left(1 - \frac{it}{\hbar}H_0 + \cdots\right)|\text{vac}\rangle = e^{-\frac{i}{\hbar}H_0 t}|\text{vac}\rangle,\end{aligned} \tag{2.5}$$

2 Generally, an acceptable interaction can also include *decay* and *oscillation* potentials. However, as we have established in Subsection **2**-1.2.3, such interactions are absent in QED. Therefore, we do not consider them here.

$$|\alpha(t)\rangle = e^{-\frac{i}{\hbar}H^d t}\alpha^\dagger|\text{vac}\rangle = \left(1 - \frac{it}{\hbar}(H_0 + V^d) + \cdots\right)\alpha^\dagger|\text{vac}\rangle$$
$$= \left(1 - \frac{it}{\hbar}H_0 + \cdots\right)\alpha^\dagger|\text{vac}\rangle = e^{-\frac{i}{\hbar}H_0 t}|\alpha\rangle, \tag{2.6}$$

as one can expect from simple physical considerations. This means that our dressed interaction (2.4) also removes any *self-interaction* in these states. All these properties look very attractive to us, so we will base our search for the dressed interaction in RQD on the following postulate.

Postulate 2.1 (the absence of self-interaction). The vacuum and one-particle states should not experience self-interaction. In particular, time evolutions of these states should be free from the influence of interaction and described by the free Hamiltonian, as in (2.5)–(2.6). In other words, the potential energy V^d must be a *phys* operator.

2.1.2 Requirements for dressed interactions

Summarizing discussions from various parts of our book, we can now formulate an (incomplete) list of conditions that any realistic interaction operator V^d must satisfy.

Postulate 2.2 (properties of dressed interactions).
(A) Poincaré invariance (Statement **1**-3.2).
(B) Instant form of dynamics (Postulate **1**-6.3).
(C) Cluster separability (Postulate **1**-6.5 and Theorem **2**-2.5).
(D) The *phys* type of the operator V^d (Postulate 2.1).
(E) Finiteness: coefficient functions of potentials in V^d tend to finite values in the limit $\Lambda \to \infty$.
(F) Asymptotic decay: Coefficient functions of V^d rapidly tend to zero as their arguments (particle momenta) move away from the energy shell.[3]
(G) Scattering equivalence: The scattering operator S^d in our dressed theory remains exactly the same as the operator S^c of the renormalized QED. This condition also implies (see Subsection **1**-7.1.6.) that energies of bound states coincide in the two theories.

Our condition (D) practically excludes from consideration all usual field-theoretic Hamiltonians (see footnote 17 on page 7). In addition, the Hamiltonian H^c of the renormalized QED diverges in the ultraviolet limit $\Lambda \to \infty$, thus violating our condition (E). The main question is whether there are nontrivial interactions having all the required

3 According to Theorem **2**-2.4, this requirement guarantees the convergence of all loop integrals involving vertices V^d. We believe that true physical potentials must have this property.

properties (A)–(G)? Our answer to this question is *yes*. In the rest of this chapter we are going to justify this answer.

2.1.3 Two approaches to dressing

Our search for the dressed Hamiltonian H^d will be based on the scattering equivalence condition (G). We have two options to proceed. One possibility is to use the fact (see Subsection **1**-7.2.1) that scattering equivalence of the Hamiltonians H^d and H^c implies a unitary connection between them. This "unitary dressing transformation" approach will be considered in Section 2.3. In Section 2.2 we will explore an alternative way, namely, a simple fit of V^d to the known S-operator.

2.2 Fitting dressed Hamiltonian to S-operator

We take into account the fact that the scattering operator S^c of the renormalized QED is well studied even in high orders of perturbation theory. This information is obtained not only from precision experiments, but also from very accurate – and now already routine – QED calculations within the Feynman diagram technique. We can take advantage of this vast amount of knowledge accumulated over many decades of theoretical research and simply fit the desired components V_i^d of the dressed Hamiltonian (2.3) to the known operators S_i^c, as we did in our toy model in Section 1.4.

2.2.1 System of equations

On the one hand, traditional QED gives us the following Dyson series for the S-operator:

$$S^c = 1 + S_2^c + S_3^c + S_4^c + \cdots = 1 + \underbrace{\Sigma_2^c} + \underbrace{\Sigma_3^c} + \underbrace{\Sigma_4^c} + \cdots. \tag{2.7}$$

On the other hand, in the dressed theory with the Hamiltonian (2.3), the S-operator has the form[4]

$$S^d = 1 + \underbrace{V_2^d} + \underbrace{V_3^d} + \underbrace{V_4^d} + \underbrace{\underline{V_2^d V_2^d}} \ldots. \tag{2.8}$$

According to condition (G), these two operators must coincide in each order, i. e.,

$$S_i^d = S_i^c, \quad i = 2, 3, 4, \ldots. \tag{2.9}$$

4 We substituted the dressed interaction V^d from (2.3) into formulas (**1**-7.14) and (**1**-7.15).

Thus, we obtain the following system of relations between V_i^d and Σ_i^c on the energy shell[5]:

$$\underbrace{V_2^d} = \underbrace{\Sigma_2^c}, \tag{2.10}$$

$$\underbrace{V_3^d} = \underbrace{\Sigma_3^c}, \tag{2.11}$$

$$\underbrace{V_4^d} = \underbrace{\Sigma_4^c} - \underbrace{\underline{V_2^d V_2^d}}, \tag{2.12}$$

$$\underbrace{V_i^d} = \underbrace{\Sigma_i^c} + \underbrace{Y_i}, \quad i > 4, \tag{2.13}$$

where Y_i denotes sums of products of dressed interactions V_j^d from lower orders ($2 \leq j \leq i-2$).[6] Equations (2.10)–(2.13) are independent of the ultraviolet cutoff Λ, so they remain valid also in the limit $\Lambda \to \infty$. In this limit, operators Σ_i^c on the right-hand sides are finite, Λ-independent and assumed to be known on the energy shell from standard QED calculations. Our goal is to solve the system of equations (2.10)–(2.13), i. e., to find operators V_i^d that satisfy conditions (A)–(F).

2.2.2 Dressed potentials in second order

According to the no-self-scattering condition in Subsection **2**-4.1.1, the operator on the right-hand side of (2.10) is *phys*. Using equations (**2**-1.68) and (**2**-3.19), we obtain

$$\begin{aligned}\underbrace{(\Sigma_2^c)^{\text{phys}}} &= \underbrace{(\Sigma_2^n)^{\text{phys}}} = \underbrace{V_2} + \underbrace{(\underline{V_1 V_1})^{\text{phys}}} = \underbrace{V_2} - \underbrace{(\underline{V_1 V_1})^{\text{phys}}} \\ &= \underbrace{V_2} - \frac{1}{2}\underbrace{[\underline{V_1, V_1}]^{\text{phys}}} = \underbrace{(\Phi_2^n)^{\text{phys}}}.\end{aligned}$$

Then from (2.10) it follows that on the energy shell

$$V_2^d = (\Phi_2^n)^{\text{phys}}. \tag{2.14}$$

Recall that the part of $(\Phi_2^n)^{\text{phys}}$ responsible for the electron–proton scattering was already calculated in Subsections **2**-3.2.1 and **2**-3.2.2. So, from (2.10) we immediately obtain the corresponding dressed potential on the energy shell. However, we cannot say anything about the behavior of coefficient functions of V_2^d outside the energy shell. We can still satisfy condition (G) by choosing any off-shell behavior that we like. Of course, when making this choice we should also satisfy conditions (A), (B), (C), (E) and (F). But, as we will see in Section 2.3, this is not a difficult task. As a result, we can

5 In (**2**-1.66) we have shown that for regular operators the "underbrace" symbol means $\underbrace{V} = -2\pi i V \circ \delta(\mathcal{E}_V)$. Thus, equations (2.10)–(2.13) have nontrivial meaning only on energy shells where $\mathcal{E}_V = 0$.

6 In our book we do not go beyond the fourth perturbation order, so we will not need explicit expressions for Y_i, where $i > 4$.

get a wide selection of operators $\{V_2^d\}$, agreeing with all our conditions. The coefficient functions of all these operators coincide on their energy shells, but their behaviors outside energy shells can vary a lot.

This means that our conditions (A)–(G) are not sufficient for finding the unique V^d. Complete determination of V^d in the entire momentum space (on and off the energy shell) requires a combination of various hard-to-get pieces of information that go beyond the elementary two-particle S-matrix: experimental data from multiparticle and multichannel scattering, wave functions of bound states, the time evolution of interacting systems, etc. We will keep this ambiguity in mind when deriving dressed interactions V^d in subsequent sections.

Even without complete knowledge of off-shell coefficient functions, we can learn a lot about the operator structure of various terms in the potential energy V^d. Some examples of potentials in V_i^d $(i = 2, 3, 4, 5, 6, \ldots)$ are shown in Table 2.1. We divide them into two groups: *elastic* potentials and *inelastic* potentials.

Elastic potentials do not change the numbers of particles in the system and their types: in these potentials, annihilation and creation operators are balanced, i. e., created particles exactly match those annihilated. Therefore, elastic potentials correspond to interactions familiar to us from ordinary nonrelativistic quantum mechanics and classical physics.[7] In Chapter 3, we will be especially interested in the proton–electron elastic potential $V_2^d[d^\dagger a^\dagger da]$.

Inelastic potentials change numbers of particles and/or their types. In the second perturbation order there are three types of inelastic processes/potentials: particle–antiparticle *pair creation*, particle–antiparticle *pair annihilation* and two types of *pair conversions*.

Table 2.1: Examples of interaction operators in RQD.

Potential	Physical meaning	Perturbation orders
	Elastic potentials	
$a^\dagger a^\dagger aa$	$e^- + e^-$ potential	$2, 4, 6, \ldots$
$d^\dagger a^\dagger da$	$e^- + p^+$ potential	$2, 4, 6, \ldots$
$a^\dagger c^\dagger ac$	$e^- + \gamma$ potential (Compton scattering)	$2, 4, 6, \ldots$
$a^\dagger a^\dagger a^\dagger aaa$	$e^- + e^- + e^-$ 3-electron potential	$4, 6, \ldots$
	Inelastic potentials	
$a^\dagger b^\dagger cc$	$e^- + e^+$ pair creation	$2, 4, 6, \ldots$
$c^\dagger c^\dagger ab$	$e^- + e^+$ annihilation	$2, 4, 6, \ldots$
$d^\dagger f^\dagger ab$	conversion of $e^- + e^+$ into $p^- + p^+$	$2, 4, 6, \ldots$
$a^\dagger b^\dagger df$	conversion of $p^- + p^+$ into $e^- + e^+$	$2, 4, 6, \ldots$
$d^\dagger a^\dagger c^\dagger da$	$e^- + p^+$ bremsstrahlung	$3, 5, \ldots$
$d^\dagger a^\dagger dac$	absorption of photon in $e^- + p^+$ scattering	$3, 5, \ldots$
$a^\dagger a^\dagger a^\dagger b^\dagger aa$	pair creation in $e^- + e^-$ scattering	$4, 6, \ldots$

7 See examples in Subsections **2**-1.2.7 and **2**-1.2.8.

2.2.3 Dressed potentials in third order

Similar to the second order discussed above, third-order interactions V_3^d near the energy shell can be found from (2.11). We have

$$V_3^d \approx (\Sigma_3^c)^{\text{phys}} = (\Sigma_3^n)^{\text{phys}}. \tag{2.15}$$

All third-order potentials are inelastic. Two of them are shown in Table 2.1. The bremsstrahlung potential $d^\dagger a^\dagger c^\dagger da$ describes the production of a photon in an electron–proton collision.[8] The Hermitian conjugate term $d^\dagger a^\dagger dac$ describes the inverse effect – the absorption of a photon by a colliding pair of charged particles.

Just as in the second order, all third-order dressed potentials remain largely undetermined outside their energy shells.

2.2.4 Dressed potentials in fourth and higher orders

Let us now turn to the fourth-order operator V_4^d in (2.12). According to the renormalized QED, the $\underbrace{\Sigma_4^c}$ term is well known on the energy shell, but what can we say about the other term, $-\underbrace{V_2^d \underline{V_2^d}}$?

As we explained in Section **2**-2.3, diagrams for the product $V_2^d \underline{V_2^d}$ should be constructed from the vertices V_2^d and $\underline{V_2^d}$ by contracting some external lines of the factors and converting them into internal lines and loops. As follows from Table 2.1, all second-order vertices in V_2^d have two incoming and two outgoing external lines. For us, only one vertex $V_2^d[d^\dagger a^\dagger da]$ is of interest. It is shown in Figure 2.1 (a). The diagram for the product $V_2^d \underline{V_2^d}$ (see Figure 2.1 (b)) has a loop with unlimited integration momentum. Therefore, this product depends on the behavior of the coefficient function of the potential $V_2^d[d^\dagger a^\dagger da]$ in the entire momentum space – on and off the energy shell. Now we can appreciate the meaning of condition (F) in Postulate 2.2. If the coef-

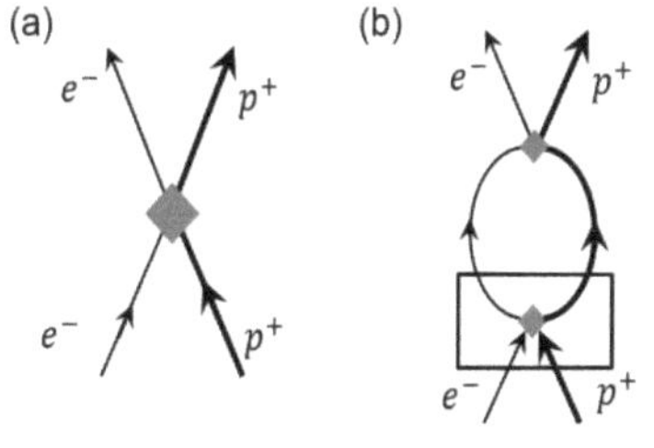

Figure 2.1: Examples of diagrams in the dressed theory. (a) Interaction vertex $V_2^d[d^\dagger a^\dagger da]$. (b) A diagram in the product $V_2^d \underline{V_2^d}$ drawn by the rules from Section **2**-2.3.

8 In the language of classical electrodynamics, this can be interpreted as radiation caused by acceleration of charged particles. A more detailed analysis of this effect is postponed to Section 5.1.

ficient functions of $V_2^d[d^\dagger a^\dagger da]$ did not have the rapidly decreasing asymptotics, then by Theorem **2**-2.4 the loop integral in $\underline{V_2^d V_2^d}$ could diverge and we would not be able to define a finite potential V_4^d.

So, due to the (supposed) convergence of the loop integrals in $\underline{V_2^d V_2^d}$,[9] we conclude that both terms on the right-hand side of (2.12) are known and finite on the energy shell. Then, the dressed potential V_4^d is well-defined and finite on the energy shell. We can repeat these arguments for solving equations (2.13) and finding V_i^d in higher orders i of perturbation theory, thus completing our construction of the dressed Hamiltonian $H^d = H_0 + V_2^d + V_3^d + V_4^d + \cdots$, which is free from ultraviolet divergences.

2.2.5 What are advantages of dressed Hamiltonian?

Summing up our qualitative reasoning in this section, it is time to mention several important differences between the initial QED interaction V^c and the dressed potential V^d. Due to the renormalizability property (see Subsection **2**-4.3.5), the complete interaction V^c (in all orders of perturbation theory) consisted of only several terms and could be represented in a closed analytical form (**2**-4.44). On the other hand, the dressed interaction V^d is not representable in a closed form. The higher the perturbation order i, the more different, increasingly complex types of interactions are contained in V_i^d. However, to the credit of RQD, all these operators are finite, do not depend on the ultraviolet cutoff parameter Λ and directly correspond to actual interactions and processes in nature.

Now, when we want to calculate scattering amplitudes, we can directly substitute our newly found potential V^d in the perturbation theory series (2.8). Of course, in these calculations we will encounter loop integrals, but they are guaranteed to converge because of the rapidly damped asymptotic behavior of the coefficient functions of the potentials V_i^d (= condition (F)). Hence, the dressed theory contains ultraviolet divergences neither in the Hamiltonian nor in the S-operator.

2.3 Unitary dressing transformation in QED

In the preceding section, we defined the dressed interaction by a straightforward fit to the known S-operator. We required that the Hamiltonians H^d and H^c have the same spectra and identical scattering matrices. Hence it is natural to conclude that these Hamiltonians are related to each other by a unitary transformation. This is the central idea of the "unitary dressing transformation" method, which we consider in the present section.

9 The diagram shown in Figure 2.1 will be evaluated approximately in Subsection 5.2.1.

2.3.1 System of equations

So, we guess that the Hamiltonians H^d and H^c are related by a *unitary dressing transformation,*

$$\begin{aligned} H^d &\equiv H_0 + V^d = e^{i\Xi} H^c e^{-i\Xi} &\quad (2.16)\\ &\equiv e^{i\Xi}(H_0 + V^c)e^{-i\Xi} \\ &= (H_0 + V^c) + i[\Xi, (H_0 + V^c)] - \frac{1}{2!}[\Xi, [\Xi, (H_0 + V^c)]] + \cdots, &\quad (2.17)\end{aligned}$$

whose generator Ξ is assumed to be Hermitian. Let us rewrite equation (2.17) as a perturbation series, with the help of expansions (2.1), (2.3) and

$$\Xi = \Xi_1 + \Xi_2 + \cdots. \tag{2.18}$$

Collecting terms of equal orders, we obtain a linked chain of equations for determination of the operators Ξ_i, i. e.,[10]

$$V_1^d = V_1^c + i[\Xi_1, H_0], \tag{2.19}$$

$$V_2^d = V_2^c + i[\Xi_2, H_0] + i[\Xi_1, V_1^c] - \frac{1}{2!}[\Xi_1, [\Xi_1, H_0]], \tag{2.20}$$

$$\begin{aligned} V_3^d &= V_3^c + i[\Xi_3, H_0] + i[\Xi_2, V_1^c] + i[\Xi_1, V_2^c], \\ &\quad - \frac{1}{2!}[\Xi_2, [\Xi_1, H_0]] - \frac{1}{2!}[\Xi_1, [\Xi_2, H_0]] - \frac{1}{2!}[\Xi_1, [\Xi_1, V_1^c]] \\ &\quad - \frac{i}{3!}[\Xi_1, [\Xi_1, [\Xi_1, H_0]]] \end{aligned} \tag{2.21}$$

$$\ldots.$$

Next, following the plan outlined in Subsection **2**-4.1.5, we introduce the infrared (λ) and ultraviolet (Λ) cutoffs, which will guarantee that the counterterms in V_i^c are not singular and finite[11] in all perturbation orders. Moreover, these cutoffs will ensure the convergence of all loop integrals in products and commutators of the potentials V_i^c. Of course, in order to get a self-consistent and accurate theory, we should get rid of the artificial cutoffs λ and Λ at the end of our calculations. Physically meaningful quantities should tend to finite values in these limits. We will analyze the ultraviolet limit $\Lambda \to \infty$ in Subsection 2.3.5. The infrared limit $\lambda \to 0$ will be handled in Chapter 5.

10 For the operator V^c we have in mind the QED interaction (**2**-3.14) + (**2**-3.15) in the Coulomb gauge plus required renormalization counterterms: $V^c = V_1 + V_2 + Q$. Strictly speaking, we have derived counterterms Q only in the Feynman gauge (see Section **2**-4.2). However, this gap is not going to stop us, because our calculations in this section will have a formal character, and the explicit form of Q will not be needed. Our choice of the Coulomb gauge is explained by the desire to have a simple proof of the relativistic invariance in Subsection 2.3.6.

11 Though they depend on the parameters λ and Λ.

Now we should solve equations (2.19)–(2.21) order by order, i. e., choose operators $\Xi_i = \Xi_i^{\text{phys}} + \Xi_i^{\text{unp}}$ so that interaction terms V_i^d on the left-hand sides satisfy conditions (C)–(G) from Postulate 2.2.[12]

2.3.2 Unitary dressing in first order

Let us start from the first-order equation (2.19). In the renormalized QED, the first-order interaction operator $V_1^c = V_1$ is *unphys* (**2**-D.9). According to condition (D) of Postulate 2.2, this term on the right-hand side of (2.19) should be completely canceled. This can be achieved if we choose[13]

$$\begin{aligned}\Xi_1^{\text{unp}} &= i\underline{V_1},\\ \Xi_1^{\text{phys}} &= 0.\end{aligned} \tag{2.22}$$

With this choice, not just the *unphys* part of V_1^d turns to zero, but the entire first-order potential vanishes, i. e., $V_1^d = 0$, so that conditions (C)–(F) are trivially satisfied.[14]

The coefficient function of V_1 is nonsingular, and this *unphys* operator does not have an energy shell. By Theorem B.2 this means that the operator Ξ_1^{unp} in (2.22) is smooth, and by Theorem B.1 the presence of this term in the dressing transformation $e^{i\Xi}$ has no effect on the S-operator, which agrees with our condition (G). So, we have managed to fulfill all requirements for the dressing in the first perturbation order.

2.3.3 Unitary dressing in second order

Now we can substitute the above operator Ξ_1 into equation (2.20) and obtain an expression for the second-order dressed potential

$$\begin{aligned}V_2^d &= V_2^c + i[\Xi_2, H_0] - [\underline{V_1}, V_1] + \frac{1}{2!}[\underline{V_1}, V_1]\\ &= V_2^c + i[\Xi_2, H_0] - \frac{1}{2}[\underline{V_1}, V_1].\end{aligned} \tag{2.23}$$

---

12 Note that in our definition we set $\Xi_i^{\text{ren}} = 0$. This is necessary, because according to Theorem B.1 any admixture of *renorm* terms in Ξ would destroy the desired scattering equivalence of H^d and H^c (= condition (G)). The relativistic invariance conditions (A) and (B) will be considered separately in Subsection 2.3.6.

13 In (2.22) we used formula (**2**-1.81) to solve the commutator equation $i[\Xi_1^{\text{unp}}, H_0] = -V_1$.

14 Generally speaking, for Ξ_1^{phys} we could choose any smooth *phys* operator that vanishes on its energy shell. Then we would have a *phys* interaction $V_1^d = i[\Xi_1^{\text{phys}}, H_0]$ not contributing to the S-operator (see Theorem B.1). This freedom of choice is directly related to the uncertainty of V^d, explained in Subsections 2.2.2–2.2.4. Nevertheless, we decided to choose $\Xi_1^{\text{phys}} = 0$ and $V_1^d = 0$, because we presume that dressed interactions should be at least quadratic in the coupling constant.

It will be convenient to write separately the *unphys*, *phys* and *renorm* parts of this equation, i. e.,

$$(V_2^d)^{\text{unp}} = (V_2^c)^{\text{unp}} + i[\Xi_2^{\text{unp}}, H_0] - \frac{1}{2}[\underline{V_1}, V_1]^{\text{unp}}, \tag{2.24}$$

$$(V_2^d)^{\text{phys}} = (V_2^c)^{\text{phys}} + i[\Xi_2^{\text{phys}}, H_0] - \frac{1}{2}[\underline{V_1}, V_1]^{\text{phys}}, \tag{2.25}$$

$$(V_2^d)^{\text{ren}} = (V_2^c)^{\text{ren}} - \frac{1}{2}[\underline{V_1}, V_1]^{\text{ren}}. \tag{2.26}$$

All components on the right-hand sides of (2.24)–(2.26) (except terms containing Ξ_2^{unp} and Ξ_2^{phys}) are already known to us: The operator $(V_2^c)^{\text{ren}}$ comes from electron and photon self-energy counterterms described in Section **2**-4.2. The operator $(V_2^c)^{\text{unp}}$ has contributions from V_2^{unp} in (**2**-D.13). The calculation of the commutator $[\underline{V_1}, V_1]^{\text{phys}}$ was explained in Subsection **2**-3.2.1. Commutators $[\underline{V_1}, V_1]^{\text{unp}}$ and $[\underline{V_1}, V_1]^{\text{ren}}$ can be obtained in a similar way.

Now, our goal is to choose operators Ξ_2^{phys} and Ξ_2^{unp}, so that dressed potentials on the left-hand sides of (2.24)–(2.26) satisfy our conditions (C)–(G).

From the condition (D) it follows that the operator $(V_2^d)^{\text{unp}}$ should turn to zero. To achieve that, we choose

$$\Xi_2^{\text{unp}} = \underline{i(V_2^c)^{\text{unp}}} - \underline{\frac{i}{2}[\underline{V_1}, V_1]^{\text{unp}}}. \tag{2.27}$$

The operators V_1 and $(V_2^c)^{\text{unp}}$ are smooth. Hence, by Theorems B.2 and **2**-2.7, the operators $\underline{V_1}$, $[\underline{V_1}, V_1]^{\text{unp}}$, $\underline{(V_2^c)^{\text{unp}}}$ and Ξ_2^{unp} are also smooth, and by Theorem B.1 the presence of the smooth Ξ_2^{unp} in the transformation $e^{i\Xi}$ does not change the S-operator. This is exactly what we need to satisfy condition (G).

The simplest solution for the *phys* part of the transformation would be to set

$$\Xi_2^{\text{phys}} = 0 \tag{2.28}$$

in the second term on the right-hand side of (2.25). Two other terms there are known smooth operators.[15] Then, due to Theorem **2**-2.5, we would decide that the second-order dressed interaction

$$(V_2^d)^{\text{phys}} = (V_2^c)^{\text{phys}} - \frac{1}{2}[\underline{V_1}, V_1]^{\text{phys}} \tag{2.29}$$

is cluster-separable and satisfies condition (C). Indeed, the choice (2.28)–(2.29) is made in many works on clothing/dressing in QFT [226, 214].

15 In QED $(V_2^c)^{\text{phys}}$ is equal to the operator V_2^{phys} from formula (**2**-D.12), because there are no *phys* counterterms in the second order.

In our work we will also make the choice (2.28). However, we would like to note that this choice is not fully justified, because it (i) is arbitrary and (ii) does not guarantee the fulfillment of (F).[16] In a more general approach, we could choose for Ξ_2^{phys} any smooth *phys* operator which vanishes on its energy shell. This would not affect the energy shell behavior of $(V_2^d)^{phys}$, still preserve the S-operator and give the necessary flexibility for fulfilling condition (F).[17] This flexibility of Ξ_2^{phys} and $(V_2^d)^{phys}$ outside the energy shell is fully consistent with the general off-shell ambiguity of V^d described in Subsections 2.2.2–2.2.4.

So, our choice of the operators Ξ_2^{unp} and Ξ_2^{phys} guarantees that the S-operator does not change in the second perturbation order $S_2^d = S_2^c$. In particular, this means that $\Phi_2^d = \Phi_2^c$ and, according to (**2**-4.5),

$$(V_2^d)^{ren} = (\Phi_2^d)^{ren} = 0.$$

This implies that two terms on the right-hand side of (2.26) must cancel each other. In other words, formula (**2**-2.27) for the mass renormalization counterterm $(V_2^c)^{ren} = \frac{1}{2}[\underline{V_1}, V_1]^{ren} = -(V_1\underline{V_1})^{ren}$ can be derived as the self-consistency condition for the dressing procedure [150].

2.3.4 Unitary dressing in higher orders

Thus, we have succeeded in proving conditions (C)–(G) up to the second perturbation order. For higher orders, $i > 2$, the logic of choosing operators Ξ_i and the proofs of conditions (C)–(G) are basically the same as in the preceding subsection. The defining equation for V_i^d can be written in the general form (compare with (2.23))

$$V_i^d = V_i^c + i[\Xi_i, H_0] + \Theta_i, \tag{2.30}$$

where Θ_i is a sum of multiple commutators and t-integrals involving potentials V_j^c from lower orders ($1 \le j < i$). This equation is solved by the following operator[18]:

$$\Xi_i = i\Xi_i^{unp} + \Xi_i^{phys},$$

16 That is, from (2.29) it does not follow that coefficient functions of $(V_2^d)^{phys}$ decay sufficiently rapidly away from the energy shell.

17 In other words, we can always adjust the off-shell behavior of Ξ_2^{phys} in such a way that coefficient functions of $(V_2^d)^{phys}$ rapidly tend to zero away from the energy shell. This will guarantee the convergence of all loop integrals involving vertices $(V_2^d)^{phys}$.

18 Compare with (2.27)–(2.28). As we explained in Subsection 2.3.3, there is a considerable freedom of choosing Ξ_i^{phys} outside the energy shell. This freedom can be used to fulfill condition (F). Therefore, we do not regard (2.31) as a rigorous equality. We also assume that (just as in the second order) all *renorm* terms on the right-hand side of (2.30) cancel automatically.

$$\Xi_i^{\text{unp}} = \underline{i\Theta_i^{\text{unp}}} + \underline{i(V_i^c)^{\text{unp}}},$$
$$\Xi_i^{\text{phys}} = 0. \tag{2.31}$$

Just as in the second order, operators Ξ_i are smooth, so that conditions (C) and (G) are valid and $(V_i^d)^{\text{ren}} = (V_i^d)^{\text{unp}} = 0$.

Solving equations (2.30) order by order, we finally obtain the complete dressed Hamiltonian,

$$H^d = e^{i\Xi} H^c e^{-i\Xi} = H_0 + V_2^d + V_3^d + V_4^d + \cdots, \tag{2.32}$$

which has all desired properties (C)–(G), as promised.

2.3.5 Limit of infinite ultraviolet cutoff

Up to now, in our calculations we used a finite cutoff parameter Λ. This allowed us to avoid ultraviolet divergences in all loop integrals. In the final theory, of course, we must get to the limit $\Lambda \to \infty$. Our approach will only make sense if we can verify that all physically meaningful dressed operators remain finite and independent of Λ in this limit.

Obviously, conditions (C), (D) and (F) do not depend on the cutoff Λ. Therefore, they remain valid in the limit $\Lambda \to \infty$. However, how can we guarantee that in this limit condition (E) is fulfilled as well, i. e., that coefficient functions of our dressed potentials V_i^d do not diverge? The simplest way to verify this is by referring to the system of equations (2.10)–(2.13), which follow from the proved preservation of the S-operator (condition (G)).

The S-operator components Σ_i^c are Λ-independent.[19] This immediately implies the Λ-independence of the potentials V_2^d and V_3^d. In the case of the potential V_4^d, we need to establish additionally the convergence of the product $V_2^d \underline{V_2^d}$. This fact follows from the rapidly damped asymptotics of V_2^d outside the energy shell (condition (F)) and Theorem **2**-2.4, which guarantees the convergence of loop integrals. Similarly, we can prove the finiteness of V_i^d in higher orders.

So, in the limit $\Lambda \to \infty$, we obtained a finite and Λ-independent dressed Hamiltonian H^d (2.32). Ultraviolet divergences inherent in H^c have migrated into the generator Ξ of the unitary dressing transformation. However, this generator does not correspond to any observable quantity, so there is nothing shameful in the fact that Ξ does not have a well-defined limit at $\Lambda \to \infty$.

As we have already mentioned, the unitary dressing described here is equivalent to the straightforward fitting of H^d to S^c explained in Section 2.2. Both methods have

19 This property is the main achievement of the traditional renormalization procedure in QED.

their advantages and disadvantages. Let us first point out the main drawback of the unitary dressing method. In accordance with condition (D), the main task of the transformation (2.17) is the extraction of *phys* components of the operator V^d. So, we have to solve individual equations for *unphys* and *phys* parts in each perturbation order. This also means that the original QED interaction V^c cannot be left in the convenient form of a polynomial in field operators (**2**-4.44), where *unphys*, *phys* and *renorms* are mixed together. For unitary dressing purposes interactions should be represented as polynomials in creation and annihilation operators, as we did in Appendix **2**-D. In high orders of perturbation theory, the number of different terms and their commutators grows very rapidly, so that calculations become extremely cumbersome. On the other hand, the S-matrix components required in the fitting approach of Section 2.2 could be evaluated in the compact field notation with the help of convenient Feynman diagrams. Thus, the unitary transformation method loses to the fitting method in terms of the simplicity of computations.

2.3.6 Poincaré invariance of dressed theory

The important advantage of the unitary transformation method is that it permits a simple proof of the relativistic invariance (= conditions (A) and (B)). For this proof we have to establish that there is a dressed operator $\boldsymbol{K}^d$, such that the 10 operators $\{\boldsymbol{P}_0, \boldsymbol{J}_0, \boldsymbol{K}^d, H^d\}$ satisfy commutators of the Poincaré Lie algebra. Having at our disposal the dressing operator $\exp(i\Xi)$ constructed above and the boost operator $\boldsymbol{K}^c$ of the original renormalized QED,[20] this problem has a simple solution [213]. If we choose the dressing transformation $\exp(i\Xi)$ commuting with $\boldsymbol{P}_0$, $\boldsymbol{J}_0$ and define $\boldsymbol{K}^d \equiv e^{i\Xi}\boldsymbol{K}^c e^{-i\Xi}$, then the complete set of dressed Poincaré generators is obtained as follows:

$$\{\boldsymbol{P}_0, \boldsymbol{J}_0, \boldsymbol{K}^d, H^d\} = e^{i\Xi}\{\boldsymbol{P}_0, \boldsymbol{J}_0, \boldsymbol{K}^c, H^c\}e^{-i\Xi}.$$

The unitarity of $\exp(i\Xi)$ guarantees the preservation of commutation relations (condition (A)), while commutators $[\Xi, \boldsymbol{P}_0] = [\Xi, \boldsymbol{J}_0] = 0$ ensure the instant form of dynamics (condition (B)).

20 Recall that in Subsection **2**-4.2.8 we decided not to construct this operator explicitly, although we assumed that such an operator exists and ensures the relativistic invariance of the renormalized theory $\{\boldsymbol{P}_0, \boldsymbol{J}_0, \boldsymbol{K}^c, H^c\}$.

3 Coulomb potential and beyond

I've got all the medals I need. What could I lose? So that's why I'm prepared to take a gamble that a young researcher wouldn't be prepared to take.
Roger Penrose

In Chapter 2 we derived general formulas of the dressed theory. Our next goal is to demonstrate that this method is really useful for solving specific problems. We will do this in the next few chapters according to the following plan. In this chapter we will derive the dressed electron–proton interaction in the second perturbation order. We are going to get the famous Coulomb–Darwin–Breit potential, which yields a rather accurate energy spectrum of the hydrogen atom. The next step is to extend the dressed theory to higher orders of perturbation theory. In particular, in the third order, we expect to see the instability of excited atomic levels. To analyze this effect, we will discuss quantum mechanics of unstable states in Chapter 4. In Chapter 5, we will apply this theory, together with fourth-order radiative corrections, to derive two classical QED results – the anomalous magnetic moment of the electron and the Lamb shift in the hydrogen atom.

3.1 Coulomb–Darwin–Breit Hamiltonian

Here we want to demonstrate how the dressed second-order interaction (2.14) can be written in a form suitable for concrete calculations, i. e., expressed through creation and annihilation operators of particles. We will see that in the $(v/c)^2$ approximation, the pair interaction between protons and electrons has the form of the *Coulomb–Darwin–Breit* potential.[1] Its main part is the usual Coulomb interaction. In addition, there are relativistic corrections responsible for the magnetic, contact, spin–orbit, spin–spin and other interactions, well known in calculations of atomic and molecular systems.

3.1.1 Electron–proton potential in momentum representation

Here we are going to apply the general dressing theory developed in Chapter 2 to the specific case of interaction between two electric charges – an electron and a proton. For this we use equation (2.14), i. e.,

$$V_2^d = (\Phi_2^n)^{\text{phys}} = \left(V_2 - \frac{1}{2}[\underline{V_1}, V_1]\right)^{\text{phys}},$$

1 Our calculations of the electron–proton potential in this section can be compared with § 83 in [14] and with [115].

https://doi.org/10.1515/9783110493221-003

and extract from this formula only terms having the operator structure $V_2^d[d^\dagger a^\dagger da]$. The corresponding expression has been calculated already in (**2**-3.24)–(**2**-3.25).

The operator $V_2^d[d^\dagger a^\dagger da]$ acts nontrivially in all sectors of the Fock space containing at least one electron and one proton. Here, for simplicity, we will restrict our attention to the subspace "1 proton + 1 electron," which we denote by the symbol $\mathscr{H}_{pe} \equiv \mathscr{H}(1,0,1,0,0)$. If $\Psi(\boldsymbol{p}\tau;\boldsymbol{q}\sigma)$ is the wave function of a two-particle state in $\mathscr{H}_{pe}$, then the potential energy operator $V_2^d[d^\dagger a^\dagger da]$ transforms $\Psi(\boldsymbol{p}\tau;\boldsymbol{q}\sigma)$ by formula (**2**-1.84), i. e.,

$$\begin{aligned}\Psi'(\boldsymbol{p}\tau,\boldsymbol{q}\sigma) &= V_2^d[d^\dagger a^\dagger da]\Psi(\boldsymbol{p}\tau,\boldsymbol{q}\sigma)\\ &= \sum_{\tau'\sigma'}\int d\boldsymbol{k}v_2[\boldsymbol{p}'\tau',\boldsymbol{q}'\sigma';(\boldsymbol{p}-\boldsymbol{k})\tau,(\boldsymbol{q}+\boldsymbol{k})\sigma]\,\Psi((\boldsymbol{p}-\boldsymbol{k})\tau',(\boldsymbol{q}+\boldsymbol{k})\sigma'),\end{aligned} \tag{3.1}$$

where v_2 is the coefficient function of $V_2^d[d^\dagger a^\dagger da] \equiv (\Phi_2^n)^{\text{phys}}[d^\dagger a^\dagger da]$. This function has been calculated in (**2**-G.2) already; we have[2]

$$\begin{aligned}v_2[\boldsymbol{p}'\tau',\boldsymbol{q}'\sigma';\boldsymbol{p}\tau,\boldsymbol{q}\sigma] &= \phi_2(\boldsymbol{p}'\tau',\boldsymbol{q}'\sigma';\boldsymbol{p}\tau,\boldsymbol{q}\sigma)\\ &= \frac{e^2\hbar^2}{(2\pi\hbar)^3}\chi_{\sigma'}^{(el)\dagger}\chi_{\tau'}^{(pr)\dagger}\bigg(-\frac{1}{k^2}+\frac{1}{8m_p^2c^2}+\frac{1}{8m_e^2c^2}+\frac{\boldsymbol{p}\cdot\boldsymbol{q}}{m_pm_ec^2k^2}\\ &\quad-\frac{(\boldsymbol{p}\cdot\boldsymbol{k})(\boldsymbol{q}\cdot\boldsymbol{k})}{m_pm_ec^2k^4}+\frac{i\boldsymbol{\sigma}_{pr}\cdot[\boldsymbol{k}\times\boldsymbol{p}]}{4m_p^2c^2k^2}-\frac{i\boldsymbol{\sigma}_{el}\cdot[\boldsymbol{k}\times\boldsymbol{q}]}{4m_e^2c^2k^2}-\frac{i\boldsymbol{\sigma}_{pr}\cdot[\boldsymbol{k}\times\boldsymbol{q}]}{2m_pm_ec^2k^2}\\ &\quad+\frac{i\boldsymbol{\sigma}_{el}\cdot[\boldsymbol{k}\times\boldsymbol{p}]}{2m_pm_ec^2k^2}+\frac{(\boldsymbol{\sigma}_{pr}\cdot\boldsymbol{\sigma}_{el})}{4m_pm_ec^2}-\frac{(\boldsymbol{\sigma}_{pr}\cdot\boldsymbol{k})(\boldsymbol{\sigma}_{el}\cdot\boldsymbol{k})}{4m_pm_ec^2k^2}\bigg)\chi_\sigma^{(el)}\chi_\tau^{(pr)}.\end{aligned} \tag{3.2}$$

3.1.2 Electron–proton potential in position representation

The physical meaning of the interaction (3.1)–(3.2) becomes more transparent in the position representation, which is obtained by (i) replacing the variables $\boldsymbol{p}$ and $\boldsymbol{q}$ by differentiation operators $\hat{\boldsymbol{p}} = -i\hbar(d/d\boldsymbol{x})$, $\hat{\boldsymbol{q}} = -i\hbar(d/d\boldsymbol{y})$ and (ii) performing the Fourier transform[3]

$$\begin{aligned}&\hat{V}_2^d[d^\dagger a^\dagger da]\Psi(\boldsymbol{x}\pi,\boldsymbol{y}\epsilon)\\ &= \sum_{\tau',\sigma'}\int d\boldsymbol{k}e^{\frac{i}{\hbar}\boldsymbol{k}\cdot(\boldsymbol{y}-\boldsymbol{x})}v_2[\hat{\boldsymbol{p}}'\tau',\hat{\boldsymbol{q}}'\sigma';(\hat{\boldsymbol{p}}-\boldsymbol{k})\tau,(\hat{\boldsymbol{q}}+\boldsymbol{k})\sigma]\Psi(\boldsymbol{x}\tau',\boldsymbol{y}\sigma')\\ &= \frac{e^2\hbar^2}{(2\pi\hbar)^3}\int d\boldsymbol{k}e^{\frac{i}{\hbar}\boldsymbol{k}\cdot(\boldsymbol{y}-\boldsymbol{x})}\bigg(-\frac{1}{k^2}+\frac{1}{8m_p^2c^2}+\frac{1}{8m_e^2c^2}+\frac{\hat{\boldsymbol{p}}\cdot\hat{\boldsymbol{q}}}{m_pm_ec^2k^2}\\ &\quad-\frac{(\hat{\boldsymbol{p}}\cdot\boldsymbol{k})(\hat{\boldsymbol{q}}\cdot\boldsymbol{k})}{m_pm_ec^2k^4}+\frac{i\boldsymbol{\sigma}_{pr}\cdot[\boldsymbol{k}\times\hat{\boldsymbol{p}}]}{4m_p^2c^2k^2}-\frac{i\boldsymbol{\sigma}_{el}\cdot[\boldsymbol{k}\times\hat{\boldsymbol{q}}]}{4m_e^2c^2k^2}-\frac{i\boldsymbol{\sigma}_{pr}\cdot[\boldsymbol{k}\times\hat{\boldsymbol{q}}]}{2m_pm_ec^2k^2}\end{aligned}$$

2 The transferred momentum is $\boldsymbol{k} \equiv \boldsymbol{q}' - \boldsymbol{q} = \boldsymbol{p} - \boldsymbol{p}'$.

3 See Subsection **2**-1.2.8. Here $\boldsymbol{x}$ is the proton position, $\boldsymbol{y}$ is the electron position and $\boldsymbol{r} \equiv \boldsymbol{y} - \boldsymbol{x}$.

$$+\frac{i\boldsymbol{\sigma}_{el}\cdot[\boldsymbol{k}\times\hat{\boldsymbol{p}}]}{2m_pm_ec^2k^2}+\frac{(\boldsymbol{\sigma}_{pr}\cdot\boldsymbol{\sigma}_{el})}{4m_pm_ec^2}-\frac{(\boldsymbol{\sigma}_{pr}\cdot\boldsymbol{k})(\boldsymbol{\sigma}_{el}\cdot\boldsymbol{k})}{4m_pm_ec^2k^2}\bigg)\Psi(\boldsymbol{x}\tau;\boldsymbol{y}\sigma).$$

Using formulas (**2**-1.89), (**1**-A.1) and (**2**-A.3)–(**2**-A.6), we obtain the following position-space potential[4]:

$$\begin{aligned}\hat{V}_2^d[d^\dagger a^\dagger da] &= -\frac{e^2}{4\pi r}+\frac{e^2\hbar^2}{8c^2}\left(\frac{1}{m_p^2}+\frac{1}{m_e^2}\right)\delta(\boldsymbol{r})\\ &+\frac{e^2}{8\pi m_pm_ec^2r}\left[\hat{\boldsymbol{p}}\cdot\hat{\boldsymbol{q}}+\frac{(\boldsymbol{r}\cdot\hat{\boldsymbol{p}})(\boldsymbol{r}\cdot\hat{\boldsymbol{q}})}{r^2}\right]-\frac{e^2[\boldsymbol{r}\times\hat{\boldsymbol{p}}]\cdot\boldsymbol{s}_{pr}}{8\pi m_p^2c^2r^3}\\ &+\frac{e^2[\boldsymbol{r}\times\hat{\boldsymbol{q}}]\cdot\boldsymbol{s}_{el}}{8\pi m_e^2c^2r^3}+\frac{e^2[\boldsymbol{r}\times\hat{\boldsymbol{q}}]\cdot\boldsymbol{s}_{pr}}{4\pi m_pm_ec^2r^3}-\frac{e^2[\boldsymbol{r}\times\hat{\boldsymbol{p}}]\cdot\boldsymbol{s}_{el}}{4\pi m_pm_ec^2r^3}\\ &+\frac{e^2}{m_pm_ec^2}\left(-\frac{\boldsymbol{s}_{pr}\cdot\boldsymbol{s}_{el}}{4\pi r^3}+3\frac{(\boldsymbol{s}_{pr}\cdot\boldsymbol{r})(\boldsymbol{s}_{el}\cdot\boldsymbol{r})}{4\pi r^5}+\frac{2}{3}(\boldsymbol{s}_{pr}\cdot\boldsymbol{s}_{el})\delta(\boldsymbol{r})\right).\end{aligned}\tag{3.3}$$

With the $(v/c)^2$ accuracy, the free Hamiltonian H_0 has the form

$$\begin{aligned}\hat{H}_0 &= \sqrt{m_p^2c^4+\hat{p}^2c^2}+\sqrt{m_e^2c^4+\hat{q}^2c^2}\\ &= m_pc^2+m_ec^2+\frac{\hat{p}^2}{2m_p}+\frac{\hat{q}^2}{2m_e}-\frac{\hat{p}^4}{8m_p^3c^2}-\frac{\hat{q}^4}{8m_e^3c^2}+O(1/c^4).\end{aligned}\tag{3.4}$$

The first two terms are simply constants (rest energies of the particles) that can be eliminated by a suitable choice of zero on the energy scale. So, finally, the second-order RQD energy of the electron + proton interacting pair takes the form of the so-called *Coulomb–Darwin–Breit* Hamiltonian, i. e.,

$$\begin{aligned}\hat{H}^d &= \hat{H}_0+\hat{V}_2^d(\hat{\boldsymbol{p}},\hat{\boldsymbol{q}},\boldsymbol{r},\hat{\boldsymbol{s}}_{el},\hat{\boldsymbol{s}}_{pr})+\cdots\\ &= \frac{\hat{p}^2}{2m_p}+\frac{\hat{q}^2}{2m_e}+\hat{V}_{\text{Coulomb}}+\hat{V}_{\text{orbit}}+\hat{V}_{\text{spin-orbit}}+\hat{V}_{\text{spin-spin}}+\cdots.\end{aligned}\tag{3.5}$$

This is similar to the familiar nonrelativistic Hamiltonian, where $\hat{p}^2/(2m_p)+\hat{q}^2/(2m_e)$ is regarded as the kinetic energy operator and $\hat{V}_{\text{Coulomb}}$ is the usual Coulomb potential between two charged particles,

$$\hat{V}_{\text{Coulomb}} = -\frac{e^2}{4\pi r}.\tag{3.6}$$

4 We noticed that Pauli matrices are proportional to spin operators (**2**-B.1): $\hat{\boldsymbol{s}}_{el}=(\hbar/2)\boldsymbol{\sigma}_{el}$, $\hat{\boldsymbol{s}}_{pr}=(\hbar/2)\boldsymbol{\sigma}_{pr}$. Some of the interaction terms are non-Hermitian due to the noncommutativity of $\boldsymbol{r}$ and $\boldsymbol{p}$, $\boldsymbol{q}$. This inconvenience can be easily overcome by symmetrizing the products of noncommuting operators, i. e., replacing $AB\to(AB+BA)/2$.

Here $\hat{V}_{\text{orbit}}$ is the spin-independent relativistic correction to the Coulomb potential

$$\begin{aligned}\hat{V}_{\text{orbit}} &= -\frac{\hat{p}^4}{8m_p^3c^2} - \frac{\hat{q}^4}{8m_e^3c^2} + \frac{e^2\hbar^2}{8c^2}\left(\frac{1}{m_p^2} + \frac{1}{m_e^2}\right)\delta(\boldsymbol{r}) \\ &\quad + \frac{e^2}{8\pi m_p m_e c^2 r}\left[\hat{\boldsymbol{p}} \cdot \hat{\boldsymbol{q}} + \frac{(\boldsymbol{r} \cdot \hat{\boldsymbol{p}})(\boldsymbol{r} \cdot \hat{\boldsymbol{q}})}{r^2}\right]. \end{aligned} \tag{3.7}$$

The first two terms should be interpreted as relativistic corrections to one-particle energies. The third (*contact interaction*) term can be ignored in the classical limit $\hbar \to 0$. Keeping the accuracy of $(v/c)^2$ and making substitutions $\hat{\boldsymbol{p}}/m_p \to \hat{\boldsymbol{v}}_{pr}$ and $\hat{\boldsymbol{q}}/m_e \to \hat{\boldsymbol{v}}_{el}$, we obtain the remaining term in the more familiar form of the *Darwin potential* [21], i. e.,

$$\hat{V}_{\text{Darwin}} = \frac{e^2}{8\pi c^2 r}\left[\hat{\boldsymbol{v}}_{pr} \cdot \hat{\boldsymbol{v}}_{el} + \frac{(\boldsymbol{r} \cdot \hat{\boldsymbol{v}}_{pr})(\boldsymbol{r} \cdot \hat{\boldsymbol{v}}_{el})}{r^2}\right], \tag{3.8}$$

which describes the velocity-dependent (magnetic) interaction between charged particles.

The two remaining terms in (3.5) are the *spin–orbit* and the *spin–spin* potentials, which depend on particle spins, i. e.,

$$\hat{V}_{\text{spin-orbit}} = -\frac{e^2[\boldsymbol{r} \times \hat{\boldsymbol{p}}] \cdot \hat{\boldsymbol{s}}_{pr}}{8\pi m_p^2 c^2 r^3} + \frac{e^2[\boldsymbol{r} \times \hat{\boldsymbol{q}}] \cdot \hat{\boldsymbol{s}}_{el}}{8\pi m_e^2 c^2 r^3} + \frac{e^2[\boldsymbol{r} \times \hat{\boldsymbol{q}}] \cdot \hat{\boldsymbol{s}}_{pr}}{4\pi m_p m_e c^2 r^3} - \frac{e^2[\boldsymbol{r} \times \hat{\boldsymbol{p}}] \cdot \hat{\boldsymbol{s}}_{el}}{4\pi m_p m_e c^2 r^3}, \tag{3.9}$$

$$\hat{V}_{\text{spin-spin}} = \frac{e^2}{m_p m_e c^2}\left(-\frac{\hat{\boldsymbol{s}}_{pr} \cdot \hat{\boldsymbol{s}}_{el}}{4\pi r^3} + \frac{3(\hat{\boldsymbol{s}}_{pr} \cdot \boldsymbol{r})(\hat{\boldsymbol{s}}_{el} \cdot \boldsymbol{r})}{4\pi r^5} - \frac{1}{3}(\hat{\boldsymbol{s}}_{pr} \cdot \hat{\boldsymbol{s}}_{el})\delta(\boldsymbol{r})\right). \tag{3.10}$$

Since our dressing transformation preserves commutation relations of the Poincaré Lie algebra (see Subsection 2.3.6), we can be sure that the Coulomb–Darwin–Breit Hamiltonian is relativistically invariant, at least up to the order $(v/c)^2$. In Appendix E, we double-checked this fact by a direct calculation.

The Coulomb–Darwin–Breit Hamiltonian (3.5) has been successfully applied to various problems in the physics of electromagnetic phenomena, such as fine and hyperfine structures of the atomic spectra [14, 21], superconductivity and properties of plasma [71, 68–70, 72]. In Chapter 6 we will see that in the classical limit this potential reproduces all basic results of classical electrodynamics very well.

3.2 Hydrogen atom

In the preceding section, we derived the electron–proton interaction potential in the second perturbation order. It can be used to study bound states of these two particles, i. e., the hydrogen atom. In particular, we are interested in energies and wave functions of these bound states. In the dressed approach, this requires solving the stationary Schrödinger equation (**1**-6.79) in the electron–proton sector $\mathscr{H}_{pe}$ of the Fock space.

In other words, we just need to diagonalize the dressed Hamiltonian (3.5). So, unlike the traditional QFT [266], our description of bound states is not different from nonrelativistic quantum mechanics.

3.2.1 Nonrelativistic Schrödinger equation

In order to solve the two-particle Schrödinger equation, we use the fact that the Hamiltonian (3.5) commutes with the total momentum operator $\boldsymbol{P}_0 = \boldsymbol{p} + \boldsymbol{q}$ and therefore leaves invariant the zero momentum subspace in $\mathscr{H}_{pe}$. Restricting ourselves to this subspace, we can set $\hat{\boldsymbol{Q}} \equiv \hat{\boldsymbol{q}} = -\hat{\boldsymbol{p}}$ on the right-hand side of equation (3.5) and regard $\hat{\boldsymbol{Q}}$ as the operator of differentiation with respect to the relative position $\boldsymbol{r}$, so we have

$$\hat{\boldsymbol{Q}} = -i\hbar\frac{\partial}{\partial \boldsymbol{r}}.$$

If we make these substitutions in (3.5), then the first two terms can be rewritten as

$$\frac{\hat{p}^2}{2m_p} + \frac{\hat{q}^2}{2m_e} = \frac{(m_e + m_p)\hat{Q}^2}{2m_e m_p} = \frac{\hat{Q}^2}{2\mu}, \tag{3.11}$$

where $\mu = m_e m_p/(m_e + m_p)$ is the *reduced mass*. The accuracy of our approach would not suffer much if we take into account the large proton mass ($m_p/m_e \approx 1\,837$) and set $\mu \approx m_e$, $\hat{Q} \approx \hat{q}$.[5]

Energies ε and wave functions $\Psi_\varepsilon(\boldsymbol{r}, \pi, \epsilon)$ of bound states of the hydrogen atom are solutions of the stationary Schrödinger equation

$$\hat{H}^d\left(-i\hbar\frac{\partial}{\partial \boldsymbol{r}}, \boldsymbol{r}, \boldsymbol{s}_{el}, \boldsymbol{s}_{pr}\right)\Psi_\varepsilon(\boldsymbol{r}, \pi, \epsilon) = \varepsilon\Psi_\varepsilon(\boldsymbol{r}, \pi, \epsilon). \tag{3.12}$$

An analytical solution is impossible. The best we can do is to apply perturbation theory from Subsection **1**-6.5.2. For that, we first divide $\hat{H}^d$ in (3.5) into the zero-order Hamiltonian $\hat{H}_{e+p}$ and the perturbation operator $\hat{V}_{\text{pert}}$, so that

$$\hat{H}^d = \hat{H}_{e+p} + \hat{V}_{\text{pert}}.$$

As the zero-order Hamiltonian we choose the first three terms in (3.5), which include the Coulomb interaction potential. We have

$$\hat{H}_{e+p} = \frac{\hat{q}^2}{2m_e} - \frac{e^2}{4\pi r}. \tag{3.13}$$

5 Thus, we simplified our problem to the level of one electron moving in an external potential.

All remaining terms in (3.5) will be regarded as perturbation, i. e.,

$$\hat{V}_{\text{pert}} \equiv \hat{V}_{\text{orbit}} + \hat{V}_{\text{spin-orbit}} + \hat{V}_{\text{spin-spin}}. \tag{3.14}$$

In the zeroth approximation, the Schrödinger equation

$$\hat{H}_{e+p}\Psi_{\varepsilon}(\boldsymbol{r},\pi,\epsilon) = \left(\frac{\hat{q}^2}{2m_e} - \frac{e^2}{4\pi r}\right)\Psi_{\varepsilon}(\boldsymbol{r},\pi,\epsilon) = \varepsilon\Psi_{\varepsilon}(\boldsymbol{r},\pi,\epsilon) \tag{3.15}$$

does not depend on spin variables π, ϵ, so wave functions can be written as products of orbital and spin parts, so we have

$$\Psi_{\varepsilon}(\boldsymbol{r},\pi,\epsilon) = \psi_{\varepsilon}(\boldsymbol{r})\chi(\pi,\epsilon).$$

Eigenvalues ε do not depend on the spin functions $\chi(\pi,\epsilon)$, which can be chosen as arbitrary sets of four complex numbers satisfying the normalization condition

$$|\chi(\uparrow\uparrow)|^2 + |\chi(\downarrow\uparrow)|^2 + |\chi(\uparrow\downarrow)|^2 + |\chi(\downarrow\downarrow)|^2 = 1.$$

The orbital parts are subject to the eigenvalue equation

$$\left(-\frac{\hbar^2}{2m_e}\frac{\partial^2}{\partial \boldsymbol{r}^2} - \frac{e^2}{4\pi r}\right)\psi_{\varepsilon}(\boldsymbol{r}) = \varepsilon\psi_{\varepsilon}(\boldsymbol{r}), \tag{3.16}$$

which is more convenient to write in spherical coordinates:

$$-\frac{\hbar^2}{2m_e}\left(\frac{1}{r^2}\frac{\partial}{\partial r}\left[r^2\frac{\partial}{\partial r}\right] + \frac{1}{r^2\sin\theta}\frac{\partial}{\partial\theta}\left[\sin\theta\frac{\partial}{\partial\theta}\right] + \frac{1}{r^2\sin^2\theta}\frac{\partial^2}{\partial\varphi^2}\right)$$
$$\times\,\psi_{\varepsilon}(r,\theta,\varphi) - \frac{e^2}{4\pi r}\psi_{\varepsilon}(r,\theta,\varphi) = \varepsilon\psi_{\varepsilon}(r,\theta,\varphi).$$

The analytical solution of this equation can be found in any quantum mechanics textbook [13, 154]. The eigenstates are marked by the *principal* ($\mathbb{n} = 1,2,3,\ldots$), *orbital* ($\mathbb{l} = 0,1,\ldots,\mathbb{n}-1$) and *magnetic* ($\mathbb{m} = -\mathbb{l},-\mathbb{l}+1,\ldots,\mathbb{l}-1,\mathbb{l}$) quantum numbers. Energy eigenvalues are degenerate with respect to $\mathbb{l}$ and $\mathbb{m}$, i. e.,

$$\varepsilon(\mathbb{n},\mathbb{l},\mathbb{m}) = -\frac{m_e c^2\alpha^2}{2\mathbb{n}^2}. \tag{3.17}$$

Several low-energy solutions are shown in Table 3.1, where

$$a_0 \equiv \frac{4\pi\hbar^2}{m_e e^2} = \frac{\hbar}{\alpha m_e c}$$

denotes the *Bohr radius* and

$$\alpha \equiv \frac{e^2}{4\pi\hbar c} \approx 1/137$$

is the *fine-structure constant*.

Table 3.1: Normalized nonrelativistic solutions for lowest levels of the hydrogen atom.

State $(\mathbb{n}, \mathbb{l}, \mathbb{m})$	**Wave function** $\psi_\varepsilon(r,\theta,\varphi)$	**Energy** (ε)
$1S(1,0,0)$	$\frac{1}{\sqrt{\pi a_0^3}} e^{-r/a_0}$	$-m_e c^2\alpha^2/2 = -13.6$ eV
$2S(2,0,0)$	$\frac{1}{4\sqrt{2\pi a_0^3}}(2-\frac{r}{a_0})e^{-r/(2a_0)}$	$-m_e c^2\alpha^2/8 = -3.4$ eV
$2P(2,1,0)$	$\frac{r}{4a_0\sqrt{2\pi a_0^3}} e^{-r/(2a_0)}\cos\theta$	$-m_e c^2\alpha^2/8 = -3.4$ eV
$2P(2,1,-1)$	$\frac{r}{8a_0\sqrt{\pi a_0^3}} e^{-r/(2a_0)}\sin\theta e^{-i\varphi}$	$-m_e c^2\alpha^2/8 = -3.4$ eV
$2P(2,1,1)$	$\frac{r}{8a_0\sqrt{\pi a_0^3}} e^{-r/(2a_0)}\sin\theta e^{i\varphi}$	$-m_e c^2\alpha^2/8 = -3.4$ eV

For further calculations, we will need expectation values of the inverse powers of r in these states. For example,

$$\begin{aligned}
\langle r^{-1}\rangle(2S) &\equiv \int d\boldsymbol{r}\psi^*_{2S}(\boldsymbol{r})\frac{1}{r}\psi_{2S}(\boldsymbol{r}) = \frac{1}{8a_0^3}\int_0^\infty drr\left(2-\frac{r}{a_0}\right)^2 e^{-r/a_0} \\
&= \frac{1}{8a_0^3}\left(4\int_0^\infty drre^{-r/a_0} - \frac{4}{a_0}\int_0^\infty drr^2e^{-r/a_0} + \frac{1}{a_0^2}\int_0^\infty drr^3e^{-r/a_0}\right) \\
&= \frac{1}{8a_0^3}(4a_0^2 - 8a_0^2 + 6a_0^2) = \frac{1}{4a_0}.
\end{aligned}$$

These results are presented in Table 3.2 together with probability densities $|\psi(\mathbf{0})|^2$ for finding the electron at the proton.

Table 3.2: Properties of nonrelativistic wave functions of the hydrogen atom.

State	$|\psi(0)|^2$	$\langle r^{-1}\rangle$	$\langle r^{-2}\rangle$	$\langle r^{-3}\rangle$
$1S$	$1/(\pi a_0^3)$	$1/(a_0)$	$2/(a_0^2)$	
$2S$	$1/(8\pi a_0^3)$	$1/(4a_0)$	$1/(4a_0^2)$	
$2P$	0	$1/(4a_0)$	$1/(12a_0^2)$	$1/(24a_0^3)$

3.2.2 Perturbation theory in hydrogen atom

In the preceding subsection, we were able to find the entire spectrum of eigenfunctions Ψ_ε and eigenvalues ε of the zero-order Hamiltonian $\hat{H}_{e+p}$. In accordance with perturbation theory, we will seek corrections to these results caused by the small perturbation $\hat{V}_{\text{pert}}$ (3.14). In the first order this perturbation does not change wave functions, but

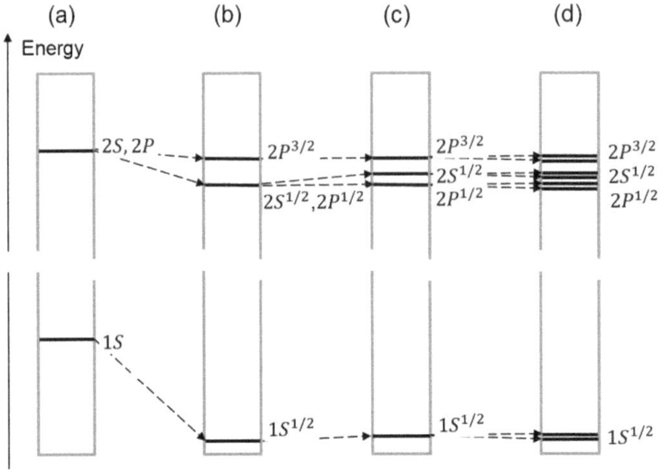

Figure 3.1: Lowest energy levels of the hydrogen atom. (a) With the nonrelativistic Coulomb potential (3.6). (b) Plus the fine structure due to *orbital* (3.7) and *spin–orbit* (3.9) corrections. (c) Plus the Lamb shift described in Section 5.4. (d) Plus the superfine structure due to *spin–spin* correction (3.10). Not to scale.

shifts energies of states according to formula (**1**-6.80), i. e.,

$$\Delta\varepsilon = \langle\Psi_\varepsilon|\hat{V}_{\text{pert}}|\Psi_\varepsilon\rangle. \tag{3.18}$$

In our approximation ($m_p \gg m_e$), only two perturbations $\hat{V}_{\text{orbit}}$ and $\hat{V}_{\text{spin-orbit}}$ are left. They are responsible for the so-called *fine structure* of the atomic spectrum (see Figure 3.1 (b)),

$$\hat{V}_{\text{pert}} \approx \hat{V}_{\text{orbit}} + \hat{V}_{\text{spin-orbit}} = \hat{V}_{\text{relat}} + \hat{V}_{\text{contact}} + \hat{V}_{\text{spin-orbit}},$$

$$\hat{V}_{\text{relat}} = -\frac{\hat{q}^4}{8m_e^3c^2}, \tag{3.19}$$

$$\hat{V}_{\text{contact}} = \frac{e^2\hbar^2}{8m_e^2c^2}\delta(\boldsymbol{r}), \tag{3.20}$$

$$\hat{V}_{\text{spin-orbit}} = \frac{e^2[\boldsymbol{r}\times\hat{\boldsymbol{q}}]\cdot\hat{\boldsymbol{s}}_{el}}{8\pi m_e^2c^2r^3} = \frac{e^2\hat{\boldsymbol{L}}\cdot\hat{\boldsymbol{s}}_{el}}{8\pi m_e^2c^2r^3}, \tag{3.21}$$

where $\hat{\boldsymbol{L}} = [\boldsymbol{r}\times\hat{\boldsymbol{q}}]$ is the electron orbital angular momentum.

3.2.3 Relativistic energy corrections (orbital)

The relativistic perturbation (3.19) gives the following energy correction:

$$\Delta\varepsilon_{\text{relat}} = -\frac{1}{8m_e^3c^2}\int d\boldsymbol{r}\psi^*\hat{q}^4\psi. \tag{3.22}$$

If ψ_ε is an eigenfunction of $\hat{H}_{e+p}$ with eigenvalue ε, then[6]

$$\hat{q}^4\psi_\varepsilon = \hat{q}^2\hat{q}^2\psi_\varepsilon = 2m_e\hat{q}^2\left(\varepsilon + \frac{e^2}{4\pi r}\right)\psi_\varepsilon = -2m_e\hbar^2\frac{\partial}{\partial\boldsymbol{r}}\cdot\left[\varepsilon\frac{\partial\psi_\varepsilon}{\partial\boldsymbol{r}} + \frac{\partial}{\partial\boldsymbol{r}}\left(\frac{e^2}{4\pi r}\psi_\varepsilon\right)\right]$$
$$= -2m_e\hbar^2\left[\varepsilon\frac{\partial^2\psi_\varepsilon}{\partial\boldsymbol{r}^2} + \frac{e^2}{4\pi}\left(\frac{\partial^2}{\partial\boldsymbol{r}^2}\frac{1}{r}\right)\psi_\varepsilon + \frac{e^2}{2\pi}\left(\frac{\partial}{\partial\boldsymbol{r}}\frac{1}{r}\cdot\frac{\partial\psi_\varepsilon}{\partial\boldsymbol{r}}\right) + \frac{e^2}{4\pi r}\frac{\partial^2\psi_\varepsilon}{\partial\boldsymbol{r}^2}\right]$$
$$= -2m_e\hbar^2\left[-\frac{2m}{\hbar^2}\left(\varepsilon + \frac{e^2}{4\pi r}\right)^2\psi_\varepsilon - e^2\delta(\boldsymbol{r})\psi_\varepsilon + \frac{e^2}{2\pi}\left(\frac{\partial}{\partial\boldsymbol{r}}\frac{1}{r}\cdot\frac{\partial\psi_\varepsilon}{\partial\boldsymbol{r}}\right)\right].$$

Substituting this result in (3.22) and using the expression for the gradient in spherical coordinates[7]

$$\frac{\partial f(r,\theta,\varphi)}{\partial\boldsymbol{r}} = \frac{\partial f}{\partial r}\hat{\boldsymbol{r}} + \frac{\hat{\boldsymbol{\theta}}}{r\sin\varphi}\frac{\partial f}{\partial\theta} + \frac{\hat{\boldsymbol{\varphi}}}{r}\frac{\partial f}{\partial\varphi},$$

we obtain

$$\Delta\varepsilon_{\text{relat}} = \frac{\hbar^2}{4m_e^2c^2}\int d\boldsymbol{r}\psi_\varepsilon^*\left[-\frac{2m_e}{\hbar^2}\left(\varepsilon + \frac{e^2}{4\pi r}\right)^2\psi_\varepsilon - e^2\delta(\boldsymbol{r})\psi_\varepsilon - \frac{e^2}{2\pi r^2}\frac{\partial\psi_\varepsilon}{\partial r}\right]. \tag{3.23}$$

The last term in square brackets,[8]

$$-\frac{e^2}{2\pi}\int d\boldsymbol{r}\psi_\varepsilon^*\frac{1}{r^2}\frac{\partial\psi_\varepsilon}{\partial r} = -\frac{e^2}{2\pi}\int_0^{2\pi}d\varphi\int_0^{\pi}\sin\theta d\theta\int_0^{\infty}dr\psi_\varepsilon^*\frac{\partial\psi_\varepsilon}{\partial r}$$
$$= -\frac{e^2}{4\pi}\int_0^{2\pi}d\varphi\int_0^{\pi}\sin\theta d\theta\int_0^{\infty}dr\frac{\partial|\psi_\varepsilon|^2}{\partial r} = \frac{e^2}{4\pi}\int_0^{2\pi}d\varphi\int_0^{\pi}\sin\theta d\theta|\psi_\varepsilon(\mathbf{0})|^2$$
$$= e^2|\psi_\varepsilon(\mathbf{0})|^2, \tag{3.24}$$

cancels with the second term in square brackets

$$-e^2\int d\boldsymbol{r}\psi_\varepsilon^*\delta(\boldsymbol{r})\psi_\varepsilon = -e^2|\psi_\varepsilon(\mathbf{0})|^2.$$

Therefore, the final contribution from the relativistic correction is

$$\Delta\varepsilon_{\text{relat}} = -\frac{1}{2m_ec^2}\int d\boldsymbol{r}\psi_\varepsilon^*\left(\varepsilon^2 + \frac{e^2\varepsilon}{2\pi r} + \frac{e^4}{16\pi^2r^2}\right)\psi_\varepsilon$$
$$= -\frac{1}{2m_ec^2}\left(\varepsilon^2 + \frac{e^2\varepsilon}{2\pi}\langle r^{-1}\rangle + \frac{e^4}{16\pi^2}\langle r^{-2}\rangle\right). \tag{3.25}$$

6 Here we used (3.15) and (**1**-A.2).

7 Here $\hat{\boldsymbol{r}} \equiv \boldsymbol{r}/r, \hat{\boldsymbol{\theta}}, \hat{\boldsymbol{\varphi}}$ are unit vectors pointing along the directions of growth of the corresponding coordinates.

8 Here we took into account that $\psi_\varepsilon(r,\theta,\varphi) \to 0$, when $r \to \infty$.

The energy correction due to the contact potential (3.20) is

$$\Delta\varepsilon_{\text{contact}} = \frac{e^2\hbar^2}{8m_e^2c^2}\int d\boldsymbol{r}\delta(\boldsymbol{r})|\psi_\varepsilon(\boldsymbol{r})|^2 = \frac{e^2\hbar^2}{8m_e^2c^2}|\psi_\varepsilon(\boldsymbol{0})|^2. \tag{3.26}$$

Using data from Table 3.2, we obtain numerical values of $\Delta\varepsilon_{\text{relat}}$ and $\Delta\varepsilon_{\text{contact}}$ for different states. They are shown in Table 3.3.

Table 3.3: Second-order corrections to the energies of lowest levels of the hydrogen atom.

Energy level	$1S^{1/2}$	$2S^{1/2}$	$2P^{1/2}$	$2P^{3/2}$
nonrelativistic energy (3.17)	$-\frac{m_ec^2\alpha^2}{2}$	$-\frac{m_ec^2\alpha^2}{8}$	$-\frac{m_ec^2\alpha^2}{8}$	$-\frac{m_ec^2\alpha^2}{8}$
Corrections:				
relativistic (3.25)	$-\frac{5m_ec^2\alpha^4}{8}$	$-\frac{13m_ec^2\alpha^4}{128}$	$-\frac{7m_ec^2\alpha^4}{384}$	$-\frac{7m_ec^2\alpha^4}{384}$
contact (3.26)	$\frac{m_ec^2\alpha^4}{2}$	$\frac{m_ec^2\alpha^4}{16}$	0	0
spin–orbit (3.27), (3.28)	0	0	$-\frac{m_ec^2\alpha^4}{48}$	$\frac{m_ec^2\alpha^4}{96}$
Full correction	$-\frac{m_ec^2\alpha^4}{8}$	$-\frac{5m_ec^2\alpha^4}{128}$	$-\frac{5m_ec^2\alpha^4}{128}$	$-\frac{m_ec^2\alpha^4}{128}$

3.2.4 Relativistic energy corrections (spin–orbit)

Now consider the energy shifts due to the spin–orbit interaction (3.21). This interaction does not affect S-states, which are eigenstates of the operator $\hat{L}^2$ with eigenvalue $\mathbb{l} = 0$. Let us then turn to $2P$-states, where $\mathbb{l} = 1$.

In total, the $2P$ level contains six states with different combinations of the magnetic quantum number $\mathbb{m} = -1, 0, 1$ and the electron spin projections $s_z = -1/2, 1/2$.[9] In these states, the total angular momentum $\hat{\boldsymbol{J}} = \hat{\boldsymbol{L}} + \hat{\boldsymbol{s}}_{el}$ can be either $j\hbar = (1-1/2)\hbar = \hbar/2$ or $j\hbar = (1+1/2)\hbar = 3\hbar/2$. Hence, the six states separate into two groups. One group of two states has $j = 1/2$; it is denoted $2P^{1/2}$. The second group of four states has $j = 3/2$ and is denoted $2P^{3/2}$. The unperturbed Hamiltonian (3.13) commutes with the orbital angular momentum $\hat{\boldsymbol{L}}$, with the electron spin $\hat{\boldsymbol{s}}_{el}$ and with the total angular momentum $\hat{\boldsymbol{J}}$. So all six $2P$ states have the same unperturbed energy. In other words, they are degenerate with respect to the Hamiltonian (3.13).

On the other hand, the full Hamiltonian (3.5) does not commute with $\hat{\boldsymbol{L}}$ and $\hat{\boldsymbol{s}}_{el}$ separately, but commutes only with their sum $\hat{\boldsymbol{J}}$. Therefore, total energies of the two groups of states $2P^{1/2}$ and $2P^{3/2}$ can differ. In our approximation, only the spin–orbit

9 We do not consider different orientations of the proton spin, because in our approximation ($m_p \gg m_e$) they have no effect on the energy. Therefore, we set $\boldsymbol{s}_{pr} = 0$.

potential (3.21) contributes to this difference. To calculate its effect on the $2P^{1/2}$ level, we use the formula

$$\hat{J}^2 = (\hat{\boldsymbol{L}} + \hat{\mathbf{s}}_{el})^2 = \hat{L}^2 + \hat{s}_{el}^2 + 2(\hat{\boldsymbol{L}} \cdot \hat{\mathbf{s}}_{el}).$$

Then

$$\begin{aligned}
\hat{J}^2 \psi_{2P^{1/2}} &= \hbar^2 j(j+1)\psi_{2P^{1/2}} = 3/4\hbar^2 \psi_{2P^{1/2}}, \\
\hat{L}^2 \psi_{2P^{1/2}} &= \hbar^2 l(l+1)\psi_{2P^{1/2}} = 2\hbar^2 \psi_{2P^{1/2}}, \\
\hat{s}_{el}^2 \psi_{2P^{1/2}} &= \hbar^2 s(s+1)\psi_{2P^{1/2}} = 3/4\hbar^2 \psi_{2P^{1/2}}, \\
(\hat{\boldsymbol{L}} \cdot \hat{\mathbf{s}}_{el})\psi_{2P^{1/2}} &= 1/2(\hat{J}^2 - \hat{L}^2 - \hat{s}_{el}^2)\psi_{2P^{1/2}} \\
&= \frac{(3/4 - 2 - 3/4)\hbar^2}{2}\psi_{2P^{1/2}} = -\hbar^2 \psi_{2P^{1/2}}.
\end{aligned}$$

Substituting this result in (3.21) and (3.18), we obtain

$$\begin{aligned}
\Delta\varepsilon_{\text{spin-orbit}}(2P^{1/2}) &= \frac{e^2}{8\pi m_e^2 c^2} \int d\boldsymbol{r} \psi^*_{2P^{1/2}} \frac{\hat{\boldsymbol{L}} \cdot \hat{\mathbf{s}}_{el}}{r^3} \psi_{2P^{1/2}} = -\frac{e^2\hbar^2}{8\pi m_e^2 c^2}\langle r^{-3}\rangle \\
&= -\frac{m_e c^2 \alpha^4}{48}.
\end{aligned} \tag{3.27}$$

A similar calculation yields the spin–orbit correction to the $2P^{3/2}$ energy, i. e.,

$$\begin{aligned}
(\hat{\boldsymbol{L}} \cdot \hat{\mathbf{s}}_{el})\psi_{2P^{3/2}} &= \frac{(15/4 - 2 - 3/4)\hbar^2}{2}\psi_{2P^{3/2}} = \frac{\hbar^2}{2}\psi_{2P^{3/2}}, \\
\Delta\varepsilon_{\text{spin-orbit}}(2P^{3/2}) &= \frac{m_e c^2 \alpha^4}{96};
\end{aligned} \tag{3.28}$$

see Table 3.3. In particular, it follows from this table that complete second-order corrections to the energies of the states $2S^{1/2}$ and $2P^{1/2}$ coincide. Hence these two levels remain degenerate in our approximation, so we have

$$\varepsilon(2S^{1/2}) = \varepsilon(2P^{1/2}). \tag{3.29}$$

In Chapter 5 we will take into account higher perturbation orders in RQD and find a tiny gap between the levels $2S^{1/2}$ and $2P^{1/2}$, which is called the *Lamb shift*.

4 Decays

Many things are incomprehensible to us not because our comprehension is weak, but because those things are not within the frames of our comprehension.
Kozma Prutkov

Our formulation of quantum theory in the Fock space with a variable number of particles makes it possible to describe not only interparticle interactions, but also processes in which particles are created and destroyed. The simplest example of such a process is the decay of an unstable particle. This is the topic of the present chapter.

The unstable particle is interesting for several reasons. First, it is a rare example of an interacting quantum system whose time evolution can be observed relatively easily. This time evolution is quite simple, because in many cases it can be described with only one time-dependent parameter – the nondecay probability $\Upsilon(t)$. Second, a mathematically rigorous description of such a system is possible in a small part of the Fock space, composed of only two sectors: the sector of the particle itself and the sector of its decay products. Therefore, a fairly accurate solution of this dynamical problem can be obtained in an analytical form.

In Sections 4.1 and 4.2 we will discuss the decay law of an unstable system at rest. These results will be useful to us in Chapter 5, when we discuss the spectrum of the hydrogen atom in higher perturbation orders. In Section 4.3 we will be interested in the decay law seen from a moving frame of reference. In Section 4.4, we will show that Einstein's "time dilation" formula, strictly speaking, is not applicable to decays, although the deviations are too small to be observable in modern experiments.

4.1 Unstable particle at rest

4.1.1 Quantum mechanics of particle decays

Mathematically, the decay of particles is described by the *nondecay probability* Υ, which has the following definition. Suppose we have a piece of radioactive material in which N unstable nuclei were prepared simultaneously at the time $t = 0$. Let $N_u(t)$ be the number of nuclei that remain undecayed at the time $t > 0$. Then at each moment of time our radioactive material can be described by the ratio $N_u(t)/N \leq 1$.

In the spirit of quantum mechanics, we will regard the N unstable nuclei as an ensemble of equally prepared systems and consider the ratio $N_u(t)/N$ as the property of an individual particle – the probability to find this particle in the initial (nondecayed) state. Then the nondecay probability $\Upsilon(t)$ is defined as the large N limit

$$\Upsilon(t) = \lim_{N\to\infty} N_u(t)/N. \tag{4.1}$$

The function $\Upsilon(t)$ is also called the *decay law* of the particle (nucleus).

https://doi.org/10.1515/9783110493221-004

Let us now turn to the description of an isolated unstable system from the point of view of quantum theory. We consider a model theory with three particles α, β and γ. The particle α is massive and unstable. To simplify calculations, we assume that this particle has zero spin and only one decay channel $\alpha \to \beta + \gamma$. The *decay products* β and γ are assumed to be stable, and their masses satisfy the inequality

$$m_\alpha > m_\beta + m_\gamma, \tag{4.2}$$

which makes the decay $\alpha \to \beta + \gamma$ energetically possible.

Observation performed on an unstable particle can lead to one of two results: either the α particle will be detected in the undecayed state, or its decay products $(\beta + \gamma)$ will be registered. Therefore, it makes sense to describe this system in only two sectors of the Fock space[1]:

$$\mathcal{H} = \mathcal{H}_\alpha \oplus \mathcal{H}_{\beta\gamma}, \tag{4.3}$$

where $\mathcal{H}_\alpha$ is the space of states of the unstable particle α and $\mathcal{H}_{\beta\gamma} \equiv \mathcal{H}_\beta \otimes \mathcal{H}_\gamma$ is the orthogonal subspace of the decay products.

Now we can introduce the Hermitian operator T corresponding to the experimental proposition “the particle α exists.” This operator is completely determined by its eigenvalues and eigensubspaces. When a measurement finds the system in a state corresponding to the particle α, then the value of T is 1. In the case when the $\beta + \gamma$ decay products are observed, the value of T is 0. Obviously, T is a projection onto the subspace $\mathcal{H}_\alpha$. For each normalized state vector $|\Psi\rangle \in \mathcal{H}$, the probability of finding the unstable particle α is given by the expectation value of this operator

$$\Upsilon = \langle\Psi|T|\Psi\rangle. \tag{4.4}$$

In other words, we can say that the nondecay probability Υ is the square of the projection $T|\Psi\rangle$, i. e.,

$$\Upsilon = \langle\Psi|TT|\Psi\rangle = \|T|\Psi\rangle\|^2, \tag{4.5}$$

where we used the property $T^2 = T$ from Theorem **1**-H.1.

Each vector $|\Psi\rangle \in \mathcal{H}_\alpha$ describes a state in which the particle α is found with 100 % certainty. We shall assume that the unstable system was prepared in one of such states $|\Psi(0)\rangle$ at the time instant $t = 0$, so that

$$\Upsilon(0) = \langle\Psi(0)|T|\Psi(0)\rangle = 1, \tag{4.6}$$

and ask how this probability changes over time.

1 This is, of course, an approximation in which the interaction between decay products β and γ (and the possibility of creation of new particles due to this interaction) is ignored. However, experience shows that this approximation is rather good.

In order to obtain the time evolution, we need to know the total Hamiltonian $H = H_0 + V$ in the Hilbert space $\mathscr{H}$. The free part of this Hamiltonian H_0 can be easily constructed by the usual rules from Subsection **2**-1.1.3. In the subspace $\mathscr{H}_\alpha$, this operator is

$$H_0|_{\mathscr{H}_\alpha} = \sqrt{m_\alpha^2 c^4 + p^2c^2}, \tag{4.7}$$

while in the decay products subspace

$$H_0|_{\mathscr{H}_{\beta\gamma}} = \sqrt{m_\beta^2 c^4 + p^2c^2} + \sqrt{m_\gamma^2 c^4 + p^2c^2}. \tag{4.8}$$

For the interaction we choose the simplest operator that can be responsible for the process $\alpha \leftrightarrow \beta + \gamma$, i. e.,[2]

$$V = \int d\boldsymbol{p}d\boldsymbol{q}(V(\boldsymbol{p},\boldsymbol{q})\alpha^\dagger_{\boldsymbol{p}+\boldsymbol{q}}\beta_{\boldsymbol{p}}\gamma_{\boldsymbol{q}} + V^*(\boldsymbol{p},\boldsymbol{q})\gamma^\dagger_{\boldsymbol{q}}\beta^\dagger_{\boldsymbol{p}}\alpha_{\boldsymbol{p}+\boldsymbol{q}}). \tag{4.9}$$

This operator leaves invariant the sector $\mathscr{H} = \mathscr{H}_\alpha \oplus \mathscr{H}_{\beta\gamma}$ of the full Fock space.

Then the time evolution of the initial state $|\Psi(0)\rangle$ is given by equation (**1**-6.88)[3]:

$$|\Psi(t)\rangle = e^{-\frac{i}{\hbar}Ht}|\Psi(0)\rangle, \tag{4.10}$$

which leads to the following decay law:

$$\Upsilon(t) = \langle\Psi(0)|e^{\frac{i}{\hbar}Ht}Te^{-\frac{i}{\hbar}Ht}|\Psi(0)\rangle. \tag{4.11}$$

Clearly, the interaction V should not commute with the projection T, i. e.,

$$[H,T] \neq 0. \tag{4.12}$$

Then the subspace $\mathscr{H}_\alpha$ of the particle α is not invariant with respect to time translations, and the decay law (4.11) is a nontrivial function of time.

Figure 4.1 is a schematic visualization of the decay time evolution in the Hilbert space. The full Hilbert space $\mathscr{H}$ is represented as the direct sum of two orthogonal subspaces $\mathscr{H}_\alpha$ and $\mathscr{H}_{\beta\gamma}$. We have assumed that the initial state vector $|\Psi(0)\rangle$ at time $t = 0$ lies entirely in the subspace $\mathscr{H}_\alpha$, so that the nondecay probability $\Upsilon(0)$ is equal

2 For Hermiticity, the interaction operator along with the term $\gamma^\dagger\beta^\dagger\alpha$, responsible for the decay, must contain also the term $\alpha^\dagger\beta\gamma$ describing the inverse process $\beta + \gamma \to \alpha$. Due to relation (4.2), these two terms have nontrivial energy surfaces. Therefore, in accordance with our classification in Subsection **2**-1.2.3, they are of the *decay* type. We know that there are no *decay*-type interaction operators in QED. Therefore, within QED, results of this chapter can be applicable only to decays of bound multiparticle states; see Section 5.1.

3 In this chapter, we are working in the Schrödinger picture.

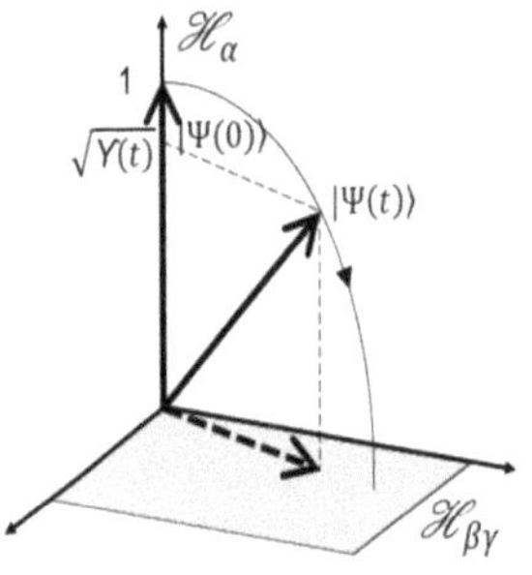

Figure 4.1: Time evolution of the state vector $|\Psi(t)\rangle$ of an unstable system.

to 1, as in (4.6). At later times, $t > 0$, the state vector $|\Psi(t)\rangle = e^{-\frac{i}{\hbar}Ht}|\Psi(0)\rangle$ acquires a component[4] lying in the subspace of decay products $\mathscr{H}_{\beta\gamma}$. So, the nondecay probability $\Upsilon(t)$ decreases monotonically with time, while the probability $(1 - \Upsilon(t))$ to find the decay products increases. We will see later that under very general assumptions the decay law has a universal exponential form $\Upsilon(t) \approx e^{-\Gamma t}$, where the decay rate Γ is proportional to the strength of the decay interaction V.

Before calculating the decay law (4.11), we have to do some preparatory work. In the next two subsections we are going to build two useful bases. The first one is the basis $|\boldsymbol{p}\rangle$ of eigenvectors of the total momentum operator $\boldsymbol{P}_0$ in the subspace $\mathscr{H}_{\alpha}$. The second one is the basis $|\boldsymbol{p}, m\rangle$ of common eigenvectors of $\boldsymbol{P}_0$ and the interacting mass $M = \sqrt{H^2 - P_0^2c^2}/c^2$ in the entire Hilbert space $\mathscr{H}$.

4.1.2 Noninteracting representation of Poincaré group

Let us first consider the simplest case, when the interaction responsible for the decay is "turned off," i. e., we set $V(\boldsymbol{p}, \boldsymbol{q}) = 0$ in (4.9). In this case the system's dynamics is governed by the *noninteracting* representation U_g^0 of the Poincaré group in $\mathscr{H}$. This representation is constructed in accordance with the Hilbert space structure (4.3) as follows:

$$U_g^0 \equiv U_g^{\alpha} \oplus (U_g^{\beta} \otimes U_g^{\gamma}), \tag{4.13}$$

where U_g^{α}, U_g^{β} and U_g^{γ} are unitary irreducible representations of the Poincaré group associated with the particles α, β and γ, respectively. We will denote the generators of this representation as $\{\boldsymbol{P}_0, \boldsymbol{J}_0, \boldsymbol{K}_0, H_0\}$. In particular, the energy operator H_0 is defined by formulas (4.7) and (4.8). In agreement with (4.2), the noninteracting *mass* operator

$$M_0 = +\frac{1}{c^2}\sqrt{H_0^2 - P_0^2c^2}$$

4 Shown by the dashed arrow in the figure.

has a continuous spectrum in the interval $[m_\beta + m_\gamma, \infty)$ and a discrete point m_α embedded in this interval.

From the definition (4.13) it is clear that subspaces $\mathscr{H}_\alpha$ and $\mathscr{H}_{\beta\gamma}$ are invariant with respect to U_g^0. Moreover, the projection T commutes with noninteracting generators, i. e.,

$$[T, \boldsymbol{P}_0] = [T, \boldsymbol{J}_0] = [T, \boldsymbol{K}_0] = [T, H_0] = 0. \tag{4.14}$$

Similar to Subsection **1**-5.1.3, we can use the noninteracting representation U_g^0 for building the basis $|\boldsymbol{p}\rangle$ of eigenvectors of the total momentum operator $\boldsymbol{P}_0$ in the subspace of the unstable particle $\mathscr{H}_\alpha$. Then any state $|\Psi\rangle \in \mathscr{H}_\alpha$ can be represented as a linear combination (integral) of these basis vectors,

$$|\Psi\rangle = \int d\boldsymbol{p}\psi(\boldsymbol{p})|\boldsymbol{p}\rangle, \tag{4.15}$$

and the projection T can be written as (**1**-5.22)

$$T = \int d\boldsymbol{p}|\boldsymbol{p}\rangle\langle\boldsymbol{p}|. \tag{4.16}$$

Denote $\boldsymbol{R}_0$ the Newton–Wigner position operator (**1**-4.31), corresponding to the representation (4.13). Then, in agreement with (**1**-5.38), the exponent $e^{\frac{i}{\hbar}\hat{\boldsymbol{R}}_0\cdot\boldsymbol{b}}$ acts as a translation operator in the momentum space, i. e., $e^{\frac{i}{\hbar}\hat{\boldsymbol{R}}_0\cdot\boldsymbol{b}}\psi(\boldsymbol{p}) = \psi(\boldsymbol{p}-\boldsymbol{b})$. This implies the following representation for the momentum eigenvectors:

$$|\boldsymbol{p}\rangle = e^{\frac{i}{\hbar}\boldsymbol{R}_0\cdot\boldsymbol{p}}|\boldsymbol{0}\rangle, \tag{4.17}$$

which will be helpful later in this chapter.

4.1.3 Normalized eigenvectors of momentum

The improper (nonnormalizable) state vectors $|\boldsymbol{p}\rangle$ are convenient for writing arbitrary states $|\Psi\rangle \in \mathscr{H}_\alpha$ in the form of integrals (4.15). However, these vectors themselves are not precise representatives of quantum states because they are not normalized. For example, the wave function of the vector $|\boldsymbol{q}\rangle$ in the momentum representation is the delta function

$$\psi_{\boldsymbol{q}}(\boldsymbol{p}) = \langle\boldsymbol{p}|\boldsymbol{q}\rangle = \delta(\boldsymbol{p}-\boldsymbol{q}) \tag{4.18}$$

and the corresponding "probability" of finding the particle is infinite, i. e.,

$$\int d\boldsymbol{p}|\psi_{\boldsymbol{q}}(\boldsymbol{p})|^2 = \int d\boldsymbol{p}|\delta(\boldsymbol{p}-\boldsymbol{q})|^2 = \infty.$$

Therefore, for reliable calculations of probabilities we need other (normalized) sharp-momentum state vectors, for which we introduce a special notation $|\boldsymbol{q}\gg$. In order to satisfy the normalization condition,

$$\int d\boldsymbol{p}|\phi_{\boldsymbol{q}}(\boldsymbol{p})|^2 = 1,$$

the wave function $\phi_{\boldsymbol{q}}(\boldsymbol{p})$ of the state $|\boldsymbol{q}\gg$ should be formally represented by a square root of the Dirac delta function,[5]

$$\phi_{\boldsymbol{q}}(\boldsymbol{p}) = \sqrt{\delta(\boldsymbol{p}-\boldsymbol{q})}, \tag{4.19}$$

$$\int d\boldsymbol{p}|\phi_{\boldsymbol{q}}(\boldsymbol{p})|^2 = \int d\boldsymbol{p}|\sqrt{\delta(\boldsymbol{p}-\boldsymbol{q})}|^2 = \int d\boldsymbol{p}\delta(\boldsymbol{p}-\boldsymbol{q}) = 1.$$

4.1.4 Interacting representation of Poincaré group

Let us now "switch on" the interaction (4.9) responsible for the decay and discuss the interacting representation U_g of the Poincaré group with generators $\{\boldsymbol{P}, \boldsymbol{J}, \boldsymbol{K}, H\}$. As usual, we prefer to work in Dirac's instant form of dynamics, where generators of space translations and rotations are interaction-free:

$$\boldsymbol{P} = \boldsymbol{P}_0,$$
$$\boldsymbol{J} = \boldsymbol{J}_0,$$

while generators of time translations and boosts contain interaction terms:

$$H = H_0 + V,$$
$$\boldsymbol{K} = \boldsymbol{K}_0 + \boldsymbol{Z}.$$

Besides, to simplify calculations, we will suppose that the interacting representation U_g belongs to the Bakamjian–Thomas form of dynamics,[6] where the mass operator

$$M \equiv \frac{1}{c^2}\sqrt{H^2 - P_0^2c^2}$$

5 As an alternative way to achieve the same goal we could try to keep momentum eigenfunctions as delta functions $\phi_{\boldsymbol{q}}(\boldsymbol{p}) = N\delta(\boldsymbol{p}-\boldsymbol{q})$, but use the normalization factor $N = (\int d\boldsymbol{p}|\phi_{\boldsymbol{q}}(\boldsymbol{p})|^2)^{-1/2}$ that is formally equal to zero. This would result in the uncertainty of the type $0 \times \infty$. Probably, such manipulations with infinitely large and infinitesimally small quantities can be justified within *nonstandard analysis* [87].

6 See Subsection **1**-6.3.2. A description of decays in other forms of dynamics will be discussed in Subsection 4.4.4.

commutes with the noninteracting Newton–Wigner position, i. e.,

$$[\boldsymbol{R}_0, M] = 0. \tag{4.20}$$

Our next task is to define the basis of the common eigenvectors of $|\boldsymbol{p}, m\rangle$ of the commuting operators $\boldsymbol{P}_0$ and M in $\mathscr{H}$.[7] These eigenvectors satisfy the conditions

$$\boldsymbol{P}_0|\boldsymbol{p}, m\rangle = \boldsymbol{p}|\boldsymbol{p}, m\rangle, \tag{4.21}$$
$$M|\boldsymbol{p}, m\rangle = m|\boldsymbol{p}, m\rangle. \tag{4.22}$$

Here we make an approximation that the decay interaction is rather weak and does not perturb the mass spectrum. So, we assume that the spectrum of M remains continuous in the interval $m \in [m_\beta + m_\gamma, \infty)$.

The vectors $|\boldsymbol{p}, m\rangle$ are also eigenvectors of the interacting Hamiltonian $H = \sqrt{M^2c^4 + P_0^2c^2}$, i. e.,

$$H|\boldsymbol{p}, m\rangle = \omega_{\boldsymbol{p}}|\boldsymbol{p}, m\rangle,$$

where $\omega_{\boldsymbol{p}} \equiv \sqrt{m^2c^4 + c^2p^2}$. The zero-momentum subspace contains a part of our basis. We have

$$\boldsymbol{P}_0|\boldsymbol{0}, m\rangle = \boldsymbol{0}.$$
$$M|\boldsymbol{0}, m\rangle = m|\boldsymbol{0}, m\rangle.$$

The basis $|\boldsymbol{p}, m\rangle$ in the entire space $\mathscr{H}$ is constructed by the formula[8]

$$|\boldsymbol{p}, m\rangle = \sqrt{\frac{mc^2}{\omega_{\boldsymbol{p}}}} e^{-\frac{ic}{\hbar}\boldsymbol{K}\cdot\boldsymbol{\vartheta}}|\boldsymbol{0}, m\rangle,$$

where the rapidity vector $\boldsymbol{\vartheta}$ is related to the momentum by the formula

$$\boldsymbol{p} = \frac{\boldsymbol{\vartheta}}{\vartheta} mc \sinh \vartheta.$$

Basis vectors are normalized to the delta function, i. e.,

$$\langle \boldsymbol{q}, m|\boldsymbol{p}, m'\rangle = \delta(\boldsymbol{q} - \boldsymbol{p})\delta(m - m'). \tag{4.23}$$

7 Complete sets of mutually commuting operators containing $\boldsymbol{P}_0$ and M have other independent observables as well, such as, for example, the square of the total angular momentum J_0^2 and its component J_{0z}. Therefore, an unambiguous labeling of basis vectors should include their eigenvalues as well, i. e., $|\boldsymbol{p}, m, j^2, j_z, \ldots\rangle$. However, these additional quantum numbers are not relevant to our description of the decay, so we omit them.

8 Compare with (**1**-5.5) and (**1**-5.29). Here $\boldsymbol{K}$ is the interacting boost operator whose explicit form will not be needed.

The action of inertial transformations on these states can be found in the same manner as in Sections **1**-5.1 and **1**-5.2. In particular, for time translations and boosts along the x-axis we have[9]

$$e^{-\frac{i}{\hbar}Ht}|\boldsymbol{p},m\rangle = e^{-\frac{i}{\hbar}\omega_{\boldsymbol{p}}t}|\boldsymbol{p},m\rangle, \tag{4.24}$$

$$e^{\frac{ic}{\hbar}K_x\theta}|\boldsymbol{p},m\rangle = \sqrt{\frac{\omega_{\boldsymbol{\theta}^{-1}\boldsymbol{p}}}{\omega_{\boldsymbol{p}}}}|\boldsymbol{\theta}^{-1}\boldsymbol{p},m\rangle, \tag{4.25}$$

$$\omega_{\boldsymbol{p}} = \sqrt{m^2c^4 + p^2c^2},$$

$$\boldsymbol{\theta}^{-1}\boldsymbol{p} \equiv \left(p_x \cosh\theta - \frac{\omega_{\boldsymbol{p}}}{c}\sinh\theta, p_y, p_z\right). \tag{4.26}$$

Next we notice that due to equations (**1**-4.25) and (4.20), vectors $e^{\frac{i}{\hbar}\boldsymbol{R}_0\cdot\boldsymbol{p}}|\boldsymbol{0},m\rangle$ also satisfy eigenvalue equations (4.21)–(4.22), because

$$\begin{aligned}
\boldsymbol{P}_0(e^{\frac{i}{\hbar}\boldsymbol{R}_0\cdot\boldsymbol{p}}|\boldsymbol{0},m\rangle) &= e^{\frac{i}{\hbar}\boldsymbol{R}_0\cdot\boldsymbol{p}}(e^{-\frac{i}{\hbar}\boldsymbol{R}_0\cdot\boldsymbol{p}}\boldsymbol{P}_0 e^{\frac{i}{\hbar}\boldsymbol{R}_0\cdot\boldsymbol{p}})|\boldsymbol{0},m\rangle \\
&= e^{\frac{i}{\hbar}\boldsymbol{R}_0\cdot\boldsymbol{p}}(\boldsymbol{P}_0 + \boldsymbol{p})|\boldsymbol{0},m\rangle = \boldsymbol{p}(e^{\frac{i}{\hbar}\boldsymbol{R}_0\cdot\boldsymbol{p}}|\boldsymbol{0},m\rangle), \\
M(e^{\frac{i}{\hbar}\boldsymbol{R}_0\cdot\boldsymbol{p}}|\boldsymbol{0},m\rangle) &= e^{\frac{i}{\hbar}\boldsymbol{R}_0\cdot\boldsymbol{p}}M|\boldsymbol{0},m\rangle = m(e^{\frac{i}{\hbar}\boldsymbol{R}_0\cdot\boldsymbol{p}}|\boldsymbol{0},m\rangle),
\end{aligned}$$

So, these vectors should be proportional to the basis vectors $|\boldsymbol{p},m\rangle$ and we can write

$$e^{\frac{i}{\hbar}\boldsymbol{R}_0\cdot\boldsymbol{p}}|\boldsymbol{0},m\rangle = \nu(\boldsymbol{p},m)|\boldsymbol{p},m\rangle, \tag{4.27}$$

where $\nu(\boldsymbol{p},m)$ is a unimodular factor, i. e.,

$$|\nu(\boldsymbol{p},m)| = 1. \tag{4.28}$$

In contrast to (4.17), here we cannot conclude that $\nu(\boldsymbol{p},m) = 1$. However, we assume that our interaction is not pathological and that the factor $\nu(\boldsymbol{p},m)$ is smooth, i. e., without rapid oscillations.

Obviously, the vector $|\boldsymbol{0}\rangle$ from the noninteracting basis (4.17) can be expressed as a linear combination of the interacting basis vectors $|\boldsymbol{0},m\rangle$ with zero momentum. So, we can write

$$|\boldsymbol{0}\rangle = \int_{m_\beta+m_\gamma}^{\infty} dm\mu(m)|\boldsymbol{0},m\rangle, \tag{4.29}$$

where $\mu(m)$ is a yet unknown function that depends on the choice of the interaction Hamiltonian V and has the following properties:

$$\mu(m) = \langle\boldsymbol{0},m|\boldsymbol{0}\rangle, \tag{4.30}$$

9 Compare with equations (**1**-5.9), (**1**-5.30) and (**1**-5.11).

$$\int_{m_\beta+m_\gamma}^{\infty} dm|\mu(m)|^2 = 1.$$

The physical meaning of $|\mu(m)|^2$ is the probability density for finding the eigenvalue m of the interacting mass M in the initial unstable state $|\mathbf{0}\rangle \in \mathcal{H}_\alpha$. We shall call this function the *mass distribution* of the unstable particle.

Next we use equations (4.17) and (4.27) to expand vectors $|\boldsymbol{p}\rangle \in \mathcal{H}_\alpha$ in the basis $|\boldsymbol{p}, m\rangle$, and obtain[10]

$$|\boldsymbol{p}\rangle = e^{\frac{i}{\hbar}\boldsymbol{R}_0\cdot\boldsymbol{p}}|\mathbf{0}\rangle = e^{\frac{i}{\hbar}\boldsymbol{R}_0\cdot\boldsymbol{p}} \int_{m_b+m_c}^{\infty} dm\mu(m)|\mathbf{0}, m\rangle = \int_{m_b+m_c}^{\infty} dm\mu(m)\nu(\boldsymbol{p}, m)|\boldsymbol{p}, m\rangle. \tag{4.31}$$

Hence, any state vector from the subspace $\mathcal{H}_\alpha$ can be written as

$$|\Psi\rangle = \int d\boldsymbol{p}\psi(\boldsymbol{p})|\boldsymbol{p}\rangle \tag{4.32}$$

$$= \int d\boldsymbol{p} \int_{m_\beta+m_\gamma}^{\infty} dm\mu(m)\nu(\boldsymbol{p}, m)\psi(\boldsymbol{p})|\boldsymbol{p}, m\rangle. \tag{4.33}$$

From (4.23) we also obtain the following useful result:

$$\langle\boldsymbol{q}|\boldsymbol{p}, m\rangle = \int_{m_\beta+m_\gamma}^{\infty} dm'\mu^*(m')\nu^*(\boldsymbol{q}, m')\langle\boldsymbol{q}, m'|\boldsymbol{p}, m\rangle = \nu^*(\boldsymbol{p}, m)\mu^*(m)\delta(\boldsymbol{q} - \boldsymbol{p}). \tag{4.34}$$

4.1.5 Decay law

Now we are ready to find the time evolution of the state vector (4.32) prepared at $t = 0$ inside the subspace $\mathcal{H}_\alpha$. Applying equations (4.10), (4.31) and (4.24), we obtain

$$\begin{aligned}
|\Psi(t)\rangle &= \int d\boldsymbol{p}\psi(\boldsymbol{p})e^{-\frac{i}{\hbar}Ht}|\boldsymbol{p}\rangle \\
&= \int d\boldsymbol{p}\psi(\boldsymbol{p}) \int_{m_\beta+m_\gamma}^{\infty} dm\mu(m)\nu(\boldsymbol{p}, m)e^{-\frac{i}{\hbar}Ht}|\boldsymbol{p}, m\rangle \\
&= \int d\boldsymbol{p}\psi(\boldsymbol{p}) \int_{m_\beta+m_\gamma}^{\infty} dm\mu(m)\nu(\boldsymbol{p}, m)e^{-\frac{i}{\hbar}\omega_{\boldsymbol{p}}t}|\boldsymbol{p}, m\rangle.
\end{aligned}$$

---

10 Note that this simple equality is possible only in the Bakamjian–Thomas dynamics, where the commutator (4.20) holds.

From (4.34) we get the following inner product of this vector with $|\boldsymbol{q}\rangle$:

$$\begin{aligned}\langle\boldsymbol{q}|\Psi(t)\rangle &= \int d\boldsymbol{p}\psi(\boldsymbol{p}) \int_{m_\beta+m_\gamma}^{\infty} dm\mu(m)\nu(\boldsymbol{p},m)e^{-\frac{i}{\hbar}\omega_{\boldsymbol{p}}t}\langle\boldsymbol{q}|\boldsymbol{p},m\rangle \\ &= \int d\boldsymbol{p}\psi(\boldsymbol{p}) \int_{m_\beta+m_\gamma}^{\infty} dm|\mu(m)|^2\nu(\boldsymbol{p},m)\nu^*(\boldsymbol{p},m)e^{-\frac{i}{\hbar}\omega_{\boldsymbol{p}}t}\delta(\boldsymbol{q}-\boldsymbol{p}) \\ &= \psi(\boldsymbol{q}) \int_{m_\beta+m_\gamma}^{\infty} dm|\mu(m)|^2 e^{-\frac{i}{\hbar}\omega_{\boldsymbol{q}}t}. \end{aligned} \tag{4.35}$$

Now, the decay law is obtained by substituting (4.16) in equation (4.11) and applying (4.35). We have

$$\begin{aligned}\Upsilon(t) &= \int d\boldsymbol{q}\langle\Psi(t)|\boldsymbol{q}\rangle\langle\boldsymbol{q}|\Psi(t)\rangle = \int d\boldsymbol{q}|\langle\boldsymbol{q}|\Psi(t)\rangle|^2 \\ &= \int d\boldsymbol{q}|\psi(\boldsymbol{q})|^2 \left|\int_{m_\beta+m_\gamma}^{\infty} dm|\mu(m)|^2 e^{-\frac{i}{\hbar}\omega_{\boldsymbol{q}}t}\right|^2. \end{aligned} \tag{4.36}$$

This formula makes sense only if the state $|\Psi\rangle \in \mathscr{H}_\alpha$ is *normalized*. For example, according to Subsection 4.1.3, a particle at rest can be described by a zero-momentum state $|\mathbf{0}\gg$, whose wave function is the “square root of the delta function.” We have

$$\begin{aligned}\psi_{\mathbf{0}}(\boldsymbol{q}) &\approx \sqrt{\delta(\boldsymbol{q})}, \\ \int d\boldsymbol{q}|\psi_{\mathbf{0}}(\boldsymbol{q})|^2 &\approx \int d\boldsymbol{q}\delta(\boldsymbol{q}) = 1. \end{aligned} \tag{4.37}$$

Then the decay law of the particle at rest,[11]

$$\Upsilon_{|\mathbf{0}\gg}(t) \approx \left|\int_{m_\beta+m_\gamma}^{\infty} dm|\mu(m)|^2 e^{-\frac{i}{\hbar}mc^2t}\right|^2, \tag{4.38}$$

is fully determined by the mass distribution $|\mu(m)|^2$. In the next section, we will consider an exactly solvable decay model, in which the functions $|\mu(m)|^2$ and $\Upsilon(t)$ can be calculated analytically.

4.2 Breit–Wigner formula

Here we would like to derive a beautiful result due to Breit and Wigner, which explains why the decay law $\Upsilon(t)$ is almost always exponential.

---

11 Compare, for example, with equation (3.8) in [85].

4.2.1 Schrödinger equation

In this section, we are dealing with the decay of a particle at rest. Therefore, it suffices to restrict ourselves to the subspace $\mathcal{H}_{\mathbf{0}} \subseteq \mathcal{H}$ of states having zero momentum. This subspace can be decomposed into a direct sum,

$$\mathcal{H}_0 = \mathcal{H}_{\alpha\mathbf{0}} \oplus \mathcal{H}_{(\beta\gamma)\mathbf{0}},$$

where

$$\mathcal{H}_{\alpha\mathbf{0}} = \mathcal{H}_{\mathbf{0}} \cap \mathcal{H}_\alpha$$

is the one-dimensional subspace that includes the zero-momentum state $|\mathbf{0}\rangle$ of the particle α and

$$\mathcal{H}_{(\beta\gamma)\mathbf{0}} = \mathcal{H}_{\mathbf{0}} \cap (\mathcal{H}_\beta \otimes \mathcal{H}_\gamma)$$

is the subspace of the decay products with vanishing total momentum $\boldsymbol{P}_0 = \boldsymbol{p}_\beta + \boldsymbol{p}_\gamma = 0$. Hence, vectors of the two-particle basis $|\boldsymbol{\rho}\rangle$ in $\mathcal{H}_{(\beta\gamma)\mathbf{0}}$ can be labeled by eigenvalues of the relative momentum,

$$\boldsymbol{\rho} = \boldsymbol{p}_\beta = -\boldsymbol{p}_\gamma,$$

and each state $|\Psi\rangle \in \mathcal{H}_{\mathbf{0}}$ can be expanded in the above basis $\{|\mathbf{0}\rangle, |\boldsymbol{\rho}\rangle\}$, i. e.,

$$|\Psi\rangle = \mu^* |\mathbf{0}\rangle + \int d\boldsymbol{\rho}\zeta(\boldsymbol{\rho})|\boldsymbol{\rho}\rangle.$$

The coefficients of this expansion will be represented as an infinite column vector, i. e.,

$$|\Psi\rangle = \begin{bmatrix} \mu^* \\ \zeta(\boldsymbol{\rho}_1) \\ \zeta(\boldsymbol{\rho}_2) \\ \zeta(\boldsymbol{\rho}_3) \\ \dots \end{bmatrix}.$$

The zeroth component is a complex number $\mu^* \equiv \langle\mathbf{0}|\Psi\rangle$, and other components represent the complex function $\zeta(\boldsymbol{\rho}) \equiv \langle\boldsymbol{\rho}|\Psi\rangle$ at different values of the relative momentum $\boldsymbol{\rho}$.[12] For brevity, we will also use the following notation:

$$|\Psi\rangle = \begin{bmatrix} \mu^* \\ \zeta(\boldsymbol{\rho}) \end{bmatrix}. \tag{4.39}$$

12 Of course, the spectrum of $\boldsymbol{\rho}$ is continuous and, strictly speaking, cannot be described by the set of discrete values $\boldsymbol{\rho}_i$. However, we can justify our approximation by the usual trick of placing the system inside a finite box. Then the momentum spectrum becomes discrete, and in the limit of infinite box size we return to the familiar situation of the continuous spectrum.

The wave function of the normalized state vector $|\Psi\rangle$ satisfies the condition

$$|\mu|^2 + \int d\boldsymbol{p}|\zeta(\boldsymbol{p})|^2 = 1. \tag{4.40}$$

The probability that a measurement performed on the state $|\Psi\rangle$ will find the unstable particle α is

$$\Upsilon = |\mu|^2.$$

According to our assumption, in the initial state

$$|\Psi(0)\rangle = |\mathbf{0}\rangle = \begin{bmatrix} 1 \\ 0 \end{bmatrix} \tag{4.41}$$

this particle is found with 100 % probability.

Next we can get representations of different operators in the basis $\{|\mathbf{0}\rangle, |\boldsymbol{p}\rangle\}$. The matrix of the free Hamiltonian is diagonal. We have

$$H_0 = \begin{bmatrix} m_\alpha c^2 & 0 & 0 & 0 & \cdots \\ 0 & \eta_{p_1} c^2 & 0 & 0 & \cdots \\ 0 & 0 & \eta_{p_2} c^2 & 0 & \cdots \\ 0 & 0 & 0 & \eta_{p_3} c^2 & \cdots \\ \cdots & \cdots & \cdots & \cdots & \ddots \end{bmatrix} \equiv \begin{bmatrix} m_\alpha c^2 & 0 \\ 0 & \eta_p c^2 \end{bmatrix},$$

where

$$\eta_p = \frac{1}{c^2}\left(\sqrt{m_\beta^2 c^4 + c^2 p^2} + \sqrt{m_\gamma^2 c^4 + c^2 p^2}\right) \tag{4.42}$$

is the mass of the two-particle system $(\beta + \gamma)$ expressed as a function of the relative momentum p. In the subspace $\mathscr{H}_0$, the interaction operator (4.9) takes the form

$$\begin{aligned} V &= \int d\boldsymbol{p}[V(\boldsymbol{p}, -\boldsymbol{p})\alpha_{\mathbf{0}}^\dagger \beta_{\boldsymbol{p}} \gamma_{-\boldsymbol{p}} + V^*(\boldsymbol{p}, -\boldsymbol{p})\gamma_{-\boldsymbol{p}}^\dagger \beta_{\boldsymbol{p}}^\dagger \alpha_{\mathbf{0}}] \\ &\equiv \int d\boldsymbol{p}[g(\boldsymbol{p})\alpha_{\mathbf{0}}^\dagger \beta_{\boldsymbol{p}} \gamma_{-\boldsymbol{p}} + g^*(\boldsymbol{p})\gamma_{-\boldsymbol{p}}^\dagger \beta_{\boldsymbol{p}}^\dagger \alpha_{\mathbf{0}}]. \end{aligned}$$

In our basis, it is represented by the following matrix[13]:

$$V = \begin{bmatrix} 0 & g(\boldsymbol{p}_1) & g(\boldsymbol{p}_2) & g(\boldsymbol{p}_3) & \cdots \\ g^*(\boldsymbol{p}_1) & 0 & 0 & 0 & \cdots \\ g^*(\boldsymbol{p}_2) & 0 & 0 & 0 & \cdots \\ g^*(\boldsymbol{p}_3) & 0 & 0 & 0 & \cdots \\ \cdots & \cdots & \cdots & \cdots & \ddots \end{bmatrix}$$

13 Here $g(\boldsymbol{p}) \equiv V(\boldsymbol{p}, -\boldsymbol{p})$, and the symbol $\int d\boldsymbol{q}g(\boldsymbol{q})\ldots$ denotes a linear functional that yields a number $\int d\boldsymbol{q}g(\boldsymbol{q})\zeta(\boldsymbol{q})$ when acting on an arbitrary test function $\zeta(\boldsymbol{q})$.

$$\equiv \begin{bmatrix} 0 & \int d\boldsymbol{q}g(\boldsymbol{q})\dots \\ g^*(\boldsymbol{p}) & 0 \end{bmatrix},$$

where $g(\boldsymbol{p})$ is the matrix element of the interaction operator taken between states $|\mathbf{0}\rangle \equiv \alpha_0^\dagger|\text{vac}\rangle$ and $|\boldsymbol{p}\rangle \equiv \gamma_{-\boldsymbol{p}}^\dagger\beta_{\boldsymbol{p}}^\dagger|\text{vac}\rangle$, i. e.,

$$g(\boldsymbol{p}) = \langle\mathbf{0}|V|\boldsymbol{p}\rangle. \tag{4.43}$$

Therefore, the full Hamiltonian $H = H_0 + V$ acts on vectors (4.39) as follows:

$$H\begin{bmatrix} \mu^* \\ \zeta(\boldsymbol{p}) \end{bmatrix} = \begin{bmatrix} m_\alpha c^2 & g(\boldsymbol{p}_1) & g(\boldsymbol{p}_2) & g(\boldsymbol{p}_3) & \cdots \\ g^*(\boldsymbol{p}_1) & \eta_{\boldsymbol{p}_1}c^2 & 0 & 0 & \cdots \\ g^*(\boldsymbol{p}_2) & 0 & \eta_{\boldsymbol{p}_2}c^2 & 0 & \cdots \\ g^*(\boldsymbol{p}_3) & 0 & 0 & \eta_{\boldsymbol{p}_3}c^2 & \cdots \\ \cdots & \cdots & \cdots & \cdots & \ddots \end{bmatrix}\begin{bmatrix} \mu^* \\ \zeta(\boldsymbol{p}_1) \\ \zeta(\boldsymbol{p}_2) \\ \zeta(\boldsymbol{p}_3) \\ \cdots \end{bmatrix}$$

$$= \begin{bmatrix} m_\alpha c^2\mu^* + \int d\boldsymbol{q}g(\boldsymbol{q})\zeta(\boldsymbol{q}) \\ g^*(\boldsymbol{p}_1)\mu^* + \eta_{\boldsymbol{p}_1}\zeta(\boldsymbol{p}_1)c^2 \\ g^*(\boldsymbol{p}_2)\mu^* + \eta_{\boldsymbol{p}_2}\zeta(\boldsymbol{p}_2)c^2 \\ g^*(\boldsymbol{p}_3)\mu^* + \eta_{\boldsymbol{p}_3}\zeta(\boldsymbol{p}_3)c^2 \\ \cdots \end{bmatrix}$$

$$\equiv \begin{bmatrix} m_\alpha c^2\mu^* + \int d\boldsymbol{q}g(\boldsymbol{q})\zeta(\boldsymbol{q}) \\ g^*(\boldsymbol{p})\mu^* + \eta_{\boldsymbol{p}}\zeta(\boldsymbol{p})c^2 \end{bmatrix}.$$

The next step is to find eigenvalues[14] and eigenvectors

$$|\mathbf{0}, \mathrm{m}_i\rangle \equiv \begin{bmatrix} \mu^*(\mathrm{m}_i) \\ \zeta_{\mathrm{m}_i}(\boldsymbol{p}) \end{bmatrix} \tag{4.44}$$

of the full Hamiltonian $H = Mc^2$, i. e.,

$$H|\mathbf{0}, \mathrm{m}_i\rangle = \mathrm{m}_i c^2|\mathbf{0}, \mathrm{m}_i\rangle. \tag{4.45}$$

In the infinite box limit, these eigenvalues coalesce into the continuous spectrum $m \in [m_\beta + m_\gamma, \infty)$ of the interacting mass operator. Then the matrix problem presented above is equivalent to solving the following system of equations:

$$m_\alpha c^2\mu^*(m) + \int d\boldsymbol{q}g(\boldsymbol{q})\zeta_m(\boldsymbol{q}) = mc^2\mu^*(m), \tag{4.46}$$

$$g^*(\boldsymbol{p})\mu^*(m) + \eta_{\boldsymbol{p}}c^2\zeta_m(\boldsymbol{p}) = mc^2\zeta_m(\boldsymbol{p}). \tag{4.47}$$

14 We are denoting the discrete eigenvalues of M by $\mathrm{m}_i c^2$, where index i runs through all integers from 1 to ∞.

Equation (4.47) implies

$$\zeta_m(\boldsymbol{\rho}) = \frac{g^*(\boldsymbol{\rho})\mu^*(m)}{mc^2 - \eta_{\boldsymbol{\rho}}c^2}. \tag{4.48}$$

Inserting this result in (4.46), we obtain a nonlinear equation, determining the spectrum of eigenvalues $\mathfrak{m}_i$. We have

$$m - m_\alpha = \frac{1}{c^4}\int d\boldsymbol{\rho}\frac{|g(\boldsymbol{\rho})|^2}{m - \eta_{\boldsymbol{\rho}}}. \tag{4.49}$$

To ensure conservation of the angular momentum, the function $|g(\boldsymbol{\rho})|$ must depend only on the modulus (ρ) of its argument. So, we rewrite equation (4.49) in the form

$$m - m_\alpha = F(m), \tag{4.50}$$

where we introduce the notation

$$F(m) \equiv \int_0^\infty d\rho\frac{G(\rho)}{m - \eta_\rho}, \tag{4.51}$$

$$G(\rho) \equiv \frac{4\pi\rho^2}{c^4}|g(\rho)|^2. \tag{4.52}$$

From the normalization condition (4.40)

$$|\mu(m)|^2 + \int d\boldsymbol{\rho}|\zeta_m(\boldsymbol{\rho})|^2 = 1 \tag{4.53}$$

and equation (4.48), we finally obtain the following formula for the mass distribution:

$$|\mu(m)|^2\left(1 + \int d\rho\frac{G(\rho)}{(m - \eta_\rho)^2}\right) = 1,$$

which implies

$$|\mu(m)|^2 = \frac{1}{1 - F'(m)}, \tag{4.54}$$

where $F'(m)$ is the derivative of (4.51). This means that in order to calculate the desired decay law (4.38), we have to find the derivatives $F'(\mathfrak{m})$ at points $\mathfrak{m}$ of the interacting mass spectrum.[15] In the next subsection, we consider in detail the solution of this nontrivial problem.

15 That is, the points that are solutions of the nonlinear equation (4.49).

4.2.2 Discrete approximation

Now return to the discrete approximation where equation (4.49) takes the form

$$m - m_\alpha = \sum_{i=1}^{\infty} \frac{G(\rho_i)}{m - m_i}. \tag{4.55}$$

Let us turn to its graphical solution in Figure 4.2. On the mass axis (m), the points of the spectrum ($m_i \equiv \eta_{\rho_i}$) of the noninteracting mass operator M_0 are shown by hollow circles. The lowest (threshold) eigenvalue is denoted by $m_1 = m_\beta + m_\gamma = \eta_0$. The function $F(m)$ on the right-hand side of (4.55) is shown by a bold solid line. It has singularities at the points m_i.

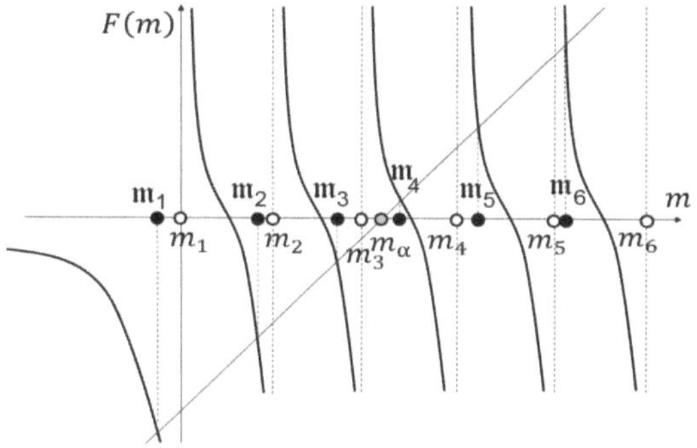

Figure 4.2: Graphical solution of equation (4.55). Hollow circles mark the spectrum of the operator $H_0/c^2 = M_0$. Filled circles mark the spectrum of the operator $H/c^2 = M$.

According to equation (4.55), the spectrum of the interacting mass consists of points where the line $m - m_\alpha$ intersects the graph of $F(m)$. These points $\mathfrak{m}_i$ ($i = 1, 2, 3, \ldots$) are shown by filled circles in the figure. This means that the derivatives required in equation (4.54) are the slopes of the function $F(m)$ at the points $\mathfrak{m}_1, \mathfrak{m}_2, \mathfrak{m}_3, \ldots$.[16]

We will solve this problem separately below the threshold $m_1 = m_\beta + m_\gamma$ and above it.

In the former region $(-\infty, m_\beta + m_\gamma)$, $F(m)$ is a continuous smooth function. It tends to zero in the limit $m \to -\infty$ and decreases monotonously with increasing m. Moreover, this functions tends to a well-defined shape in the large box limit. Thus, the graphical solution of equation (4.55) in this region is obtained as the intersection of the line $m - m_\alpha$ and the graph of the function $F(m)$. The corresponding value $\mathfrak{m}_1 < m_\beta + m_\gamma$ is a discrete eigenvalue of the interacting mass operator. The related eigenstate is a superposition of the unstable particle α and its decay products $\beta + \gamma$.

16 Note that these derivatives are negative. This is consistent with the fact that on the right-hand side of (4.54) we expect to have a strictly positive quantity.

In the latter region $(m_\beta + m_\gamma, +\infty)$, the situation is less certain, because in the large box limit, the distances between the neighboring poles m_i (and points $\mathfrak{m}_i$) tend to zero, the function $F(m)$ oscillates wildly and its derivative tends to (minus) infinity everywhere.

To overcome this difficulty, let us first return temporarily to the continuous spectrum of p and change the integration variable in (4.51). We start with

$$z = \eta_p \equiv \frac{1}{c^2}\left(\sqrt{m_\beta^2 c^4 + p^2c^2} + \sqrt{m_\gamma^2 c^4 + p^2c^2}\right),$$

so that the inverse function

$$p = \eta^{-1}(z)$$

expresses the relative momentum p as a function of the total mass z of the decay products. Then, denoting

$$\Gamma(z) \equiv 2\pi \frac{d\eta^{-1}(z)}{dz} G(\eta^{-1}(z)) \tag{4.56}$$

and choosing a point of interest $\overline{m}$ from the interval $(m_\beta + m_\gamma, \infty)$, we obtain from (4.51) the desired function near this point. We have

$$\begin{aligned} F(m) &= \int_{m_\beta+m_\gamma}^{\infty} dz \left|\frac{dp}{dz}\right| \frac{G(\eta^{-1}(z))}{m - z} = \int_{m_\beta+m_\gamma}^{\infty} dz \frac{\Gamma(z)}{2\pi(m-z)} \\ &= \int_{m_\beta+m_\gamma}^{\overline{m}-\Delta} dz \frac{\Gamma(z)}{2\pi(m-z)} + \int_{\overline{m}-\Delta}^{\overline{m}+\Delta} dz \frac{\Gamma(z)}{2\pi(m-z)} + \int_{\overline{m}+\Delta}^{\infty} dz \frac{\Gamma(z)}{2\pi(m-z)}. \end{aligned} \tag{4.57}$$

Here we divided the integration interval $[m_\beta + m_\gamma, +\infty)$ into three segments. When $\Delta \to 0$, the sum of the first and the third terms on the right-hand side of (4.57) gives the *principal value* integral that we denoted $\mathcal{P}(m)$. We have

$$\int_{m_\beta+m_\gamma}^{\overline{m}-\Delta} dz \frac{\Gamma(z)}{2\pi(m-z)} + \int_{\overline{m}+\Delta}^{\infty} dz \frac{\Gamma(z)}{2\pi(m-z)} \xrightarrow{\Delta\to 0} P \int_{m_\beta+m_\gamma}^{\infty} dz \frac{\Gamma(z)}{2\pi(m-z)} \equiv \mathcal{P}(m). \tag{4.58}$$

Next we take a closer look at the second integral on the right-hand side of (4.57). If the parameter Δ is sufficiently small,[17] then the function $\Gamma(z)$ can be regarded as being constant, i. e., $\Gamma(z) = \Gamma(\overline{m})$, in the interval $[\overline{m} - \Delta, \overline{m} + \Delta]$.

In the discrete approximation, the density of the spectrum points m_j in the interval $[\overline{m} - \Delta, \overline{m} + \Delta]$ also can be regarded as a constant. So, the poles of the function $F(m)$

17 But still much larger than the separation between neighboring points m_j and m_{j+1} of the discrete mass spectrum.

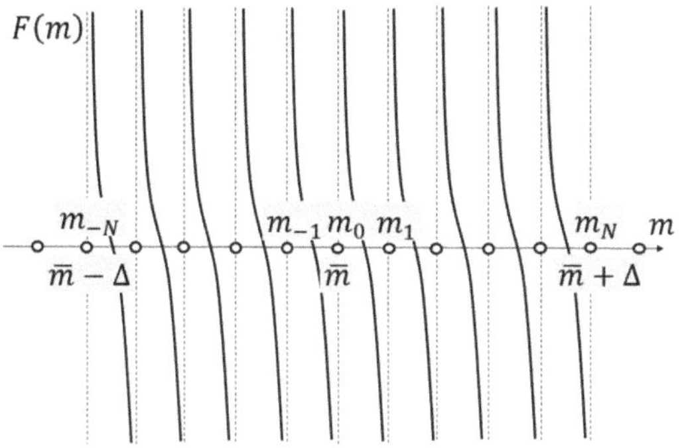

Figure 4.3: To the derivation of formula (4.60).

divide this interval into $2N$ small equal segments (see Figure 4.3), i. e.,

$$m_j = m_0 + j\frac{\Delta}{N}, \tag{4.59}$$

where $m_0 = \overline{m}$, the index j runs from $-N$ to N and the integral is approximated by the *partial sum*

$$\int_{\overline{m}-\Delta}^{\overline{m}+\Delta} dz \frac{\Gamma(z)}{2\pi(m-z)} \approx \frac{\Gamma(\overline{m})}{2\pi} \int_{\overline{m}-\Delta}^{\overline{m}+\Delta} dz \frac{1}{m-z} \approx \frac{\Gamma(\overline{m})}{2\pi} \sum_{j=-N}^{N} \frac{\Delta/N}{m - m_0 - j\frac{\Delta}{N}}. \tag{4.60}$$

In the limits $\Delta \to 0$ and $N \to \infty$, the index j runs through all values from $-\infty$ to ∞. Then the right-hand side of (4.60) defines some analytical function whose poles are spread uniformly (4.59) on the real axis and have equal residues

$$\frac{\Gamma(\overline{m})\Delta}{2\pi N}. \tag{4.61}$$

Since any analytical function is determined uniquely by its poles and residues, we conclude that the integral (4.60) has the following representation:

$$\int_{\overline{m}-\Delta}^{\overline{m}+\Delta} dz \frac{\Gamma(z)}{2\pi(m-z)} \approx \frac{\Gamma(\overline{m})}{2} \cot\left(\frac{\pi N}{\Delta}(m - m_0)\right). \tag{4.62}$$

Indeed, the cotangent function on the right-hand side of (4.62) also has poles located at (4.59) with the same residues (4.61). Now we can combine our results (4.58) and (4.62) to write for all values of $m \in (m_\beta + m_\gamma, +\infty)$

$$F(m) = \mathcal{P}(m) + \frac{\Gamma(m)}{2} \cot\left(\frac{\pi N m}{\Delta}\right).$$

Next, using

$$\cot(ax)' = -a(1 + \cot^2(ax))$$

and neglecting derivatives of smooth functions $\mathcal{P}(m)$ and $\Gamma(m)$, we obtain

$$F'(m) \approx -\frac{\pi\Gamma(m)N}{2\Delta}\left(1+\cot^2\left(\frac{\pi Nm}{\Delta}\right)\right). \tag{4.63}$$

As we said already, in the desired limit $N \to \infty$ the density of the poles grows, and the function $F(m)$ oscillates wildly between $-\infty$ and ∞. However, our task is simplified by the fact that for (4.54) we only need derivatives $F'(m)$ at very specific points, which are solutions of the equation

$$F(m) = m - m_\alpha.$$

At these points we have

$$m - m_\alpha = \mathcal{P}(m) + \frac{\Gamma(m)}{2}\cot\left(\frac{\pi Nm}{\Delta}\right),$$
$$\cot\left(\frac{\pi Nm}{\Delta}\right) = \frac{2(m - m_\alpha - \mathcal{P}(m))}{\Gamma(m)},$$
$$\cot^2\left(\frac{\pi Nm}{\Delta}\right) = \frac{4(m - m_\alpha - \mathcal{P}(m))^2}{\Gamma^2(m)}.$$

Substituting this result in equations (4.63) and (4.54), we get the desired answer:

$$F'(m) = -\pi\frac{\Gamma(m)N}{2\Delta}\left(1+\frac{4(m-m_\alpha-\mathcal{P}(m))^2}{\Gamma^2(m)}\right),$$
$$|\mu(m)|^2 = \frac{1}{1+\pi\Gamma(m)N/(2\Delta)(1+4(m_\alpha+\mathcal{P}(m)-m)^2/\Gamma^2(m))} \tag{4.64}$$
$$\approx \frac{\Gamma(m)\Delta/(2\pi N)}{\Gamma^2(m)/4+(m_\alpha+\mathcal{P}(m)-m)^2}, \tag{4.65}$$

where in the denominator of (4.64) we neglected the unity in comparison with the much greater term $\propto N/\Delta$.

4.2.3 Mass distribution

Equation (4.65) gives the probability of finding the particle α at each point of the discrete mass spectrum $\mathfrak{m}_2, \mathfrak{m}_3, \mathfrak{m}_4, \ldots$. Naturally, these probabilities tend to zero when the box's size is growing and the density of mass eigenvalues $N\Delta^{-1}$ goes to infinity. However, in this continuous limit, we are not interested in the (vanishing) probability at each point of the spectrum. The more important quantity is the *probability density*, which can be obtained by multiplying the right-hand side of (4.65) by the number ($N\Delta^{-1}$) of mass eigenvalues in the unit interval. Then, in the continuous limit, the mass distribution of the unstable particle converges to the famous *Breit–Wigner* formula

$$|\mu(m)|^2 = \frac{\Gamma(m)/(2\pi)}{\Gamma^2(m)/4+(m_\alpha+\mathcal{P}(m)-m)^2}. \tag{4.66}$$

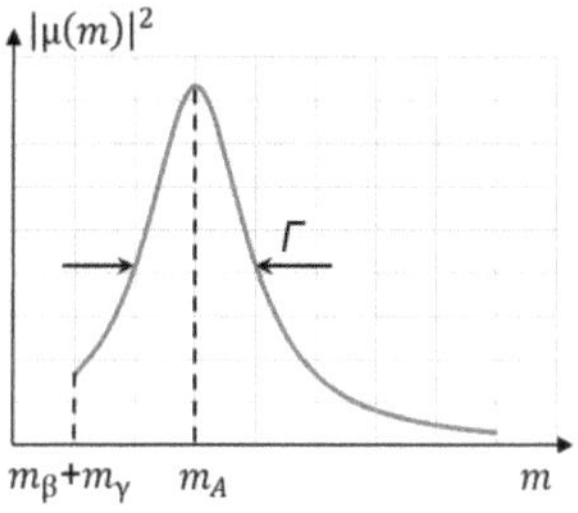

Figure 4.4: Mass distribution of a typical unstable particle.

This resonance distribution is shown in Figure 4.4. It describes an unstable particle with the expectation value of mass (= center of the resonance) $m_A \equiv m_\alpha + \mathcal{P}(m_A)$ and the width of the resonance $\Delta m \approx \Gamma(m_A)$.

For unstable particle whose decays are sufficiently slow, so that their time dependence can be observed in experiments, the resonance in Figure 4.4 is very narrow, so we can replace functions $\Gamma(m)$ and $\mathcal{P}(m)$ by their values (constants) at the center of the resonance: $\Gamma \equiv \Gamma(m_A)$ and $\mathcal{P} \equiv \mathcal{P}(m_A)$. We will also neglect a small contribution from the isolated point $\mathfrak{m}_1 < m_\beta + m_\gamma$ of the mass spectrum.[18] Then the mass distribution vanishes below the threshold ($m < m_\beta + m_\gamma$), while above the threshold ($m > m_\beta + m_\gamma$) we get our final formula:

$$|\mu(m)|^2 \approx \frac{\Gamma/(2\pi)}{\Gamma^2/4 + (m_A - m)^2}. \tag{4.67}$$

4.2.4 Exponential decay law

To conclude our discussion of the unstable particle at rest, we are going to calculate its decay law. The dynamics of the initial state (4.41) (= pure particle α) is described by the time evolution operator

$$|\mathbf{0}, t\rangle = e^{-\frac{i}{\hbar}Ht}|\mathbf{0}\rangle.$$

To evaluate this expression, it will be convenient to use expansion (4.29) for the vector $|\mathbf{0}\rangle$. Then, from (4.44), (4.45) and (4.48) we obtain

$$\begin{aligned} e^{-\frac{i}{\hbar}Ht}|\mathbf{0}\rangle &= \int\limits_{m_\beta+m_\gamma}^{\infty} dm\mu(m)e^{-\frac{i}{\hbar}Ht}|\mathbf{0}, m\rangle = \int\limits_{m_\beta+m_\gamma}^{\infty} dm\mu(m)e^{-\frac{i}{\hbar}mc^2t}|\mathbf{0}, m\rangle \\ &= \int\limits_{m_\beta+m_\gamma}^{\infty} dme^{-\frac{i}{\hbar}mc^2t}|\mu(m)|^2 \begin{bmatrix} 1 \\ g^*(\boldsymbol{\rho})/(mc^2 - \eta_{\boldsymbol{\rho}}c^2) \end{bmatrix} \\ &\equiv \begin{bmatrix} I(t) \\ J(\boldsymbol{\rho}, t) \end{bmatrix}. \end{aligned} \tag{4.68}$$

18 See the beginning of Subsection 4.2.2.

In accordance with (4.38), the integral $I(t)$ defines the probability of finding the particle α at time t, i. e., the decay law of this particle.[19] Inserting (4.67) in the integrand, we obtain

$$\Upsilon_{|\mathbf{0}\gg}(t) = |I(t)|^2 \approx \frac{1}{4\pi^2}\left|\int_{m_\beta+m_\gamma}^{\infty} dm \frac{\Gamma e^{-\frac{i}{\hbar}mc^2t}}{\Gamma^2/4 + (m_A - m)^2}\right|^2. \tag{4.69}$$

For most unstable particles,[20]

$$\Gamma \ll m_A - (m_\beta + m_\gamma), \tag{4.70}$$

so the integrand is well localized near the value $m \approx m_A$, and we can introduce an additional approximation by setting the low integration limit in (4.69) equal to $-\infty$. Then, the integral is calculated by the method from Appendix C (by setting $p = 0$ there), and the decay law takes the familiar exponential form

$$\Upsilon(t) \approx \frac{1}{4\pi^2}\left|-2\pi e^{-\frac{i}{\hbar}m_Ac^2t}\exp\left(-\frac{\Gamma c^2 t}{2\hbar}\right)\right|^2 = \exp\left(-\frac{\Gamma c^2 t}{\hbar}\right) = \exp\left(-\frac{t}{\tau_0}\right), \tag{4.71}$$

where

$$\tau_0 = \frac{\hbar}{\Gamma c^2} \tag{4.72}$$

is the *lifetime* of the unstable particle at rest. The nondecay probability drops from 1 to $1/e$ during the lifetime.

With the help of equations (4.43), (4.52) and (4.56) we also see that the *decay rate*

$$\begin{aligned}\frac{1}{\tau_0} &= \frac{\Gamma c^2}{\hbar} = \frac{2\pi c^2}{\hbar} G(\eta^{-1}(z)) \frac{d\eta^{-1}(z)}{dz}\bigg|_{z=m_A} \\ &= \frac{8\pi^2\eta^2(z)}{c^2\hbar} |\langle \mathbf{0}|V|\eta^{-1}(z)\rangle|^2 \frac{d\eta^{-1}(z)}{dz}\bigg|_{z=m_A}\end{aligned} \tag{4.73}$$

is proportional to the square of the matrix element of perturbation V, taken between the initial and final states of the system. It is also proportional to the kinematical factor $d\eta^{-1}(z)/dz|_{z=m_A}$, which is fully determined by the three involved masses m_A, m_β and m_γ (see equation (4.42)).

---

19 The second integral $J(\boldsymbol{p}, t)$ in (4.68) describes the evolution of the wave function of the decay products β and γ. We are not going to discuss it here.

20 Approximation (4.70) can be violated for particles (or *resonances*) decaying due to strong nuclear forces. In these cases, the parameter Γ is comparable with the particle mass and the lifetime is very short ($\tau_0 \approx 10^{-23}$ s), so that the time dependence of the decay law cannot be studied experimentally. Such short-lived particles are identified by resonance behavior of scattering cross sections.

Formulas (4.67), (4.71) and (4.73) are important, because they were derived from very general premises. In our derivation we did not employ the perturbation theory. In fact, the only significant approximation was the assumption about the weakness of the decay interaction V. This condition is satisfied for most known decays, so it is not surprising that the exponential decay law is applicable to diverse unstable systems.

4.3 Decay law of moving particle

Formula (4.36) is the decay law $\Upsilon(0,t)$, observed from the reference frame O at rest. In this section, we will be interested in the decay law $\Upsilon(\theta,t')$ seen from the moving frame O'.[21] Particular cases of this formula for states with sharp momentum or velocity will be considered in Subsections 4.3.2 and 4.3.4, respectively.

4.3.1 General formula for decay law

Suppose that an observer O at rest describes the initial state (at $t = 0$) by the vector $|\Psi\rangle$ in the Hilbert space. Then the observer O' moving along the axis x describes the same state by the vector[22]

$$|\Psi(\theta,0)\rangle = e^{\frac{ic}{\hbar}K_x\theta}|\Psi\rangle.$$

The time dependence of this state is given by the formula[23]

$$|\Psi(\theta,t')\rangle = e^{-\frac{i}{\hbar}Ht'}e^{\frac{ic}{\hbar}K_x\theta}|\Psi\rangle. \tag{4.74}$$

According to our result (4.5), the decay law from the point of view of O' is

$$\begin{aligned}\Upsilon(\theta,t') &= \langle\Psi(\theta,t')|T|\Psi(\theta,t')\rangle \qquad (4.75)\\ &= \||T|\Psi(\theta,t')\rangle\|^2\\ &= \langle\Psi|e^{-\frac{ic}{\hbar}K_x\theta}e^{\frac{i}{\hbar}Ht'}Te^{-\frac{i}{\hbar}Ht'}e^{\frac{ic}{\hbar}K_x\theta}|\Psi\rangle. \qquad (4.76)\end{aligned}$$

Next we expand vector $|\Psi\rangle$ in the basis (4.33) and use equations (4.74), (4.25) and (4.24) to obtain

$$|\Psi(\theta,t')\rangle = \int d\boldsymbol{p}\psi(\boldsymbol{p})e^{-\frac{i}{\hbar}Ht'}e^{\frac{ic}{\hbar}K_x\theta}|\boldsymbol{p}\rangle$$

---

21 As usual, θ is the observer's rapidity, which is related to its speed v by the formula $v = c\tanh\theta$, and t' is time measured by the moving observer's clock.

22 See formula (1-5.54) for passive transformations of state vectors. We work in the Schrödinger representation, where operators of observables are the same in all frames of reference.

23 Here t' is time measured by the moving observer O'.

$$= \int d\mathbf{p}\psi(\mathbf{p}) \int\limits_{m_\beta+m_\gamma}^{\infty} dm\mu(m)\nu(\mathbf{p},m)e^{-\frac{i}{\hbar}Ht'}e^{\frac{ic}{\hbar}K_x\theta}|\mathbf{p},m\rangle$$

$$= \int d\mathbf{p}\psi(\mathbf{p}) \int\limits_{m_\beta+m_\gamma}^{\infty} dm\mu(m)\nu(\mathbf{p},m)e^{-\frac{i}{\hbar}\omega_{\boldsymbol{\theta}^{-1}\mathbf{p}}t'}\sqrt{\frac{\omega_{\boldsymbol{\theta}^{-1}\mathbf{p}}}{\omega_{\mathbf{p}}}}|\boldsymbol{\theta}^{-1}\mathbf{p},m\rangle.$$

The inner product of this vector with $|\mathbf{q}\rangle$ can be found with the help of (4.34), (**1**-5.31) and the introduction of a new integration variable $\mathbf{y} = \boldsymbol{\theta}^{-1}\mathbf{p} \equiv (p_x\cosh\theta - (\omega_{\mathbf{p}}/c)\sinh\theta, p_y, p_z)$. We have

$$\begin{aligned}
\langle \mathbf{q}|\Psi(\theta,t')\rangle &= \int d\mathbf{p}\psi(\mathbf{p}) \int\limits_{m_\beta+m_\gamma}^{\infty} dm\mu(m)\nu(\mathbf{p},m)e^{-\frac{i}{\hbar}\omega_{\boldsymbol{\theta}^{-1}\mathbf{p}}t'}\langle \mathbf{q}|\boldsymbol{\theta}^{-1}\mathbf{p},m\rangle\sqrt{\frac{\omega_{\boldsymbol{\theta}^{-1}\mathbf{p}}}{\omega_{\mathbf{p}}}} \\
&= \int d\mathbf{p}\psi(\mathbf{p}) \int\limits_{m_\beta+m_\gamma}^{\infty} dm|\mu(m)|^2\nu(\mathbf{p},m)\nu^*(\boldsymbol{\theta}^{-1}\mathbf{p},m)e^{-\frac{i}{\hbar}\omega_{\boldsymbol{\theta}^{-1}\mathbf{p}}t'} \\
&\quad \times \delta(\mathbf{q}-\boldsymbol{\theta}^{-1}\mathbf{p})\sqrt{\frac{\omega_{\boldsymbol{\theta}^{-1}\mathbf{p}}}{\omega_{\mathbf{p}}}} \\
&= \int\limits_{m_\beta+m_\gamma}^{\infty} dm \int d\mathbf{y}\frac{\omega_{\boldsymbol{\theta}\mathbf{y}}}{\omega_{\mathbf{y}}}\sqrt{\frac{\omega_{\mathbf{y}}}{\omega_{\boldsymbol{\theta}\mathbf{y}}}}\psi(\boldsymbol{\theta}\mathbf{y})\nu(\boldsymbol{\theta}\mathbf{y},m)\nu^*(\mathbf{y},m)|\mu(m)|^2e^{-\frac{i}{\hbar}\omega_{\mathbf{y}}t'}\delta(\mathbf{q}-\mathbf{y}) \\
&= \int\limits_{m_\beta+m_\gamma}^{\infty} dm\sqrt{\frac{\omega_{\boldsymbol{\theta}\mathbf{q}}}{\omega_{\mathbf{q}}}}\psi(\boldsymbol{\theta}\mathbf{q})\nu(\boldsymbol{\theta}\mathbf{q},m)\nu^*(\mathbf{q},m)|\mu(m)|^2e^{-\frac{i}{\hbar}\omega_{\mathbf{q}}t'}.
\end{aligned}$$

The decay law in the frame O' for all values of θ and t' is now obtained by substituting (4.16) into equation (4.75). We have

$$\begin{aligned}
\Upsilon(\theta,t') &= \int d\mathbf{q}\langle\Psi(\theta,t')|\mathbf{q}\rangle\langle\mathbf{q}|\Psi(\theta,t')\rangle = \int d\mathbf{q}|\langle\mathbf{q}|\Psi(\theta,t')\rangle|^2 \\
&= \int d\mathbf{q}\left|\int\limits_{m_\beta+m_\gamma}^{\infty} dm\sqrt{\frac{\omega_{\boldsymbol{\theta}\mathbf{q}}}{\omega_{\mathbf{q}}}}\psi(\boldsymbol{\theta}\mathbf{q})\nu(\boldsymbol{\theta}\mathbf{q},m)\nu^*(\mathbf{q},m)|\mu(m)|^2e^{-\frac{i}{\hbar}\omega_{\mathbf{q}}t'}\right|^2. \qquad (4.77)
\end{aligned}$$

This general and exact expression is not very convenient in actual calculations. So, in the following subsections we will consider specific situations where equation (4.77) can be simplified.

4.3.2 Decays of states with definite momentum

In the rest frame ($\theta = 0$), formula (4.77) coincides exactly with our previous result (4.36):

$$\Upsilon(0,t) = \int d\mathbf{q}|\psi(\mathbf{q})|^2\left|\int\limits_{m_\beta+m_\gamma}^{\infty} dm|\mu(m)|^2e^{-\frac{i}{\hbar}\omega_{\mathbf{q}}t}\right|^2. \qquad (4.78)$$

In Section 4.2 we used this formula to calculate the decay law for a particle with zero momentum. Here we are going to consider an unstable particle with nonzero momentum $\boldsymbol{p}$, whose state is described by the normalized vector $|\boldsymbol{p}\gg \in \mathscr{H}_\alpha$ and whose wave function is given by the square root of the delta function (4.37),

$$\psi(\boldsymbol{q}) = \sqrt{\delta(\boldsymbol{q}-\boldsymbol{p})}. \tag{4.79}$$

Substitution into (4.78) yields the following decay law:

$$\Upsilon_{|\boldsymbol{p}\gg}(0,t) = \left| \int_{m_\beta+m_\gamma}^{\infty} dm |\mu(m)|^2 e^{-\frac{i}{\hbar}\omega_{\boldsymbol{p}}t} \right|^2. \tag{4.80}$$

In a number of works [233, 131, 221, 253, 92] it was noticed that this result does not agree with Einstein's "time dilation" formula. Indeed, if we interpret $|\boldsymbol{p}\gg$ as a state of a particle moving with definite velocity

$$v = \frac{c^2 p}{\sqrt{m_A^2 c^4 + p^2 c^2}} = c \tanh\theta,$$

then the decay law (4.80) is *not* connected with the decay law at rest (4.38) by Einstein's formula (A.11), i. e.,

$$\Upsilon_{|\boldsymbol{p}\gg}(0,t) \neq \Upsilon_{|\boldsymbol{0}\gg}(0,t/\cosh\theta) \tag{4.81}$$

This observation caused some doubts [233, 131, 221] regarding the applicability of special relativity to particle decays. However, at a closer inspection it appears that this result by itself does not contradict the "time dilation" formula (A.11). Equation (4.81) compares decay laws of two momentum eigenstates $|\boldsymbol{0}\gg$ and $|\boldsymbol{p}\gg$, observed from the same reference frame. This is not the same as formula (A.11), which compares observations made on the same particle by two different observers.

If observer O sees a particle α described by the zero-momentum, zero-velocity state vector $|\boldsymbol{0}\gg$, then from the point of view of O', this particle is described by a state

$$e^{\frac{ic}{\hbar}\boldsymbol{K}\cdot\boldsymbol{\theta}}|\boldsymbol{0}\gg, \tag{4.82}$$

which is *not* an eigenstate of the total momentum $\boldsymbol{P}_0$. So, strictly speaking, equation (4.80) is not applicable to this state. However, one can easily see that (4.82) is an eigenstate of the velocity operator [222]. Indeed, taking into account $V_x|\boldsymbol{0}\gg = \boldsymbol{0}$ and equation (**1**-4.7), we see that

$$\begin{aligned} V_x(e^{\frac{ic}{\hbar}K_x\theta}|\boldsymbol{0}\gg) &= e^{\frac{ic}{\hbar}K_x\theta}e^{-\frac{ic}{\hbar}K_x\theta}V_x e^{\frac{ic}{\hbar}K_x\theta}|\boldsymbol{0}\gg = e^{\frac{ic}{\hbar}K_x\theta}\frac{V_x - c\tanh\theta}{1-\frac{V_x\tanh\theta}{c}}|\boldsymbol{0}\gg \\ &\approx -c\tanh\theta(e^{\frac{ic}{\hbar}K_x\theta}|\boldsymbol{0}\gg). \end{aligned} \tag{4.83}$$

Therefore, a fair comparison with Einstein's formula (A.11) requires consideration of unstable states having sharp velocity values for both observers. We will do this in Subsection 4.3.4.

4.3.3 Approximate decay law

Before we consider decay laws for states with definite velocity, let us introduce a few realistic approximations and simplify a bit our general formula (4.77). First, we can notice that in real situations the initial state of the decaying particle $|\Psi(0)\rangle \in \mathcal{H}_\alpha$ is not an exact eigenstate of the total momentum operator: the wave function of the unstable particle cannot be localized at one point of the momentum space[24]; the wave function should have a spread (or uncertainty) of momentum $|\Delta \boldsymbol{p}|$, which corresponds to the position uncertainty $|\Delta \boldsymbol{r}| \approx \hbar/|\Delta \boldsymbol{p}|$. Second, the state $|\Psi(0)\rangle \in \mathcal{H}_\alpha$ is also not an eigenstate of the mass operator M. Real states of unstable particles are characterized by the mass uncertainty Γ (see Figure 4.4), which is related to the particle lifetime τ_0 by equation (4.72). It is important to note that in all cases of practical interest the mentioned uncertainties are related by the following inequalities

$$|\Delta \boldsymbol{p}| \gg \Gamma c, \tag{4.84}$$

$$|\Delta \boldsymbol{r}| \approx \frac{\hbar}{|\Delta \boldsymbol{p}|} \ll c\tau_0. \tag{4.85}$$

In particular, the last inequality means that position uncertainty is much smaller than the distance traveled by light during particle lifetime.[25] Thus, we can safely assume that the factor $|\mu(m)|^2$ in (4.77) has a sharp peak near $m = m_A$. The factor $\sqrt{\omega_{\theta \boldsymbol{q}}\omega_{\boldsymbol{q}}}\psi(\boldsymbol{\theta q})\nu(\boldsymbol{\theta q}, m)\nu^*(\boldsymbol{q}, m)$ there is relatively smooth,[26] so we can move its value (constant) at the point $m = m_A$ outside the integral sign. Using also (**1**-5.31), we obtain

$$\begin{aligned}
\Upsilon(\theta, t') &\approx \int d\boldsymbol{q} \left| \sqrt{\frac{\Omega_{\vartheta \boldsymbol{q}}}{\Omega_{\boldsymbol{q}}}} \psi(\boldsymbol{\vartheta q})\nu(\boldsymbol{\vartheta q}, m_A)\nu^*(\boldsymbol{q}, m_A) \right|^2 \left| \int_{m_\beta+m_\gamma}^{\infty} dm |\mu(m)|^2 e^{-\frac{i}{\hbar}\omega_{\boldsymbol{q}} t'} \right|^2 \\
&= \int d\boldsymbol{q} \frac{\Omega_{\vartheta \boldsymbol{q}}}{\Omega_{\boldsymbol{q}}} |\psi(\boldsymbol{\vartheta q})|^2 \left| \int_{m_\beta+m_\gamma}^{\infty} dm |\mu(m)|^2 e^{-\frac{i}{\hbar}\omega_{\boldsymbol{q}} t'} \right|^2 \\
&= \int d\boldsymbol{p} |\psi(\boldsymbol{p})|^2 \left| \int_{m_\beta+m_\gamma}^{\infty} dm |\mu(m)|^2 e^{-\frac{i}{\hbar}\omega_{\vartheta^{-1}\boldsymbol{p}} t'} \right|^2 .
\end{aligned} \tag{4.86}$$

Here we use the notation

$$\omega_{\boldsymbol{p}} \equiv \sqrt{m^2c^4 + p^2c^2},$$

24 As we assumed, for example, in (4.79).

25 For example, in the case of a muon $\tau_0 \approx 2.2 \times 10^{-6}$ s, so, according to (4.85), the spread of the position wave function must be less than 600 m, which seems realistic. This approximation can be violated for particles (resonances) decaying due to strong nuclear forces. For them $\tau_0 \approx 10^{-23}$ s and the wave function should be localized better than $|\Delta \boldsymbol{r}| \approx 3 \times 10^{-15}$ m, which seems problematic for our approach. However, in these cases, experimental measurements of such short lifetimes are impossible.

26 See the discussion after equation (4.27).

$$\Omega_{\boldsymbol{p}} \equiv \sqrt{m_A^2 c^4 + p^2 c^2},$$
$$\theta \boldsymbol{p} \equiv \left(p_x \cosh\theta + \frac{\omega_{\boldsymbol{p}}}{c} \sinh\theta, p_y, p_z \right),$$
$$\vartheta \boldsymbol{p} \equiv \left(p_x \cosh\theta + \frac{\Omega_{\boldsymbol{p}}}{c} \sinh\theta, p_y, p_z \right),$$
$$\omega_{\vartheta^{-1}\boldsymbol{p}} = \sqrt{m^2 c^4 + (\vartheta^{-1}\boldsymbol{p})^2 c^2}.$$

4.3.4 Decays of states with definite velocity

Next we consider an initial state in which the particle α is at rest with respect to the observer O. The wave function of this state is localized near zero momentum $\boldsymbol{p} = \boldsymbol{0}$. Therefore, in equation (4.86) we can set[27]

$$|\psi(\boldsymbol{p})|^2 \approx \delta(\boldsymbol{p}), \quad (4.87)$$
$$\vartheta^{-1}\boldsymbol{p} = (-m_A c \sinh\theta, 0, 0),$$
$$\omega_{\vartheta^{-1}\boldsymbol{p}} = \sqrt{m^2 c^4 + m_A^2 c^4 \sinh^2\theta}$$

and obtain the decay law of this particle seen by the moving observer[28]:

$$\Upsilon_{|0\gg}(\theta, t') \approx \left| \int_{m_\beta + m_\gamma}^{\infty} dm |\mu(m)|^2 e^{-\frac{it'}{\hbar}\sqrt{m^2c^4 + m_A^2 c^4 \sinh^2\theta}} \right|^2. \quad (4.88)$$

If we approximately identify $-m_A c \sinh\theta$ with the particle momentum p_x measured by the moving observer O', then we get

$$\Upsilon_{|0\gg}(\theta, t') \approx \left| \int_{m_\beta + m_\gamma}^{\infty} dm |\mu(m)|^2 e^{-\frac{i}{\hbar}\omega_{\boldsymbol{p}} t'} \right|^2. \quad (4.89)$$

In this approximation, the decay law (4.89) in the frame O' with velocity $c \tanh\theta$ is exactly the same as the decay law (4.80) of a particle having momentum $p_x = -m_A c \sinh\theta$ with respect to the stationary observer O.[29] This means that deviations from Einstein's "time dilation" formula, mentioned in Subsection 4.3.2, are a real effect, which requires serious attention.

27 As we noted in Subsection 4.3.3, real particle states are not eigenstates of the momentum (or velocity). However, their wave functions are localized in the $\boldsymbol{p}$-space much better than the slowly varying second factor in the integrand (4.86). Therefore, approximation (4.87) is justified.

28 Or, equivalently, the decay law of a particle moving relative to the observer with the speed $-c \tanh\vartheta$.

29 Note that this result is quite different from the work [222], which predicted acceleration(!) of the decay in the moving frame.

4.4 “Time dilation” in decays

In this section, we will consider a specific example of a moving unstable particle, perform numerical calculations of its decay law and explicitly demonstrate deviations from the prediction (A.11) of special relativity (see also [253]).

In addition, we will be interested in decays caused by boosts (Subsection 4.4.3) and decays in non-Bakamjian–Thomas forms of dynamics (Subsection 4.4.4).

4.4.1 Decays in moving frame

Let us now perform an approximate calculation of the decay law in a moving frame by formula (4.89). For simplicity, we assume that the inequality (4.70) holds:

$$\Gamma \ll m_A, \ (m_\beta + m_\gamma), \ (m_A - m_\beta - m_\gamma), \tag{4.90}$$

and that the mass distribution $|\mu(m)|^2$ of the unstable particle has the Breit–Wigner form (4.67) on the entire mass axis $m \in (-\infty, \infty)$. As we saw in Subsection 4.2.4, these fairly accurate approximations guarantee the pure exponential character of the decay law. Then[30]

$$\Upsilon_{|\mathbf{0}\gg}(\theta, t) = \frac{\Gamma^2 |I(\theta, t)|^2}{4\pi^2}, \tag{4.91}$$

where the integral I is calculated in (C.2) for high values of θ:

$$\begin{aligned} I(\theta, t) &= \int\limits_{-\infty}^{\infty} \frac{dm}{\Gamma^2/4 + (m - m_A)^2} e^{-\frac{it}{\hbar}\sqrt{m^2c^4 + p^2c^2}} \\ &\approx \frac{2\pi}{\Gamma} e^{-\frac{it}{\hbar}\sqrt{p^2c^2 + (m_A - i\Gamma/2)^2 c^4}} + \frac{\pi c}{p} J_1\left(\frac{pct}{\hbar}\right). \end{aligned} \tag{4.92}$$

Here $p \equiv m_A c \sinh\theta \gg m_A c$ is the particle momentum seen by the moving observer O'. Using the smallness of the parameter Γ and the relation $\sqrt{p^2c^2 + m_A^2c^4}/(m_Ac^2) = \cosh\theta$, we rewrite the square root in the exponent

$$\begin{aligned} \sqrt{p^2c^2 + (m_A - i\Gamma/2)^2c^4} &\approx \sqrt{p^2c^2 + m_A^2c^4}\sqrt{1 - \frac{im_A\Gamma c^4}{p^2c^2 + m_A^2c^4}} \\ &\approx \sqrt{p^2c^2 + m_A^2c^4} - \frac{i\Gamma c^2}{2\cosh\theta}. \end{aligned}$$

30 For brevity, we drop the prime that usually marks time t' measured by the clock of the moving observer O'.

Substituting these results in equation (4.91), we obtain the decay law[31]

$$\begin{aligned}\Upsilon_{|\mathbf{0}\gg}(\theta,t) &\approx \frac{\Gamma^2}{4\pi^2}\left|\frac{2\pi}{\Gamma}e^{-\frac{it}{\hbar}\sqrt{p^2c^2+m_A^2c^4}-t\Gamma c^2/(2\hbar\cosh\theta)}+\frac{\pi c}{p}J_1\left(\frac{pct}{\hbar}\right)\right|^2\\ &\approx e^{-t/(\tau_0\cosh\theta)}+\frac{\Gamma c}{p}e^{-t/(2\tau_0\cosh\theta)}J_1\left(\frac{pct}{\hbar}\right)\cos\left(\frac{tm_Ac^2\cosh\theta}{\hbar}\right),\end{aligned} \tag{4.93}$$

where $\tau_0 \equiv \hbar/(\Gamma c^2)$ is the lifetime (4.72) of the particle at rest.

4.4.2 Numerical results

In this subsection we would like to calculate the difference between the quantum mechanical result (4.93) and the special-relativistic formula (A.11), which in our case takes a particularly simple form,

$$\Upsilon^{SR}(\theta,t) = \Upsilon\left(0,\frac{t}{\cosh\theta}\right) = e^{-t/(\tau_0\cosh\theta)}. \tag{4.94}$$

The desired difference is

$$\begin{aligned}\Delta\Upsilon &\equiv \Upsilon_{|\mathbf{0}\gg}(\theta,t)-\Upsilon^{SR}(\theta,t)\\ &\approx \frac{\Gamma}{m_A\sinh\theta}e^{-t/(2\tau_0\cosh\theta)}J_1\left(\frac{tm_Ac^2\sinh\theta}{\hbar}\right)\cos\left(\frac{tm_Ac^2\cosh\theta}{\hbar}\right).\end{aligned} \tag{4.95}$$

It is convenient to rewrite this expression using dimensionless parameters $\varepsilon \equiv \Gamma/m_A$ and $\chi \equiv t/\tau_0$, i. e.,

$$\Delta\Upsilon \approx \frac{\varepsilon}{\sinh\theta}e^{-\chi/(2\cosh\theta)}J_1\left(\frac{\chi\sinh\theta}{\varepsilon}\right)\cos\left(\frac{\chi\cosh\theta}{\varepsilon}\right). \tag{4.96}$$

This correction is plotted in Figure 4.5 for three combinations of the parameters ε and θ. The bold line shows the case $\varepsilon = 0.02$, $\theta = 2$. The maximal value of the correction is $\approx 0.2\%$, which, seemingly, corresponds to the best accuracy obtained in measurements of decays of relativistic muons [11, 75].

However, we have no hope that this prediction will be verified in experiments anytime soon. Fact is that our selected value $\varepsilon = 0.02$ is unrealistically large for observed unstable particles and states.[32] For example, muons have parameters $\Gamma \approx$

31 The term containing the Bessel function is a small correction to the exponential term, so we neglected the square of this function.

32 This large value is characteristic for particles and resonances that decay due to strong nuclear forces. But such systems are very short-lived ($\tau_0 \approx 10^{-23}$ s), and their time-dependent decay laws cannot be measured.

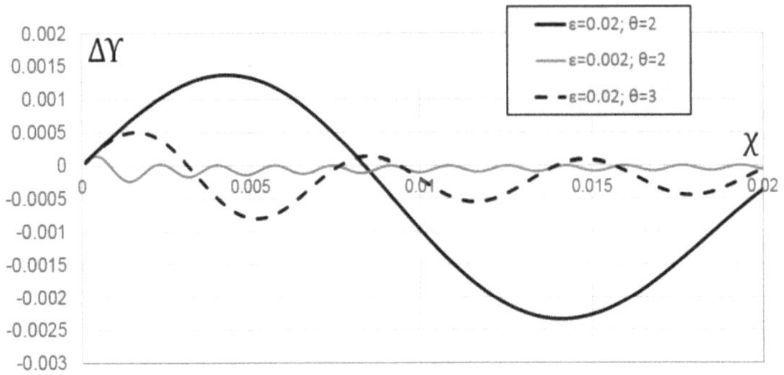

Figure 4.5: Corrections $\Delta\Upsilon$ to Einstein's formula (4.94) for the decay law of a particle moving with velocity $v = c\tanh\theta$. Parameter χ is time measured in the units of the lifetime τ_0; $\varepsilon = \Gamma/m_A$.

2×10^{-9} eV/c^2, $m_A \approx 105$ MeV/c^2, which correspond to the value $\varepsilon = \Gamma/m_A \approx 0.02\times10^{-15}$ that is many orders of magnitude smaller than ours. The maximum value of the correction $\Delta\Upsilon$ is roughly proportional to ε [221]. Therefore, its measurement in muons, just as in any other realistic unstable system, is practically impossible.

As the observer's speed increases, the magnitude of the correction decreases, as shown by the broken line in Figure 4.5 for $\varepsilon = 0.02$, $\theta = 3$. So, would deviations from the Einstein decay law become more noticeable at low speeds? Unfortunately, this is not true. At low observer speeds, we have to use equation (C.3), which leads to the correction

$$\Delta\Upsilon \approx -\varepsilon \sinh\theta e^{-\chi(2\cosh\theta)} J_1\left(\frac{\chi\sinh\theta}{\varepsilon}\right)\cos\left(\frac{\chi\cosh\theta}{\varepsilon}\right)$$

that tends to zero for $\theta \to 0$.

Thus, our decay theory, based on unitary representations of the Poincaré group in the instant form of dynamics, predicts not only the slowdown of the decay observed from moving frames, but also deviations from Einstein's formula (4.94) commonly used in such situations. Unfortunately, in real unstable systems these deviations are so small that their experimental detection is not possible at the present stage.

4.4.3 Decays caused by boosts

According to Postulate **1**-6.3, we are working in Dirac's instant form of dynamics, where spatial translations and rotations are considered kinematical (= interaction-independent), while time translations and boosts are dynamical (= interaction-dependent). As we know (see also Section 8.3), kinematical transformations trivially change the appearance of an object without affecting its internal state. Description of kinematical transformations is a purely geometric exercise that does not require knowledge of interactions in the physical system. This conclusion is supported by observations of unstable particles: for two observers at different positions or with different orientations in space, the nondecay probability appears the same.

On the other hand, dynamical transformations depend on interactions and influence the internal structure of the observed system. The dynamical effect of time translations is obvious: the unstable particle decays over time. But what about boost transformations? Does the nondecay probability depend on the speed of the observer? The special theory of relativity replies: “No. There is no such a relationship” (see Appendix A.3), and this answer is often regarded as self-evident:

> Any event that is “seen” in one inertial system is “seen” in all others. For example if observer in one system “sees” an explosion on a rocket then so do all other observers – R. Polishchuk [187].

Applying this (dubious) statement to decaying particles, we could expect that at the time $t = 0$

$$\Upsilon(\theta, 0) = 1 \tag{4.97}$$

for all θ. Here we would like to prove that this expectation is wrong.

Suppose that equation (4.97) is true, i. e., for any $|\Psi\rangle \in \mathscr{H}_\alpha$ and any $\theta > 0$, the boost leaves the state vector of the particle inside its own subspace:

$$e^{\frac{ic}{\hbar}K_x\theta}|\Psi\rangle \in \mathscr{H}_\alpha.$$

Hence it can be concluded that the subspace $\mathscr{H}_\alpha$ is invariant with respect to interacting boosts and that the operator K_x commutes with the projection T (4.16) on this subspace. Then commutators (**1**-3.54), $[T, P_{0x}] = 0$ and the Jacobi identity imply

$$[T, H] = -\frac{ic^2}{\hbar}[T, [K_x, P_{0x}]] = \frac{ic^2}{\hbar}[K_x, [P_{0x}, T]] + \frac{ic^2}{\hbar}[P_{0x}, [T, K_x]] = 0,$$

which contradicts the fundamental property (4.12) of unstable particles. To resolve this contradiction, we have to admit that the state $e^{\frac{ic}{\hbar}K_x\theta}|\Psi\rangle$ *does not* correspond to the particle α with 100 % probability. This state must contain an admixture of decay products even at the initial time $t = 0$, so we have

$$e^{\frac{ic}{\hbar}K_x\theta}|\Psi\rangle \notin \mathscr{H}_\alpha, \tag{4.98}$$

$$\Upsilon(\theta, 0) < 1, \quad \text{for } \theta \neq 0. \tag{4.99}$$

This is the “decay caused by boost,” which means, among other things, that special-relativistic formulas (4.94) and (4.97) are inaccurate and that boosts have a nontrivial effect on the internal state of the unstable particle. In other words, the composition of an unstable system is not a relativistic invariant.

In spite of its fundamental importance, this effect is rather weak. For example, in our rather good approximation (4.86), boosts did not cause decays. Indeed, at $t' = 0$ this formula predicted

$$\Upsilon(\theta, 0) = \int d\boldsymbol{p}|\psi(\boldsymbol{p})|^2 \left| \int_{m_\beta+m_\gamma}^{\infty} dm|\mu(m)|^2 \right|^2 = 1$$

instead of the expected (4.99).

4.4.4 Particle decays in different forms of dynamics

Throughout this chapter, we assumed that the interaction responsible for the decay belongs to the Bakamjian–Thomas instant form of dynamics. However, as we saw in Subsection **1**-6.4.4, this form does not tolerate cluster-separable interactions, so most likely it is not realized in nature. It would be interesting to calculate decay laws of moving particles also in non-Bakamjian–Thomas instant forms of dynamics. Although such calculations are not known to us, we can say with certainty that there is no such form of decay interaction in which Einstein's "time dilation" formula (4.94) would be exactly correct. This follows, for example, from the fact that in any instant form representation boost operators contain interaction terms. Therefore, decays caused by boosts (4.98)–(4.99) are always present.

And what if the decay interaction does not belong to the instant form of dynamics? Is there a form where Einstein's formula (4.94) is valid? Our answer to this question is "*no.*" Consider, for example, the point form (whose particular case was discussed in Subsection **1**-7.2.2). In this case, the subspace $\mathscr{H}_\alpha$ of the unstable particle is invariant with respect to boosts ($[K_{0x}, T] = 0$), so that decays caused by boosts are absent. However, we obtain the following relationship between decay laws of a particle observed from frames in relative motion[33]:

$$\begin{aligned}
\Upsilon(\theta, t) &= \langle \mathbf{0}|e^{-\frac{ic}{\hbar}K_{0x}\theta}e^{\frac{i}{\hbar}Ht}Te^{-\frac{i}{\hbar}Ht}e^{\frac{ic}{\hbar}K_{0x}\theta}|\mathbf{0}\rangle \\
&= \langle \mathbf{0}|e^{-\frac{ic}{\hbar}K_{0x}\theta}e^{\frac{i}{\hbar}Ht}e^{\frac{ic}{\hbar}K_{0x}\theta}e^{-\frac{ic}{\hbar}K_{0x}\theta}Te^{\frac{ic}{\hbar}K_{0x}\theta}e^{-\frac{ic}{\hbar}K_{0x}\theta}e^{-\frac{i}{\hbar}Ht}e^{\frac{ic}{\hbar}K_{0x}\theta}|\mathbf{0}\rangle \\
&= \langle \mathbf{0}|e^{-\frac{ic}{\hbar}K_{0x}\theta}e^{\frac{i}{\hbar}Ht}e^{\frac{ic}{\hbar}K_{0x}\theta}Te^{-\frac{ic}{\hbar}K_{0x}\theta}e^{-\frac{i}{\hbar}Ht}e^{\frac{ic}{\hbar}K_{0x}\theta}|\mathbf{0}\rangle \\
&= \langle \mathbf{0}|e^{\frac{it}{\hbar}(H\cosh\theta - cP_x\sinh\theta)}Te^{-\frac{it}{\hbar}(H\cosh\theta - cP_x\sinh\theta)}|\mathbf{0}\rangle \\
&= \langle \mathbf{0}|e^{\frac{it}{\hbar}H\cosh\theta}Te^{-\frac{it}{\hbar}H\cosh\theta}|\mathbf{0}\rangle \\
&= \Upsilon(0, t\cosh\theta),
\end{aligned}$$

33 Here we used (4.76), (**1**-4.4) and assumed that the initial state $|\mathbf{0}\rangle$ has zero interacting momentum, i. e., $\boldsymbol{P}|\mathbf{0}\rangle = \mathbf{0}$.

where the last equality follows from comparison with the decay law at rest (4.11). This result means that the decay rate in the moving frame is $\cosh\theta$ times *faster*(!) than in the frame at rest. This is in direct contradiction with the experiment.[34]

The point form of the decay interaction is unacceptable for yet another reason. Indeed, since the momentum operator $\boldsymbol{P}$ is interaction-dependent, in the point form we should see decays caused by ordinary spatial translations, i. e.,

$$e^{\frac{i}{\hbar}P_x a}|\Psi\rangle \notin \mathcal{H}_\alpha, \quad \text{for } a \neq 0 \text{ and } |\Psi\rangle \in \mathcal{H}_\alpha. \tag{4.100}$$

Decays caused by spatial translations and/or rotations are expected in all forms of dynamics, except for the instant one. Such decays contradict our experience, telling that these kinematical transformations cannot affect the internal makeup of the unstable particle. Therefore, only the instant form is suitable for describing decays. This conclusion gives additional support to our instant-form Postulate **1**-6.3; see also Section 8.3.

---

34 Note that decay laws of moving particles are different in the instant and point forms, despite the scattering equivalence of these forms (see Subsection **1**-7.2.4). As was shown in [233], these two statements do not contradict each other.

5 RQD in higher orders

Let a hundred flowers bloom, let a hundred schools of thought contend.
Mao Zedong

In Chapter 2 we analyzed formula (2.10) for the dressed interaction potential V_2^d in the second perturbation order. In this chapter we are going to extend this approach to higher orders. For example, in the third and fourth orders, the dressed potentials V_3^d and V_4^d near their energy shells are given by (see equations (2.11)–(2.12))

$$V_3^d \approx (\Sigma_3^c)^{\text{phys}}, \tag{5.1}$$

$$V_4^d \approx (\Sigma_4^c)^{\text{phys}} - V_2^d \underline{V_2^d}. \tag{5.2}$$

In Section 5.1 we will use equation (5.1) to calculate the dressed interaction potential V_3^d. This interaction is responsible for lifetimes and energy shifts of hydrogen atom levels. In Section 5.2 we will use (5.2) to derive the fourth-order electron–proton potential V_4^d. From these calculations we will get the famous QED radiative corrections: the Lamb shift and the electron anomalous magnetic moment.

5.1 Spontaneous radiative transitions

In Chapter 4, we discussed only general properties of the decay process. In particular, we were not interested in the composition of the unstable system and in the precise expression for the decay interaction V. So we left the formula for the decay rate in an unprocessed form (4.73). In this section, we want to fill this gap and perform a full calculation of the decay rate in a realistic system – an excited state of the hydrogen atom.

The simplest interaction responsible for the spontaneous emission of a photon from the excited state of the hydrogen atom has the structure $d^\dagger a^\dagger c^\dagger da$.[1] In the dressed Hamiltonian H^d, such terms first appear in the third order of perturbation theory (see Table 2.1). Therefore, our plan in this section will be as follows. In Subsection 5.1.1 we are going to find a third-order contribution to the S-operator of QED, which has the desired structure $d^\dagger a^\dagger c^\dagger da$. Then, in Subsection 5.1.2 we will get the dressed potential $V_3^d[d^\dagger a^\dagger c^\dagger da]$ near the energy shell. In Subsections 5.1.3–5.1.4, we will apply methods developed in Chapter 4 to calculate the rates of radiative transitions between hydrogen levels.

1 The two annihilation operators da destroy the electron and the proton in the initial state. The creation operators $d^\dagger a^\dagger c^\dagger$ restore the two charges and add a photon in the final state.

https://doi.org/10.1515/9783110493221-005

5.1.1 Bremsstrahlung

When two charged particles collide, there is a probability of a photon creation. The photon carries away a part of the energy of the colliding charges, so the charges are decelerated. In RQD, this *bremsstrahlung* effect is described, for example, by the interaction operator $d^\dagger a^\dagger c^\dagger da$.

In order to find the bremsstrahlung part of the S-operator in the third order, we use the time-ordered perturbation series (**1**-7.17) with the interaction operator V_1 from (**2**-3.30). The renormalized QED Hamiltonian H^c does not have *phys* counterterms in the third order.[2] This means that $(\Sigma_3^c)^{\text{phys}} = (\Sigma_3^n)^{\text{phys}}$ and[3]

$$\begin{aligned}
S_3^c &\equiv \underbrace{(\Sigma_3^c)^{\text{phys}}} = \underbrace{(\Sigma_3^n)^{\text{phys}}} \\
&= \frac{i}{3!\hbar^3}\int_{-\infty}^{+\infty} dt_1 dt_2 dt_3 T[V_1(t_1)V_1(t_2)V_1(t_3)] \\
&= \frac{i}{3!\hbar^3}\int d^4x_1 d^4x_2 d^4x_3 T[V_1(\tilde{x}_1)V_1(\tilde{x}_2)V_1(\tilde{x}_3)] \\
&= \frac{i}{3!\hbar^3}\int d^4x_1 d^4x_2 d^4x_3 T[(\mathrm{j}_{ep}^\mu(\tilde{x}_1)\mathcal{A}_\mu(\tilde{x}_1) + \mathrm{j}_{pa}^\mu(\tilde{x}_1)\mathcal{A}_\mu(\tilde{x}_1)) \\
&\quad \times (\mathrm{j}_{ep}^\nu(\tilde{x}_2)\mathcal{A}_\nu(\tilde{x}_2) + \mathrm{j}_{pa}^\nu(\tilde{x}_2)\mathcal{A}_\nu(\tilde{x}_2))(\mathrm{j}_{ep}^\lambda(\tilde{x}_3)\mathcal{A}_\lambda(\tilde{x}_3) + \mathrm{j}_{pa}^\lambda(\tilde{x}_3)\mathcal{A}_\lambda(\tilde{x}_3))],
\end{aligned} \tag{5.3}$$

where j_{ep}^μ and j_{pa}^μ are operators of the electron–positron and proton–antiproton currents, respectively. They were defined in Appendix **2**-D.1. Expanding the three parentheses, we obtain eight terms under the T-ordering sign. Not all these terms are of interest to us. The term $\mathrm{j}_{ep}\mathrm{j}_{ep}\mathrm{j}_{ep}$ cannot contribute to the electron–proton bremsstrahlung, because it lacks the proton component. Similarly, the term $\mathrm{j}_{pa}\mathrm{j}_{pa}\mathrm{j}_{pa}$ does not contribute and should be dropped. Now consider the sum $\mathrm{j}_{ep}\mathrm{j}_{ep}\mathrm{j}_{pa} + \mathrm{j}_{ep}\mathrm{j}_{pa}\mathrm{j}_{ep} + \mathrm{j}_{pa}\mathrm{j}_{ep}\mathrm{j}_{ep}$. The order of factors under the T-ordering sign is irrelevant, so these three terms are equal. Their contribution to the coefficient function of the S-operator is

$$\begin{aligned}
&s_3^{eep}(\boldsymbol{p}'\sigma', \boldsymbol{q}'\tau', \boldsymbol{s}\kappa; \boldsymbol{p}\sigma, \boldsymbol{q}\tau) \\
&= \frac{3i}{3!\hbar^3}\int d^4x_1 d^4x_2 d^4x_3 \\
&\quad \times \langle \text{vac}| a_{\boldsymbol{q}\tau} d_{\boldsymbol{p}\sigma} T[\mathrm{j}_{ep}^\mu(\tilde{x}_1)\mathcal{A}_\mu(\tilde{x}_1)\mathrm{j}_{ep}^\nu(\tilde{x}_2)\mathcal{A}_\nu(\tilde{x}_2)\mathrm{j}_{pa}^\lambda(\tilde{x}_3)\mathcal{A}_\lambda(\tilde{x}_3)] d_{\boldsymbol{p}'\sigma'}^\dagger a_{\boldsymbol{q}'\tau'}^\dagger c_{\boldsymbol{s}\kappa}^\dagger |\text{vac}\rangle \\
&= \frac{ie^3}{2\hbar^3}\int d^4x_1 d^4x_2 d^4x_3 \\
&\quad \times \langle \text{vac}| a_{\boldsymbol{q}\tau} d_{\boldsymbol{p}\sigma} T[(\overline{\psi}(\tilde{x}_1)\gamma^\mu\psi(\tilde{x}_1)\mathcal{A}_\mu(\tilde{x}_1))(\overline{\psi}(\tilde{x}_2)\gamma^\nu\psi(\tilde{x}_2)\mathcal{A}_\nu(\tilde{x}_2)) \\
&\quad \times (\overline{\Psi}(\tilde{x}_3)\gamma^\lambda\Psi(\tilde{x}_3)\mathcal{A}_\lambda(\tilde{x}_3))] d_{\boldsymbol{p}'\sigma'}^\dagger a_{\boldsymbol{q}'\tau'}^\dagger c_{\boldsymbol{s}\kappa}^\dagger |\text{vac}\rangle.
\end{aligned}$$

---

2 The only third-order counterterm (**2**-4.42) is *unphys*.

3 We assumed summation over repeated indices $\mu, \nu, \lambda = 0, 1, 2, 3$.

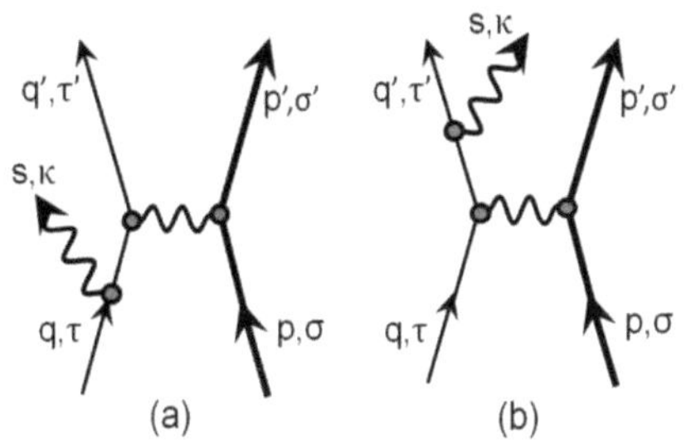

Figure 5.1: Third-order Feynman diagrams describing the photon emission in an electron–proton collision.

Next, the product of operators sandwiched between vacuum vectors $\langle \mathrm{vac}|\ldots|\mathrm{vac}\rangle$ should be brought to the normal order. Only the purely numerical term contributes to the matrix element. The easiest way to extract this term is by building two Feynman diagrams shown in Figure 5.1 and translating them into algebraic form according to the rules from Subsection **2**-3.2.4. We have

$$
\begin{aligned}
s_3^{eep}(\mathbf{p}'\sigma',\mathbf{q}'\tau',\mathbf{s}\kappa;\mathbf{p}\sigma,\mathbf{q}\tau) \\
&= \frac{ie^3c^{5/2}}{4\pi^2(2\pi\hbar)^{3/2}\sqrt{2s}}\frac{m_pm_ec^4}{\sqrt{\omega_{\mathbf{q}}\omega_{\mathbf{q}'}\Omega_{\mathbf{p}}\Omega_{\mathbf{p}'}}}\delta^4(\tilde{q}+\tilde{p}-\tilde{p}'-\tilde{q}'-\tilde{s})\frac{1}{(\tilde{p}-\tilde{p}')^2} \\
&\quad\times\overline{u}_a(\mathbf{q}',\tau')\Bigg(\not{e}_{ab}(\mathbf{s},\kappa)\frac{(\not{q}'+\not{s}+m_ec^2)_{bc}}{(\tilde{q}'+\tilde{s})^2-m_e^2c^4}\mathcal{W}_{cd}(\mathbf{p}'\sigma',\mathbf{p}\sigma) \\
&\quad+\mathcal{W}_{ab}(\mathbf{p}'\sigma',\mathbf{p}\sigma)\frac{(\not{q}-\not{s}+m_ec^2)_{bc}}{(\tilde{q}-\tilde{s})^2-m_e^2c^4}\not{e}_{cd}(\mathbf{s},\kappa)\Bigg)u_d(\mathbf{q},\tau).
\end{aligned}
$$

Similarly, one can build two diagrams, corresponding to the terms $\mathrm{j}_{ep}\mathrm{j}_{pa}\mathrm{j}_{pa}+\mathrm{j}_{pa}\mathrm{j}_{ep}\mathrm{j}_{pa}+\mathrm{j}_{pa}\mathrm{j}_{pa}\mathrm{j}_{ep}$ in equation (5.3). In contrast to Figure 5.1, these two diagrams have external photon lines originating from the proton lines. In our approximation ($m_e \ll m_p$), these terms can be neglected.

Now we assume that both the proton and the electron are nonrelativistic and simplify the expression for s_3^{eep}. We recall that all involved particles are on their mass shells, i. e.,

$$
\begin{aligned}
\tilde{q}^2 &= (\tilde{q}')^2 = m_e^2c^4, \\
\tilde{p}^2 &= (\tilde{p}')^2 = m_p^2c^4, \\
\tilde{s}^2 &= 0.
\end{aligned}
$$

Using approximations from Appendix **2**-B.9 and the formulas

$$
\begin{aligned}
(\tilde{p}-\tilde{p}')^2 &\approx -c^2(\mathbf{p}-\mathbf{p}')^2, \\
(\tilde{s}+\tilde{q}')^2-m_e^2c^4 &= 2\tilde{s}\cdot\tilde{q}', \\
(\tilde{q}-\tilde{s})^2-m_e^2c^4 &= -2\tilde{s}\cdot\tilde{q},
\end{aligned}
$$

$$\mathcal{W}^0(\boldsymbol{p}'\sigma',\boldsymbol{p}\sigma) \approx \delta_{\sigma\sigma'},$$
$$\mathcal{W}(\boldsymbol{p}'\sigma',\boldsymbol{p}\sigma) \approx 0,$$
$$\sqrt{\omega_{\boldsymbol{q}}\omega_{\boldsymbol{q}'}\Omega_{\boldsymbol{p}}\Omega_{\boldsymbol{p}'}} \approx m_p m_e c^4,$$

we obtain

$$\begin{aligned}
&s_3^{eep}(\boldsymbol{p}'\sigma',\boldsymbol{q}'\tau',\boldsymbol{s}\kappa;\boldsymbol{p}\sigma,\boldsymbol{q}\tau)\\
&\approx -\frac{ie^3c^{1/2}}{4\pi^2(2\pi\hbar)^{3/2}\sqrt{2s}}\frac{\delta^4(\tilde{q}+\tilde{p}-\tilde{p}'-\tilde{q}'-\tilde{s})\delta_{\sigma\sigma'}}{(\boldsymbol{p}-\boldsymbol{p}')^2}\\
&\times \overline{u}_a(\boldsymbol{q}',\tau')\left(\not{e}_{ab}(\boldsymbol{s},\kappa)\frac{(\not{q}'+\not{s}+m_ec^2)_{bc}}{2\tilde{q}'\cdot\tilde{s}}\gamma^0_{cd}-\gamma^0_{ab}\frac{(\not{q}-\not{s}+m_ec^2)_{bc}}{2\tilde{q}\cdot\tilde{s}}\not{e}_{cd}(\boldsymbol{s},\kappa)\right)u_d(\boldsymbol{q},\tau).
\end{aligned}$$

Next we use (**2**-B.22), the approximation $cs \ll m_ec^2$ and the Dirac equations (**2**-B.93), (**2**-B.94) to write

$$\begin{aligned}
&\overline{u}(\boldsymbol{q}',\tau')\not{e}(\not{q}'+\not{s}+m_ec^2)\gamma^0u(\boldsymbol{q},\tau)\\
&\approx \overline{u}(\boldsymbol{q}',\tau')(\not{e}\not{q}'+\not{e}m_ec^2)\gamma^0u(\boldsymbol{q}',\tau')\\
&= \overline{u}(\boldsymbol{q}',\tau')(-\not{q}'\not{e}+2(\tilde{q}'\cdot\tilde{e})+\not{e}m_ec^2)\gamma^0u(\boldsymbol{q},\tau)\\
&= 2\overline{u}(\boldsymbol{q}',\tau')(\tilde{q}'\cdot\tilde{e})\gamma^0u(\boldsymbol{q},\tau) = 2\mathcal{U}^0(\boldsymbol{q}'\tau',\boldsymbol{q}\tau)(\tilde{q}'\cdot\tilde{e}) \approx 2\delta_{\tau\tau'}(\tilde{q}'\cdot\tilde{e}),
\end{aligned}$$

$$\overline{u}(\boldsymbol{q}',\tau')\gamma^0(\not{q}-\not{s}+m_ec^2)\not{e}u(\boldsymbol{q},\tau) \approx 2\mathcal{U}^0(\boldsymbol{q}'\tau',\boldsymbol{q}\tau)(\tilde{q}\cdot\tilde{e}) = 2\delta_{\tau\tau'}(\tilde{q}\cdot\tilde{e}).$$

In the nonrelativistic limit $q, q' \ll m_ec^2$, we have

$$\begin{aligned}
\tilde{q}'\cdot\tilde{s} &= cs\omega_{\boldsymbol{q}'} - c^2(\boldsymbol{s}\cdot\boldsymbol{q}') \approx m_ec^3s,\\
\tilde{q}\cdot\tilde{s} &\approx m_ec^3s,\\
\tilde{q}\cdot\tilde{e} &= -c(\boldsymbol{q}\cdot\boldsymbol{e}),\\
\tilde{q}'\cdot\tilde{e} &= -c(\boldsymbol{q}'\cdot\boldsymbol{e}).
\end{aligned}$$

Therefore,

$$\begin{aligned}
&s_3^{eep}(\boldsymbol{p}'\sigma',\boldsymbol{q}'\tau',\boldsymbol{s}\kappa;\boldsymbol{p}\sigma,\boldsymbol{q}\tau)\\
&\approx -\frac{ie^3\sqrt{c}\delta^4(\tilde{q}+\tilde{p}-\tilde{p}'-\tilde{q}'-\tilde{s})}{4\pi^2(2\pi\hbar)^{3/2}\sqrt{2s}}\cdot\frac{\delta_{\sigma\sigma'}\delta_{\tau\tau'}}{(\boldsymbol{p}-\boldsymbol{p}')^2}\left(\frac{\tilde{q}'\cdot\tilde{e}}{\tilde{q}'\cdot\tilde{s}}-\frac{\tilde{q}\cdot\tilde{e}}{\tilde{q}\cdot\tilde{s}}\right)\\
&\approx \frac{ie^3\delta^4(\tilde{q}+\tilde{p}-\tilde{p}'-\tilde{q}'-\tilde{s})}{4\pi^2m_e(2\pi\hbar)^{3/2}\sqrt{2(cs)^3}}\cdot\frac{\delta_{\sigma\sigma'}\delta_{\tau\tau'}(\boldsymbol{q}'-\boldsymbol{q})\cdot\boldsymbol{e}(\boldsymbol{s},\kappa)}{(\boldsymbol{q}'-\boldsymbol{q})^2}.
\end{aligned}\tag{5.4}$$

This is our final expression for bremsstrahlung terms in the scattering operator. It can be compared with equations (7.57)–(7.58) in [15].

5.1.2 Bremsstrahlung potential in third order

Now, from equation (5.1) we can obtain the following third-order contribution $V_3^d[d^\dagger a^\dagger c^\dagger da]$ to the dressed Hamiltonian (2.3):

$$V_3^d[d^\dagger a^\dagger c^\dagger da] = \sum_{\sigma\tau\sigma'\tau'\kappa} \int d\boldsymbol{p} d\boldsymbol{q} d\boldsymbol{p}' d\boldsymbol{q}' d\boldsymbol{s} v_3^{eep}(\boldsymbol{p}'\sigma', \boldsymbol{q}'\tau', \boldsymbol{s}\kappa; \boldsymbol{p}\sigma, \boldsymbol{q}\tau) d^\dagger_{\boldsymbol{p}'\sigma'} a^\dagger_{\boldsymbol{q}'\tau'} c^\dagger_{\boldsymbol{s}\kappa} d_{\boldsymbol{p}\sigma} a_{\boldsymbol{q}\tau}, \quad (5.5)$$

whose coefficient function is[4]

$$v_3^{eep}(\boldsymbol{p}'\sigma', \boldsymbol{q}'\tau', \boldsymbol{s}\kappa; \boldsymbol{p}\sigma, \boldsymbol{q}\tau) \approx -\frac{e^3}{8\pi^3 m_e (2\pi\hbar)^{3/2}\sqrt{2(cs)^3}} \delta(\boldsymbol{q}+\boldsymbol{p}-\boldsymbol{p}'-\boldsymbol{q}'-\boldsymbol{s}) \frac{\delta_{\sigma\sigma'}\delta_{\tau\tau'}(\boldsymbol{q}'-\boldsymbol{q})\cdot \boldsymbol{e}(\boldsymbol{s},\kappa)}{(\boldsymbol{q}'-\boldsymbol{q})^2}. \quad (5.6)$$

Acting by the operator (5.5) on a general two-particle (electron + proton) state

$$|\Psi_i\rangle \equiv \sum_{\lambda\nu} \int d\boldsymbol{p}'' d\boldsymbol{q}'' \Psi(\boldsymbol{p}''\lambda, \boldsymbol{q}''\nu) d^\dagger_{\boldsymbol{p}''\lambda} a^\dagger_{\boldsymbol{q}''\nu} |\text{vac}\rangle$$

gives the following result:

$$\begin{aligned} V_3^d[d^\dagger a^\dagger c^\dagger da]|\Psi_i\rangle &= \sum_{\sigma\tau\sigma'\tau'\kappa} \sum_{\lambda\nu} \int d\boldsymbol{p}'' d\boldsymbol{q}'' \int d\boldsymbol{p} d\boldsymbol{q} d\boldsymbol{p}' d\boldsymbol{q}' d\boldsymbol{s} v_3^{eep}(\boldsymbol{p}'\sigma', \boldsymbol{q}'\tau', \boldsymbol{s}\kappa; \boldsymbol{p}\sigma, \boldsymbol{q}\tau) \\ &\quad \times \Psi(\boldsymbol{p}''\lambda, \boldsymbol{q}''\nu) d^\dagger_{\boldsymbol{p}'\sigma'} a^\dagger_{\boldsymbol{q}'\tau'} c^\dagger_{\boldsymbol{s}\kappa} d_{\boldsymbol{p}\sigma} a_{\boldsymbol{q}\tau} d^\dagger_{\boldsymbol{p}''\lambda} a^\dagger_{\boldsymbol{q}''\nu} |\text{vac}\rangle \\ &= \sum_{\sigma\tau\sigma'\tau'\kappa} \sum_{\lambda\nu} \int d\boldsymbol{p}'' d\boldsymbol{q}'' \int d\boldsymbol{p} d\boldsymbol{q} d\boldsymbol{p}' d\boldsymbol{q}' d\boldsymbol{s} v_3^{eep}(\boldsymbol{p}'\sigma', \boldsymbol{q}'\tau', \boldsymbol{s}\kappa; \boldsymbol{p}\sigma, \boldsymbol{q}\tau) \\ &\quad \times \Psi(\boldsymbol{p}''\lambda, \boldsymbol{q}''\nu) d^\dagger_{\boldsymbol{p}'\sigma'} a^\dagger_{\boldsymbol{q}'\tau'} c^\dagger_{\boldsymbol{s}\kappa} \delta(\boldsymbol{q}-\boldsymbol{q}'')\delta_{\tau\nu}\delta(\boldsymbol{p}-\boldsymbol{p}'')\delta_{\sigma\lambda} |\text{vac}\rangle \\ &= \sum_{\sigma'\tau'\kappa} \int d\boldsymbol{p}' d\boldsymbol{q}' d\boldsymbol{s} \\ &\quad \times \left(\sum_{\sigma\tau} \int d\boldsymbol{p} d\boldsymbol{q} v_3^{eep}(\boldsymbol{p}'\sigma', \boldsymbol{q}'\tau', \boldsymbol{s}\kappa; \boldsymbol{p}\sigma, \boldsymbol{q}\tau) \Psi(\boldsymbol{p}\sigma, \boldsymbol{q}\tau) \right) d^\dagger_{\boldsymbol{p}'\sigma'} a^\dagger_{\boldsymbol{q}'\tau'} c^\dagger_{\boldsymbol{s}\kappa} |\text{vac}\rangle, \end{aligned}$$

where the expression in big parentheses is the wave function of the new three-particle system (electron + proton + photon). With the help of (5.6) we can rewrite this wave function[5]

4 According to the rule from Section 1.4, to obtain this formula we divided the matrix element (5.4) by $(-2\pi i)$ and dropped the energy delta function $\delta(\omega_q + \Omega_p - \omega_{q'} - \Omega_{p'} - cs)$.

5 Here we neglect the photon momentum $\boldsymbol{s}$ in comparison with the momenta $(\boldsymbol{p}, \boldsymbol{p}', \boldsymbol{q}, \boldsymbol{q}')$ of massive particles.

$$\begin{aligned}
&\Psi'(\boldsymbol{p}'\sigma',\boldsymbol{q}'\tau',\boldsymbol{s}\kappa)\\
&= -\sum_{\sigma\tau}\int d\boldsymbol{p}d\boldsymbol{q}\frac{e^3\delta(\boldsymbol{q}+\boldsymbol{p}-\boldsymbol{p}'-\boldsymbol{q}'-\boldsymbol{s})\delta_{\sigma\sigma'}\delta_{\tau\tau'}(\boldsymbol{q}'-\boldsymbol{q})\cdot\boldsymbol{e}(\boldsymbol{s},\kappa)}{8\pi^3 m_e(2\pi\hbar)^{3/2}\sqrt{2(cs)^3}(\boldsymbol{q}'-\boldsymbol{q})^2}\Psi(\boldsymbol{p}\sigma,\boldsymbol{q}\tau)\\
&= -\frac{e^3}{8\pi^3 m_e(2\pi\hbar)^{3/2}\sqrt{2(cs)^3}}\int d\boldsymbol{k}\frac{\boldsymbol{k}\cdot\boldsymbol{e}(\boldsymbol{s},\kappa)}{k^2}\Psi((\boldsymbol{p}'+\boldsymbol{k})\sigma',(\boldsymbol{q}'-\boldsymbol{k})\tau').
\end{aligned}$$

Performing the Fourier transformation (**1**-5.49) and using (**2**-A.4), we switch to the position representation for fermions, i. e.,[6]

$$\begin{aligned}
\Psi'(\boldsymbol{x}\sigma',\boldsymbol{y}\tau',\boldsymbol{s}\kappa) &= -\frac{e^3}{8\pi^3 m_e(2\pi\hbar)^{9/2}\sqrt{2(cs)^3}}\int d\boldsymbol{p}'d\boldsymbol{q}'e^{\frac{i}{\hbar}\boldsymbol{p}'\cdot\boldsymbol{x}+\frac{i}{\hbar}\boldsymbol{q}'\cdot\boldsymbol{y}}\int d\boldsymbol{k}\frac{\boldsymbol{k}\cdot\boldsymbol{e}(\boldsymbol{s},\kappa)}{k^2}\\
&\quad\times\Psi((\boldsymbol{p}'+\boldsymbol{k})\sigma',(\boldsymbol{q}'-\boldsymbol{k})\tau')\\
&= -\frac{e^3}{8\pi^3 m_e(2\pi\hbar)^{9/2}\sqrt{2(cs)^3}}\int d\boldsymbol{p}'d\boldsymbol{q}'e^{\frac{i}{\hbar}(\boldsymbol{p}'-\boldsymbol{k})\cdot\boldsymbol{x}+\frac{i}{\hbar}(\boldsymbol{q}'+\boldsymbol{k})\cdot\boldsymbol{y}}\\
&\quad\times\int d\boldsymbol{k}\frac{\boldsymbol{k}\cdot\boldsymbol{e}(\boldsymbol{s},\kappa)}{k^2}\Psi(\boldsymbol{p}'\sigma',\boldsymbol{q}'\tau')\\
&= -\frac{e^3}{8\pi^3 m_e(2\pi\hbar)^{3/2}\sqrt{2(cs)^3}}\int d\boldsymbol{k}e^{\frac{i}{\hbar}\boldsymbol{k}\cdot(\boldsymbol{y}-\boldsymbol{x})}\frac{\boldsymbol{k}\cdot\boldsymbol{e}(\boldsymbol{s},\kappa)}{k^2}\\
&\quad\times\left(\frac{1}{(2\pi\hbar)^3}\int d\boldsymbol{p}'d\boldsymbol{q}'e^{\frac{i}{\hbar}\boldsymbol{p}'\cdot\boldsymbol{x}+\frac{i}{\hbar}\boldsymbol{q}'\cdot\boldsymbol{y}}\Psi(\boldsymbol{p}'\sigma',\boldsymbol{q}'\tau')\right)\\
&= -\frac{e^3\hbar^{1/2}}{m_e(2\pi)^{3/2}\sqrt{2(cs)^3}}\frac{i(\boldsymbol{y}-\boldsymbol{x})\cdot\boldsymbol{e}(\boldsymbol{s},\kappa)}{4\pi|\boldsymbol{y}-\boldsymbol{x}|^3}\Psi(\boldsymbol{x}\sigma',\boldsymbol{y}\tau').
\end{aligned}$$

This means that in the position space the third-order electron–proton bremsstrahlung potential has the form

$$V_3^d(\boldsymbol{r},\boldsymbol{s},\kappa) = -\frac{i\sqrt{\hbar}e^3(\boldsymbol{r}\cdot\boldsymbol{e}(\boldsymbol{s},\kappa))}{4\pi m_e\sqrt{2(2\pi cs)^3}r^3}c^\dagger_{\boldsymbol{s}\kappa} + \frac{i\sqrt{\hbar}e^3(\boldsymbol{r}\cdot\boldsymbol{e}^*(\boldsymbol{s},\kappa))}{4\pi m_e\sqrt{2(2\pi cs)^3}r^3}c_{\boldsymbol{s}\kappa}. \tag{5.7}$$

The first term is responsible for the emission of photons with momentum **s** and helicity κ by an electron moving in the Coulomb field of a massive proton.[7] We can expect that probability of such emission is proportional to the square of the matrix element of this operator between appropriate initial and final states. Since the potential (5.7) is proportional to the electron's acceleration in the proton's field $\boldsymbol{a} \approx e^2\boldsymbol{r}/(4\pi m_e r^3)$ (see equation (6.15)), we conclude that the energy emitted in the unit of time (the radiated power) is proportional to the acceleration squared a^2. This agrees with the classical

6 Here $\boldsymbol{x}$ and $\boldsymbol{y}$ are radius-vectors of the proton and the electron, respectively; $\boldsymbol{r} \equiv \boldsymbol{y} - \boldsymbol{x}$.

7 The second term in (5.7) is needed to preserve the Hermiticity of the Hamiltonian. The corresponding potential $V_3^d[d^\dagger a^\dagger dac]$ is Hermitian-conjugated to $V_3^d[d^\dagger a^\dagger c^\dagger da]$ and describes the absorption of photons by the interacting system electron + proton.

Larmor formula and confirms that the bremsstrahlung interaction V_3^d is directly responsible for the "radiation reaction" effect [165, 180, 181].

For Lamb shift calculations in Section 5.4, it will be convenient to assume that the photon has a small nonzero mass $\lambda > 0$.[8] Then its energy becomes $\sqrt{\lambda^2c^4 + c^2s^2}$. Using this fact in (5.7), we get our final expression for the third order potential:

$$V_3^d(\boldsymbol{r}, \boldsymbol{s}, \kappa) = -\frac{i\sqrt{\hbar}e^3\boldsymbol{r} \cdot \boldsymbol{e}(\boldsymbol{s}, \kappa)}{4\pi m_e\sqrt{2(2\pi)^3}(\lambda^2c^4 + c^2s^2)^{3/4}r^3}c_{\boldsymbol{s}\kappa}^\dagger + \frac{i\sqrt{\hbar}e^3\boldsymbol{r} \cdot \boldsymbol{e}^*(\boldsymbol{s}, \kappa)}{4\pi m_e\sqrt{2(2\pi)^3}(\lambda^2c^4 + c^2s^2)^{3/4}r^3}c_{\boldsymbol{s}\kappa}. \quad (5.8)$$

5.1.3 Instability of excited atomic states

The bremsstrahlung interaction V_3^d derived in the previous subsection is also responsible for spontaneous *radiative transitions* between energy levels in atoms and other bound systems of charges. As an example, consider two stationary states $|2P^{1/2}\rangle$ and $|1S^{1/2}\rangle$ of the hydrogen atom (see Section 3.2) and a transition between them, $|2P^{1/2}\rangle \to |1S^{1/2}\rangle + \gamma$, accompanied by the emission of a photon γ with energy $E = \epsilon(2P^{1/2}) - \epsilon(1S^{1/2}) \approx 10.2\,\text{eV}$.[9] The initial and final states of the atom will be denoted $|\Psi_i\rangle \equiv |2P^{1/2}\rangle$ and $|\Psi_f\rangle \equiv |1S^{1/2}\rangle$, respectively. They are eigenstates of the two-particle electron–proton Hamiltonian (3.13)

$$H_{e+p} = \frac{p_e^2}{2m_e} + \frac{p_p^2}{2m_p} - \frac{e^2}{4\pi r}, \quad (5.9)$$

with eigenvalues ϵ_i and ϵ_f.[10] We have

$$H_{e+p}|\Psi_i\rangle = \epsilon_i|\Psi_i\rangle,$$

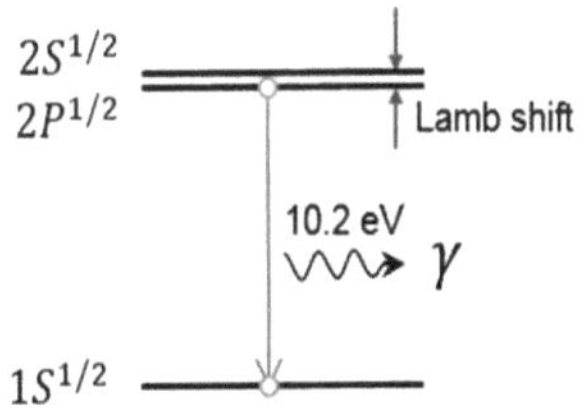

Figure 5.2: Radiative transition between $2P^{1/2}$ and $1S^{1/2}$ levels of the hydrogen atom.

8 Recall that the infrared cutoff λ was introduced in (**2**-F.20) for regularization of loop integrals. This regularization is very similar to the introduction of the photon mass λ. Of course, in the final result we should take the limit $\lambda \to 0$.

9 See Figure 5.2. The detailed kinematics of the radiative transition was described in Subsection **1**-6.5.3.

10 The eigenvalue problem for the Hamiltonian (5.9) was solved in Subsection 3.2.1.

$$H_{e+p}|\Psi_f\rangle = \varepsilon_f|\Psi_f\rangle,$$
$$\varepsilon_i = \varepsilon_f + 10.2\,\text{eV}.$$

If we add the interaction potential V_3^d to the Hamiltonian H_{e+p}, then the state $|\Psi_i\rangle$ is no longer stationary. Indeed, the two-particle subspace $\mathscr{H}_{pe}$ is not invariant with respect to this potential. The operator V_3^d has a nonzero matrix element between the stationary (excited) state $|2P^{1/2}\rangle$ of the hydrogen atom and the state $|2S^{1/2}\rangle + \gamma$, which contains the ground state of the atom plus one emitted photon γ. Therefore, the atom prepared initially in the state $|\Psi_i\rangle = |2P^{1/2}\rangle$ spontaneously evolves over time into two products: the atom in the state $|\Psi_f\rangle = |1S^{1/2}\rangle$ plus one photon. This is exactly the situation of decay discussed in Chapter 4. The particles α, β from Section 4.1 can be regarded as analogs of our multiparticle states $|\Psi_i\rangle$ and $|\Psi_f\rangle$, respectively. So, we can apply formulas derived there to the decay of the atomic state $|\Psi_i\rangle$.

5.1.4 Rate of radiative transition

According to Sections 4.1–4.2, the presence of the perturbation V_3^d should result in the decay of the level $|2P^{1/2}\rangle$ and its broadening.[11] Our next step is to calculate the probability of the spontaneous transition $|2P^{1/2}\rangle \to |2S^{1/2}\rangle + \gamma$ per unit time, i. e., the *transition rate* $\Gamma c^2/\hbar$, which is the inverse of the $|2P^{1/2}\rangle$ state's lifetime.

From equation (4.73) we get the probability density for the radiative transition $|\Psi_f\rangle \to |\Psi_i\rangle$ with the emission of one photon having momentum $\mathbf{s}$:

$$\Gamma(\mathbf{s},\kappa) = \frac{8\pi^2 s^2}{c^4}|\langle\Psi_i|V_3^d(\mathbf{r},\mathbf{s},\kappa)|\Psi_f\rangle|^2 \frac{d\eta^{-1}(z)}{dz}\bigg|_{z=m(2P^{1/2})}. \tag{5.10}$$

Using (5.7) and equality

$$[\boldsymbol{q}, H_{e+p}] = \left[\boldsymbol{q}, \frac{e^2}{4\pi r}\right] = i\hbar e^2 \frac{\boldsymbol{r}}{4\pi r^3} \tag{5.11}$$

and denoting $E \equiv \varepsilon_i - \varepsilon_f = cs$ the energy of the emitted photon, the matrix element in (5.10) can be transformed as follows:

$$\begin{aligned}\langle\Psi_i|V_3^d(\mathbf{r},\mathbf{s},\kappa)|\Psi_f\rangle &= -\frac{i\sqrt{\hbar}e^3}{m_e\sqrt{2(2\pi cs)^3}}\langle\Psi_i|\frac{\boldsymbol{r}\cdot\mathbf{e}(\mathbf{s},\kappa)}{4\pi r^3}|\Psi_f\rangle \\ &= -\frac{e}{m_e\sqrt{2\hbar(2\pi E)^3}}\langle\Psi_i|[(\boldsymbol{q}\cdot\mathbf{e})H_{e+p} - H_{e+p}(\boldsymbol{q}\cdot\mathbf{e})]|\Psi_f\rangle\end{aligned}$$

11 Notice that the broadening effect does not pertain to the ground state $|1S^{1/2}\rangle$ of the atom. This state can not spontaneously decay, since there are no available levels with lower energy. Hence, even in the presence of the bremsstrahlung potential, the ground state $|1S^{1/2}\rangle$ remains a true eigenvector of the complete Hamiltonian $H_{e+p} + V_3^d$ with a well-defined energy.

$$= \frac{eE}{m_e\sqrt{2\hbar(2\pi E)^3}} \langle \Psi_i | (\boldsymbol{q} \cdot \boldsymbol{e}) | \Psi_f \rangle.$$

Next we use

$$-\frac{im_e}{\hbar}[\boldsymbol{r}, H_{e+p}] = -\frac{im_e}{\hbar}[\boldsymbol{y}, H_{e+p}] + \frac{im_e}{\hbar}[\boldsymbol{x}, H_{e+p}] \approx \boldsymbol{q} - \frac{m_e}{m_p}\boldsymbol{p} \approx \boldsymbol{q}$$

to obtain

$$\begin{aligned}
\langle \Psi_i | V_3^d(\boldsymbol{r}, \boldsymbol{s}, \kappa) | \Psi_f \rangle &= -\frac{ieE}{\sqrt{2(2\pi\hbar E)^3}} \langle \Psi_i | [(\boldsymbol{r} \cdot \boldsymbol{e}) H_{e+p} - H_{e+p}(\boldsymbol{r} \cdot \boldsymbol{e})] | \Psi_f \rangle \\
&= \frac{ieE^2}{\sqrt{2(2\pi\hbar E)^3}} \langle \Psi_i | (\boldsymbol{r} \cdot \boldsymbol{e}) | \Psi_f \rangle \\
&= \frac{ie\sqrt{E}}{\sqrt{2(2\pi\hbar)^3}} \langle \Psi_i | (\boldsymbol{r} \cdot \boldsymbol{e}) | \Psi_f \rangle.
\end{aligned} \tag{5.12}$$

To find the function $\eta^{-1}(z)$ in (5.10) we turn to the definition (4.42). In the case considered here, one decay product is massless ($m_y = 0$). Its energy is much lower than the total energy of atomic states: $cs \ll m_i c^2 \approx m_f c^2$. Hence, we can use approximations

$$\begin{aligned}
z = \eta_s &= \frac{1}{c^2}\left(\sqrt{m_f^2 c^4 + c^2 s^2} + cs\right) \approx m_f + \frac{s}{c}, \\
s &= \eta^{-1}(z) \approx (z - m_f)c
\end{aligned}$$

and

$$\left. \frac{d\eta^{-1}(z)}{dz} \right|_{z=m_i} \approx c.$$

Substituting these results and (5.12) in (5.10), we obtain

$$\begin{aligned}
\Gamma(\boldsymbol{s}, \kappa) &\approx \frac{8\pi^2 E^2}{c^5} \left(\frac{e\sqrt{E}}{\sqrt{2(2\pi\hbar)^3}} \right)^2 |\langle \Psi_i | (\boldsymbol{r} \cdot \boldsymbol{e}) | \Psi_f \rangle|^2 \\
&= \frac{E^3}{2\pi\hbar^3 c^5} |\langle \Psi_i | (\boldsymbol{d} \cdot \boldsymbol{e}(\boldsymbol{s}, \kappa)) | \Psi_f \rangle|^2,
\end{aligned} \tag{5.13}$$

where $\boldsymbol{d} \equiv -e\boldsymbol{r} = -e\boldsymbol{y} + e\boldsymbol{x}$ is the operator of the atom's dipole moment.

The full transition probability[12] is obtained by summing the probability density (5.13) over the two photon polarizations $\kappa = \pm 1$ and integrating over all possible photon directions. We have[13]

$$\frac{1}{\tau_0} = \frac{c^2}{\hbar} \sum_{\kappa=-1}^{+1} \int d\Omega \Gamma(\boldsymbol{s}, \kappa) = \frac{E^3}{2\pi\hbar^4 c^3} \sum_{\kappa=-1}^{+1} \int d\Omega |\langle \Psi_i | (\boldsymbol{d} \cdot \boldsymbol{e}(\boldsymbol{s}, \kappa)) | \Psi_f \rangle|^2$$

---

12 It is proportional to the brightness of the corresponding spectral line.

13 Here $\int \ldots d\Omega = \int_0^{\pi} \cos\theta d\theta \int_0^{2\pi} d\varphi \ldots$ denotes the integral over orientation angles of $\boldsymbol{s}$ and $\boldsymbol{d}_{if} \equiv \langle \Psi_i | \boldsymbol{d} | \Psi_f \rangle$ is the matrix element of the dipole moment calculated between the atomic states $|\Psi_i\rangle$ and $|\Psi_f\rangle$.

$$= \frac{E^3}{2\pi\hbar^4 c^3} \sum_{\kappa=-1}^{+1} \int d\Omega |(\boldsymbol{d}_{if} \cdot \boldsymbol{e}(\mathbf{s}, \kappa))|^2.$$

Without loss of generality, we assume that the vector $\boldsymbol{d}_{if}$ is directed along the z-axis. Then, with the help of (**2**-C.15) we obtain

$$\sum_{\kappa=-1}^{+1} |(\boldsymbol{d}_{if} \cdot \boldsymbol{e}(\mathbf{s}, \kappa))|^2 = |d_{if}|^2 (|e_z(\mathbf{s}, -1)|^2 + |e_z(\mathbf{s}, +1)|^2)$$

$$= \frac{|d_{if}|^2}{2s^2} (|-s_x + is_y|^2 + |-s_x - is_y|^2) = \frac{|d_{if}|^2 (s_x^2 + s_y^2)}{s^2}$$

and the full transition rate (measured in Hz) is expressed by the following well-known formula[14]:

$$\frac{1}{\tau_0} = \frac{|d_{if}|^2 E^3}{2\pi\hbar^4 c^3} \int_0^{\pi} \sin\theta d\theta \int_0^{2\pi} d\varphi \frac{(s_x^2 + s_y^2)}{s^2} = \frac{|d_{if}|^2 E^3}{\hbar^4 c^3} \int_0^{\pi} \sin^3\theta d\theta = \frac{4|d_{if}|^2 E^3}{3\hbar^4 c^3}.$$

5.2 Radiative corrections to interaction potential

In this section, we will apply equation (5.2) to derive the fourth-order radiative correction to the electron–proton interaction potential.

5.2.1 Product of potentials in (5.2)

Let us calculate the product $-V_2^d \underline{V_2^d}$ on the right-hand side of (5.2).[15] In our approximation, for V_2^d we can take the usual nonrelativistic Coulomb potential[16]

$$V_2^d(t) = -\frac{e^2\hbar^2}{(2\pi\hbar)^3} \int d\boldsymbol{p} d\boldsymbol{q} d\boldsymbol{p}' d\boldsymbol{q}' \frac{\delta(\boldsymbol{p} + \boldsymbol{q} - \boldsymbol{p}' - \boldsymbol{q}') e^{\frac{i}{\hbar}(\Omega_p + \omega_q - \Omega'_p - \omega'_q)t}}{(\boldsymbol{q} - \boldsymbol{q}')^2 + \lambda^2 c^2} d_{\boldsymbol{p}}^\dagger a_{\boldsymbol{q}}^\dagger d_{\boldsymbol{p}'} a_{\boldsymbol{q}'}$$

and $\underline{V_2^d}$ is obtained by (**2**-1.65)

$$\underline{V_2^d(t)} = \frac{e^2\hbar^2}{(2\pi\hbar)^3} \int d\boldsymbol{t} d\boldsymbol{s} d\boldsymbol{t}' d\boldsymbol{s}' \frac{\delta(\boldsymbol{t} + \boldsymbol{s} - \boldsymbol{t}' - \boldsymbol{s}') e^{\frac{i}{\hbar}(\Omega_t + \omega_s - \Omega'_t - \omega'_s)t}}{[(\boldsymbol{s} - \boldsymbol{s}')^2 + \lambda^2 c^2][\omega_{\boldsymbol{s}} - \omega_{\boldsymbol{s}'} + \Omega_{\boldsymbol{t}} - \Omega_{\boldsymbol{t}'}]} d_{\boldsymbol{t}}^\dagger a_{\boldsymbol{s}}^\dagger d_{\boldsymbol{t}'} a_{\boldsymbol{s}'}.$$

---

14 See (45.7 a) in [14] or (19.85) in [13].

15 This product corresponds to the diagram in Figure 2.1 (b).

16 This is the largest term in the Coulomb–Darwin–Breit interaction (3.2), modified due to the presence of (fictitious) photon mass λ. We are interested only in the dominant infrared-divergent contribution to the product $V_2^d \underline{V_2^d}$, so we will be working in the nonrelativistic approximation from Appendix **2**-B.9, where particle momenta are assumed to be much less than $m_e c$. Correspondingly, we drop all terms with positive powers of $\boldsymbol{p}, \boldsymbol{p}', \boldsymbol{q}, \boldsymbol{q}'$ and/or $\boldsymbol{k} \equiv \boldsymbol{q}' - \boldsymbol{q}$. In this approximation, the product $-V_2^d \underline{V_2^d}$ does not depend on spins. So, we drop the spin labels as well.

Since we are interested only in the term $d^\dagger a^\dagger da$, we use the equality

$$
\begin{aligned}
& d^\dagger_{\boldsymbol p} a^\dagger_{\boldsymbol q} d_{\boldsymbol p'} a_{\boldsymbol q'} d^\dagger_{\boldsymbol t} a^\dagger_{\boldsymbol s} d_{\boldsymbol t'} a_{\boldsymbol s'} \\
&= (d^\dagger_{\boldsymbol p} d_{\boldsymbol p'} d^\dagger_{\boldsymbol t} d_{\boldsymbol t'})(a^\dagger_{\boldsymbol q} a_{\boldsymbol q'} a^\dagger_{\boldsymbol s} a_{\boldsymbol s'}) \\
&= (-d^\dagger_{\boldsymbol p} d^\dagger_{\boldsymbol t} d_{\boldsymbol p'} d_{\boldsymbol t'} + d^\dagger_{\boldsymbol p} d_{\boldsymbol t'} \delta(\boldsymbol t - \boldsymbol p'))(-a^\dagger_{\boldsymbol q} a^\dagger_{\boldsymbol s} a_{\boldsymbol q'} a_{\boldsymbol s'} + a^\dagger_{\boldsymbol q} a_{\boldsymbol s'} \delta(\boldsymbol s - \boldsymbol q')) \\
&= d^\dagger_{\boldsymbol p} a^\dagger_{\boldsymbol q} d_{\boldsymbol t'} a_{\boldsymbol s'} \delta(\boldsymbol t - \boldsymbol p') \delta(\boldsymbol s - \boldsymbol q') + \cdots
\end{aligned}
\tag{5.14}
$$

and obtain[17]

$$
\begin{aligned}
-V_2^d \underline{V_2^d} &= \frac{e^4\hbar^4}{(2\pi\hbar)^6} \int d\boldsymbol p d\boldsymbol q d\boldsymbol p' d\boldsymbol q' d\boldsymbol s d\boldsymbol t d\boldsymbol s' d\boldsymbol t' d^\dagger_{\boldsymbol p} a^\dagger_{\boldsymbol q} d_{\boldsymbol p'} a_{\boldsymbol q'} d^\dagger_{\boldsymbol t} a^\dagger_{\boldsymbol s} d_{\boldsymbol t'} a_{\boldsymbol s'} \\
&\quad \times \frac{\delta(\boldsymbol p + \boldsymbol q - \boldsymbol p' - \boldsymbol q') \delta(\boldsymbol s + \boldsymbol t - \boldsymbol s' - \boldsymbol t')}{[(\boldsymbol q - \boldsymbol q')^2 + \lambda^2 c^2][\omega_{\boldsymbol s} - \omega_{\boldsymbol s'} + \Omega_{\boldsymbol t} - \Omega_{\boldsymbol t'}][(\boldsymbol s - \boldsymbol s')^2 + \lambda^2 c^2]} \qquad (5.15) \\
&= \frac{e^4\hbar^4}{(2\pi\hbar)^6} \int d\boldsymbol p d\boldsymbol q d\boldsymbol p' d\boldsymbol q' d\boldsymbol s d\boldsymbol t d\boldsymbol s' d\boldsymbol t' d^\dagger_{\boldsymbol p} a^\dagger_{\boldsymbol q} d_{\boldsymbol t'} a_{\boldsymbol s'} \\
&\quad \times \frac{\delta(\boldsymbol p + \boldsymbol q - \boldsymbol p' - \boldsymbol q') \delta(\boldsymbol s + \boldsymbol t - \boldsymbol s' - \boldsymbol t') \delta(\boldsymbol q' - \boldsymbol s) \delta(\boldsymbol p' - \boldsymbol t)}{[(\boldsymbol q - \boldsymbol q')^2 + \lambda^2 c^2][\omega_{\boldsymbol s} - \omega_{\boldsymbol s'} + \Omega_{\boldsymbol t} - \Omega_{\boldsymbol t'}][(\boldsymbol s - \boldsymbol s')^2 + \lambda^2 c^2]} \\
&= \frac{e^4\hbar^4}{(2\pi\hbar)^6} \int d\boldsymbol q d\boldsymbol p d\boldsymbol p' d\boldsymbol q' \delta(\boldsymbol q + \boldsymbol p - \boldsymbol p' - \boldsymbol q') d^\dagger_{\boldsymbol p} a^\dagger_{\boldsymbol q} d_{\boldsymbol p'} a_{\boldsymbol q'} \\
&\quad \times \int \frac{d\boldsymbol s}{[(\boldsymbol q - \boldsymbol s)^2 + \lambda^2 c^2][\omega_{\boldsymbol s} - \omega_{\boldsymbol q'} + \Omega_{\boldsymbol q + \boldsymbol p - \boldsymbol s} - \Omega_{\boldsymbol p'}][(\boldsymbol q' - \boldsymbol s)^2 + \lambda^2 c^2]}.
\end{aligned}
$$

Next we introduce nonrelativistic approximations $\omega_{\boldsymbol q} \approx m_e c^2 + q^2/(2m_e)$, $\Omega_{\boldsymbol p} \approx \Omega_{\boldsymbol p'} \approx m_p c^2$ and also choose the center-of-mass reference frame where the massive proton remains fixed and the electron scattering is elastic, so we have $q = q'$. Then the desired operator has the form

$$
-V_2^d \underline{V_2^d} = \int d\boldsymbol p d\boldsymbol q d\boldsymbol p' d\boldsymbol q' \delta(\boldsymbol p + \boldsymbol q - \boldsymbol p' - \boldsymbol q') w(\boldsymbol q, \boldsymbol q') d^\dagger_{\boldsymbol p} a^\dagger_{\boldsymbol q} d_{\boldsymbol p'} a_{\boldsymbol q'}
$$

with the coefficient function

$$
\begin{aligned}
w(\boldsymbol q, \boldsymbol q') &\approx \frac{e^4\hbar^4}{(2\pi\hbar)^6} \int \frac{d\boldsymbol s}{[(\boldsymbol q - \boldsymbol s)^2 + \lambda^2 c^2][\omega_{\boldsymbol s} - \omega_{\boldsymbol q'}][(\boldsymbol q' - \boldsymbol s)^2 + \lambda^2 c^2]}. \\
&\approx \frac{2e^4\hbar^4 m_e}{(2\pi\hbar)^6} \int \frac{d\boldsymbol s}{[(\boldsymbol q - \boldsymbol s)^2 + \lambda^2 c^2][s^2 - q^2][(\boldsymbol q' - \boldsymbol s)^2 + \lambda^2 c^2]} \qquad (5.16) \\
&\approx \frac{\alpha^2 m_e c^2}{(-2\pi i)\pi q k^2} \ln\left(\frac{k^2}{\lambda^2 c^2}\right), \qquad (5.17)
\end{aligned}
$$

where the last equality follows from equation (D.2).

17 As we already mentioned in Section 2.2, the operators V_2^d and $\underline{V_2^d}$ are well-defined only on their energy shells. However, momentum integrations in (5.15) are carried out beyond the energy shells. Nevertheless, this uncertainty does not seem significant, because we are only interested in the dominant part of this integral, which comes from a small region near the singularity $\boldsymbol k = 0$ located on the energy shell.

5.2.2 Radiative corrections to Coulomb potential

So, according to equation (5.2), the electron–proton interaction potential in the fourth perturbation order is composed of two terms. The first term on the right-hand side is obtained from (**2**-4.60) by dropping the energy delta function and dividing by $(-2\pi i)$. The second term was calculated in (5.17). Then, for the full coefficient function of V_4^d we obtain

$$\begin{aligned}
&v_4^d(\boldsymbol{p},\boldsymbol{q},\boldsymbol{k};\tau,\sigma,\tau'\sigma') \\
&\quad \equiv \langle 0|a_{\boldsymbol{q}\sigma}d_{\boldsymbol{p}\tau}V_4^d d^\dagger_{\boldsymbol{p}'\tau'}a^\dagger_{\boldsymbol{q}'\sigma'}|0\rangle \frac{1}{\delta(\boldsymbol{p}+\boldsymbol{q}-\boldsymbol{p}'-\boldsymbol{q}')} \\
&\quad \approx \frac{\delta_{\tau\tau'}}{(-2\pi i)}\left[\frac{i\alpha^2}{15\pi^2 m_e^2 c}\delta_{\sigma\sigma'} + \frac{i\alpha^2}{3\pi^2 m_e^2 c}\ln\left(\frac{\lambda}{m_e}\right)\delta_{\sigma\sigma'}\right. \\
&\qquad \left. - \frac{\alpha^2\chi_\sigma^\dagger(\boldsymbol{\sigma}_{el}\cdot[\boldsymbol{k}\times\boldsymbol{q}])\chi_{\sigma'}}{4\pi^2 m_e^2 ck^2} - \frac{\alpha^2 m_e c^2}{\pi qk^2}\ln\left(\frac{k^2}{\lambda^2 c^2}\right)\delta_{\sigma\sigma'} + \frac{\alpha^2 m_e c^2}{\pi qk^2}\ln\left(\frac{k^2}{\lambda^2 c^2}\right)\delta_{\sigma\sigma'}\right] \\
&\quad = \frac{\delta_{\tau\tau'}}{(-2\pi i)}\left[\frac{i\alpha^2}{15\pi^2 m_e^2 c}\delta_{\sigma\sigma'} + \frac{i\alpha^2}{3\pi^2 m_e^2 c}\ln\left(\frac{\lambda}{m_e}\right)\delta_{\sigma\sigma'} + \frac{\alpha^2\chi_\sigma^\dagger(\boldsymbol{\sigma}_{el}\cdot[\boldsymbol{k}\times\boldsymbol{q}])\chi_{\sigma'}}{4\pi^2 m_e^2 ck^2}\right],
\end{aligned} \tag{5.18}$$

where we notice that the sum (**2**-4.59) of the ladder and cross-ladder diagrams has canceled with the term $-\underline{V_2^d V_2^d}$ from (5.17).[18]

As usual, the corresponding position–space potential is obtained by the Fourier transform with respect to the transferred momentum $\boldsymbol{k} \equiv \boldsymbol{q}' - \boldsymbol{q}$. Using formulas from Appendix **2**-A and the relationship $\boldsymbol{s}_{el} = \hbar\boldsymbol{\sigma}_{el}/2$, we obtain this potential as a sum of three terms, i. e.,

$$V_4^d(\boldsymbol{q},\boldsymbol{r},\boldsymbol{s}_{el}) = V_4^{\text{Uehling}} + V_4^{\text{vertex}} + V_4^{s-o}. \tag{5.19}$$

The first term in (5.19) is the so-called *Uehling potential*,[19]

$$V_4^{\text{Uehling}} = -\frac{4\hbar^3\alpha^2}{15m_e^2 c}\delta(\boldsymbol{r}), \tag{5.20}$$

originating from diagrams in Figures **2**-4.1 (d) and (k). The second and third terms in (5.18) are the vertex correction

$$V_4^{\text{vertex}} = -\frac{4\hbar^3\alpha^2}{3m_e^2 c}\ln\left(\frac{\lambda}{m_e}\right)\delta(\boldsymbol{r}) \tag{5.21}$$

---

18 This cancellation explains why the mentioned terms are often omitted in textbook Lamb shift calculations.

19 Compare with formula (8.27) in [15]. This interaction is sometimes regarded as an effect of *vacuum polarization*. However, for us the vacuum is nothing but empty space, and its "polarization" has no meaning. In RQD we explain higher-order effects – such as (5.20) – exclusively in terms of small corrections to interparticle potentials, without any reference to "virtual particles," "vacuum polarization" or other field-theoretical terminology.

and the fourth-order correction to the spin–orbit interaction

$$V_4^{s-o} = \frac{e^2[\boldsymbol{r} \times \boldsymbol{q}] \cdot \boldsymbol{s}_{el}}{8\pi m_e^2 c^2 r^3}\left(\frac{\alpha}{\pi}\right). \tag{5.22}$$

They originate from the diagrams in Figures **2**-4.1 (e) and (h).

To get the full radiative correction to the second-order electron–proton interaction (3.6)–(3.10), the above fourth-order potential should be added to the third-order bremsstrahlung potential (5.8).[20] Then we obtain

$$V_{3+4}^d = V_3^d(\boldsymbol{r}, \boldsymbol{s}, \kappa) + \frac{e^2[\boldsymbol{r} \times \boldsymbol{q}] \cdot \boldsymbol{s}_{el}}{8\pi m_e^2 c^2 r^3}\left(\frac{\alpha}{\pi}\right) - \frac{4\hbar^3\alpha^2}{15 m_e^2 c}\delta(\boldsymbol{r}) - \frac{4\hbar^3\alpha^2}{3 m_e^2 c}\ln\left(\frac{\lambda}{m_e}\right)\delta(\boldsymbol{r}). \tag{5.23}$$

5.3 Electron anomalous magnetic moment

As we discussed in Chapter **2**-4, high perturbation orders have no effect on the electron charge, because this is forbidden by the charge renormalization condition **2**-4.2. However, there are no restrictions on the electron magnetic moment. Schwinger's calculation of the fourth-order correction to this important property in 1948 was an important achievement of QED. Here we are going to reproduce Schwinger's result in the framework of our RQD.

5.3.1 Electron magnetic moment in second order

Electron *magnetic moment* reveals itself by the particle's dynamics in external "magnetic fields".[21] In our electron–proton system, the role of the magnetic field source is played by the proton. So far we worked in the infinite proton mass approximation, which assumes that the proton is at rest, so its magnetic field was absent. In order to have a nontrivial model of the electron–magnet interaction, let us now take into account the finite proton mass $m_p < \infty$. In the second perturbation order, the fourth term in the potential (3.9) is relevant. We have

$$V_{s-o} = -\frac{e^2[\boldsymbol{r} \times \hat{\boldsymbol{p}}] \cdot \hat{\boldsymbol{s}}_{el}}{4\pi m_p m_e c^2 r^3}, \tag{5.24}$$

where $\hat{\boldsymbol{p}}$ is the proton's momentum operator. This potential describes the interaction between the electron's spin $\hat{\boldsymbol{s}}_{el}$ and the proton's orbital angular momentum $[\boldsymbol{r} \times \hat{\boldsymbol{p}}]$.

20 The operator V_3^d changes the number of particles, so, strictly speaking, it cannot be called electron–proton potential.

21 Experimental manifestations of particle magnetic moments will be discussed in Chapter 6. We will also explain there why we prefer to put the term "magnetic field" in quotes.

Usually the electron's magnetic moment $\hat{\boldsymbol{\mu}}_{el}$ is related to its spin by the formula[22]

$$\hat{\boldsymbol{\mu}}_{el} \equiv -\frac{ge\hat{\mathbf{s}}_{el}}{2m_e}, \tag{5.25}$$

where g is the so-called *gyromagnetic ratio* or, simply, the *g-factor*. In our second perturbation order, $g = 2$ and the spin–orbit interaction operator is

$$V_{s-o} = \frac{e[\boldsymbol{r} \times \hat{\boldsymbol{p}}] \cdot \hat{\boldsymbol{\mu}}_{el}}{4\pi m_p c^2 r^3}. \tag{5.26}$$

In the next subsection we will analyze the fourth-order correction to this operator.

5.3.2 Fourth-order correction

The second-order contribution (5.24) to the interaction between the electron's magnetic moment and the proton corresponds to the ninth term in the S-operator formula (**2**-G.2). The term (**2**-G.5) has a similar structure and provides a fourth-order radiative correction to this interaction. This correction differs from the second-order term only by the factor $\alpha/(2\pi)$.[23] Therefore, in our approximation, the g-factor of the electron is given by the famous Schwinger formula

$$g = 2\left(1 + \frac{\alpha}{2\pi}\right) \approx 2.0023.$$

5.4 Lamb shift

In this section, our task is to understand how the correction (5.23) to the Coulomb potential affects the energy spectrum of the hydrogen atom. Within our dressed theory, we are going to calculate the famous Lamb shift.

5.4.1 Experimental data

As we saw in (3.29), in the second-order theory, hydrogen levels $|2S^{1/2}\rangle$ and $|2P^{1/2}\rangle$ have identical energies. However, in 1947, experimentalists Lamb and Retherford discovered a tiny gap between these two levels, which is now known as the *Lamb shift*. The presence of such a level splitting was very surprising to quantum physicists in the late 1940s. The attempts to resolve this contradiction played a very important role in

22 See equation (11.100) in [118].

23 When comparing (**2**-G.2) and (**2**-G.5), note that the former is the scattering phase while the latter is the scattering matrix, so they differ by the additional factor of $(-2\pi i)$.

the development of the renormalized QED. A successful calculation of the Lamb shift was a major triumph of this theory. The modern experimental value for this quantity is [267]

$$\begin{aligned} \varepsilon(2S^{1/2}) - \varepsilon(2P^{1/2}) &= 4.37 \times 10^{-6}\,\text{eV} \\ &= 2\pi\hbar \times (1\,057.8\,\text{MHz}). \end{aligned} \tag{5.27}$$

In this section we are going to calculate RQD corrections to atomic energies. The role of perturbation will be played by the third- and fourth-order potential (5.23), and the unperturbed system is an eigenstate $|n\rangle$ of the Hamiltonian H_{e+p} (5.9) with eigenvalue ε_n. It appears that even this simple approximation can reproduce the experimental result (5.27) rather well.

5.4.2 Contribution from potential V_3^d

First we calculate the energy shift of the atomic state $|n\rangle$ due to the potential V_3^d (5.8). This operator changes the number of particles, so $\langle n|V_3^d|n\rangle = 0$, and by formula (**1**-6.80) the energy correction vanishes in the first-order perturbation theory. Thus, we turn to the second-order correction (**1**-6.81), which changes the energy of the state $|n\rangle$ by[24]

$$\begin{aligned} \Delta\varepsilon_n &= \int d\boldsymbol{s} \sum_l \sum_\kappa \frac{\langle n|V_3^d|l;\boldsymbol{s},\kappa\rangle\langle l;\boldsymbol{s},\kappa|V_3^d|n\rangle}{\varepsilon_n - \varepsilon_l - cs} \\ &= \frac{\hbar e^6}{2m_e^2(2\pi)^3} \sum_l \sum_{\kappa=-1}^{+1} \int d\boldsymbol{s} \langle n|\frac{(\boldsymbol{r}\cdot\boldsymbol{e}^*(\boldsymbol{s},\kappa))}{4\pi r^3}|l\rangle \\ &\quad \times \langle l|\frac{(\boldsymbol{r}\cdot\boldsymbol{e}(\boldsymbol{s},\kappa))}{4\pi r^3}|n\rangle \frac{1}{(\lambda^2c^4 + c^2s^2)^{3/2}(\varepsilon_n - \varepsilon_l - cs)}, \end{aligned}$$

where $|l;\boldsymbol{s},\kappa\rangle \equiv |l\rangle|\boldsymbol{s},\kappa\rangle$ is the basis vector in which the atom is in the stationary state $|l\rangle$ and a free photon is in the state $|\boldsymbol{s},\kappa\rangle$ with momentum $\boldsymbol{s}$ and helicity κ. Taking into account (**2**-C.20) and

$$\sum_{\kappa=-1}^{+1} e_i^*(\boldsymbol{s},\kappa)e_j(\boldsymbol{s},\kappa) = \delta_{ij} - \frac{s_is_j}{s^2}$$

(where $i,j = x,y,z$), we obtain

$$\Delta\varepsilon_n = \frac{\hbar e^6}{2m_e^2(2\pi)^3} \sum_{lij} \langle n|\frac{r_i}{4\pi r^3}|l\rangle\langle l|\frac{r_j}{4\pi r^3}|n\rangle$$

24 This energy shift has the same nature as the mass shift $\mathcal{P}(m_A)$ from formula (4.58).

$$\times \int d\mathbf{s}\left(\delta_{ij} - \frac{s_i s_j}{s^2}\right)\frac{1}{(\lambda^2c^4 + c^2s^2)^{3/2}(\varepsilon_n - \varepsilon_l - cs)}.$$

Denote $\varepsilon_{nl} \equiv \varepsilon_n - \varepsilon_l$ and

$$I_{nl} \equiv \int_0^\infty \frac{s^2 ds}{(\lambda^2c^4 + c^2s^2)^{3/2}(\varepsilon_{nl} - cs)}.$$

Then

$$\begin{aligned}\int d\mathbf{s}\frac{s_i s_j}{s^2(\lambda^2c^4 + c^2s^2)^{3/2}(\varepsilon_{nl} - cs)} &= \delta_{ij}\int d\mathbf{s}\frac{s_z^2}{s^2(\lambda^2c^4 + c^2s^2)^{3/2}(\varepsilon_{nl} - cs)}\\ &= 2\pi\delta_{ij}\int_0^\pi \sin\theta d\theta \int_0^\infty ds\frac{s^2\cos^2\theta}{(\lambda^2c^4 + c^2s^2)^{3/2}(\varepsilon_{nl} - cs)}\\ &= 2\pi\delta_{ij}\int_{-1}^1 dt\int_0^\infty ds\frac{s^2t^2}{(\lambda^2c^4 + c^2s^2)^{3/2}(\varepsilon_{nl} - cs)} = \frac{4\pi\delta_{ij}}{3}I_{nl}\end{aligned}$$

and we can write

$$\begin{aligned}\Delta\varepsilon_n &\approx \frac{\hbar e^6}{2m_e^2(2\pi)^3}\sum_{lij}\langle n|\frac{r_i}{4\pi r^3}|l\rangle\langle l|\frac{r_j}{4\pi r^3}|n\rangle\left(4\pi\delta_{ij}I_{nl} - \frac{4\pi}{3}\delta_{ij}I_{nl}\right)\\ &= \frac{4\pi\hbar e^6}{3m_e^2(2\pi)^3}\sum_{li}\langle n|\frac{r_i}{4\pi r^3}|l\rangle\langle l|\frac{r_i}{4\pi r^3}|n\rangle I_{nl}.\end{aligned}$$

The integral I_{nl} is calculated as follows:

$$\begin{aligned}I_{nl} &= \frac{1}{c^3}\left[\frac{\lambda^2c^4 - \varepsilon_{nl}cs}{(\lambda^2c^4 + \varepsilon_{nl}^2)\sqrt{\lambda^2c^4 + c^2s^2}} + \frac{\varepsilon_{nl}^2}{(\lambda^2c^4 + \varepsilon_{nl}^2)^{3/2}}\right.\\ &\quad\left.\times\ln\left(\frac{\sqrt{\lambda^2c^4 + \varepsilon_{nl}^2}\sqrt{\lambda^2c^4 + c^2s^2} + \lambda^2c^2 + \varepsilon_{nl}cs}{\varepsilon_{nl} - cs}\right)\right]\Bigg|_{s=0}^{s=\infty}\\ &\approx \frac{1}{c^3}\left[-\frac{1}{\varepsilon_{nl}} + \frac{1}{|\varepsilon_{nl}|}\ln(-|\varepsilon_{nl}| - \lambda^2c^4|\varepsilon_{nl}|^{-1}/2 - \varepsilon_{nl}) - \frac{1}{|\varepsilon_{nl}|}\ln\left(\frac{|\varepsilon_{nl}|\lambda c^2}{\varepsilon_{nl}}\right)\right]\\ &\approx \frac{1}{c^3|\varepsilon_{nl}|}\ln\left(\frac{-[|\varepsilon_{nl}| + \lambda^2c^4|\varepsilon_{nl}|^{-1}/2 + \varepsilon_{nl}]\varepsilon_{nl}}{|\varepsilon_{nl}|\lambda c^2}\right).\end{aligned}$$

If $\varepsilon_{nl} > 0$, then[25]

$$I_{nl} = \frac{1}{c^3\varepsilon_{nl}}\ln\left(-\frac{2\varepsilon_{nl}}{\lambda c^2}\right).$$

25 We used the inequality $\lambda^2 \ll \lambda$.

If $\varepsilon_{nl} < 0$, then

$$I_{nl} = -\frac{1}{c^3\varepsilon_{nl}} \ln\left(-\frac{[\lambda^2 c^4 \varepsilon_{nl}^{-1}/2]\varepsilon_{nl}}{\varepsilon_{nl}\lambda c^2}\right) = \frac{1}{c^3\varepsilon_{nl}} \ln\left(-\frac{2\varepsilon_{nl}}{\lambda c^2}\right).$$

Taking into account that $\ln(-1) \ll \ln(2|\varepsilon_{nl}|/(\lambda c^2))$, we can write for all values of ε_{nl}

$$I_{nl} \approx \frac{1}{c^3\varepsilon_{nl}} \ln\left(\frac{2|\varepsilon_{nl}|}{\lambda c^2}\right),$$

which means that

$$\Delta\varepsilon_n \approx \frac{4\pi\hbar e^6}{3m_e^2(2\pi)^3c^3} \sum_{li} \left|\langle n|\frac{r_i}{4\pi r^3}|l\rangle\right|^2 \frac{1}{\varepsilon_{nl}} \ln\left(\frac{2|\varepsilon_{nl}|}{\lambda c^2}\right).$$

Next we use equation (5.11) and

$$\begin{aligned}
\frac{r_i}{4\pi r^3} &= \frac{1}{i\hbar e^2}[q_i, H_{e+p}], \\
\left|\langle n|\frac{r_i}{4\pi r^3}|l\rangle\right|^2 &= -\frac{1}{\hbar^2 e^4}\langle n|(q_i H_{e+p} - H_{e+p}q_i)|l\rangle\langle l|(q_i H_{e+p} - H_{e+p}q_i)|n\rangle \\
&= \frac{\varepsilon_{nl}^2}{\hbar^2 e^4}\langle n|q_i|l\rangle\langle l|q_i|n\rangle
\end{aligned}$$

to obtain

$$\Delta\varepsilon_n \approx -\frac{e^2}{6m_e^2\pi^2\hbar c^3} \sum_{li} |\langle n|q_i|l\rangle|^2 \varepsilon_{nl} \ln\left(\frac{\lambda c^2}{2|\varepsilon_{nl}|}\right).$$

Our next step is to use the so-called *Bethe logarithm* $\overline{\varepsilon}_n$, defined for *S*-states as follows[26]:

$$\sum_{li} |\langle n|q_i|l\rangle|^2 \varepsilon_{nl} \ln\left(\frac{\lambda c^2}{2|\varepsilon_{nl}|}\right) \equiv \ln\left(\frac{\lambda c^2}{2\overline{\varepsilon}_n}\right) \sum_{li} |\langle n|q_i|l\rangle|^2 \varepsilon_{nl}.$$

Then[27]

$$\begin{aligned}
\Delta\varepsilon_n &\approx -\frac{2\alpha}{3\pi m_e^2 c^2} \ln\left(\frac{\lambda c^2}{2\overline{\varepsilon}_n}\right) \sum_{l} |\langle n|q_i|l\rangle|^2 \varepsilon_{nl} \\
&= -\frac{\alpha}{3\pi m_e^2 c^2} \ln\left(\frac{\lambda c^2}{2\overline{\varepsilon}_n}\right) \sum_{l} [\langle n|(H_{e+p}q_i - q_i H_{e+p})|l\rangle\langle l|q_i|n\rangle
\end{aligned}$$

---

26 See, for example, equations (8.87) in [15] and (14.3.51) in [266]. Energy shifts of *P*-states are much smaller, and we ignore them.

27 Here we use equation (**1**-A.2).

$$
\begin{aligned}
&- \langle n|q_i|l\rangle\langle l|(H_{e+p}q_i - q_iH_{e+p})|n\rangle]\\
&= -\frac{\alpha}{3\pi m_e^2c^2}\ln\left(\frac{\lambda c^2}{2\overline{\varepsilon}_n}\right)[\langle n|(H_{e+p}q_i - q_iH_{e+p})q_i|n\rangle - \langle n|q_i(H_{e+p}q_i - q_iH_{e+p})|n\rangle]\\
&= -\frac{\alpha}{3\pi m_e^2c^2}\ln\left(\frac{\lambda c^2}{2\overline{\varepsilon}_n}\right)\langle n|([H_{e+p}, q_i]q_i - q_i[H_{e+p}, q_i])|n\rangle\\
&= -\frac{\alpha}{3\pi m_e^2c^2}\ln\left(\frac{\lambda c^2}{2\overline{\varepsilon}_n}\right)\langle n|[q_i, [q_i, H_{e+p}]]|n\rangle\\
&= -\frac{\hbar^2\alpha}{3\pi m_e^2c^2}\ln\left(\frac{\lambda c^2}{2\overline{\varepsilon}_n}\right)\langle n|\frac{\partial^2}{\partial\mathbf{r}^2}\frac{e^2}{4\pi r}|n\rangle = \frac{4\hbar^3\alpha^2}{3m_e^2c}\ln\left(\frac{\lambda c^2}{2\overline{\varepsilon}_n}\right)\langle n|\delta(\mathbf{r})|n\rangle\\
&= \frac{4\hbar^3\alpha^2}{3m_e^2c}\ln\left(\frac{\lambda c^2}{2\overline{\varepsilon}_n}\right)|\psi_n(0)|^2.
\end{aligned}
\tag{5.28}
$$

Using Table 3.2 and the well-known result[28] $\overline{\varepsilon}(2S^{1/2}) = 8.32m_ec^2\alpha^2$ we can calculate the shift of the level $2S^{1/2}$ due to the spontaneous photon emission. We have

$$
\Delta\varepsilon^{se}(2S^{1/2}) = \frac{4\hbar^3\alpha^2}{3m_e^2c}\ln\left(\frac{\lambda}{16.64m_e\alpha^2}\right)|\psi_{2S}(0)|^2 = \frac{m_ec^2\alpha^5}{6\pi}\ln\left(\frac{\lambda}{16.64m_e\alpha^2}\right). \tag{5.29}
$$

In the limit of vanishing photon mass ($\lambda \to 0$) this correction becomes infinite. This is one example of *infrared divergence.*

5.4.3 Contribution from potential V_4^d

In our next step we evaluate the effect of the three last terms in (5.23) on the energies of the states $|2S^{1/2}\rangle$ and $|2P^{1/2}\rangle$. We will use the first-order perturbation theory formula (**1**-6.80).

The contact potential in the third term in (5.23) shifts only energies of S-states, whose wave functions do not turn to zero in the origin, i. e.,

$$
\Delta\varepsilon^{\text{Uehling}}(2S^{1/2}) = -\frac{4\hbar^3\alpha^2}{15m_e^2c}|\psi_{2S}(\mathbf{0})|^2 = -\frac{m_ec^2\alpha^5}{30\pi}.
$$

Note that the last term (5.23) is quite problematic, because it diverges in the infrared limit $\lambda \to 0$. Form the point of view of classical electrodynamics (see Chapter 6), this divergence should not be a cause of concern, because trajectories of point charges never intersect and such a short-range potential has no chance to affect their dynamics. However, this potential gives divergent contributions to the energies of atomic S-states, i. e.,

$$
\Delta\varepsilon^{\text{vertex}}(2S^{1/2}) = -\frac{4\hbar^3\alpha^2}{3m_e^2c}\ln\left(\frac{\lambda}{m_e}\right)|\psi_{2S}(\mathbf{0})|^2 = -\frac{m_ec^2\alpha^5}{6\pi}\ln\left(\frac{\lambda}{m_e}\right). \tag{5.30}
$$

28 See Section 14.3 in [266].

Fortunately, this divergence is compensated by the energy shift (5.29), so the infrared divergences cancel out, and the total energy correction $\Delta\varepsilon(2S^{1/2})$ becomes finite. We have

$$\begin{aligned}\Delta\varepsilon(2S^{1/2}) &= \Delta\varepsilon^{\text{Uehling}}(2S^{1/2}) + \Delta\varepsilon^{\text{vertex}}(2S^{1/2}) + \Delta\varepsilon^{se}(2S^{1/2}) \\ &= -\frac{m_e c^2\alpha^5}{30\pi}[1 + 5\ln(16.64\alpha^2)].\end{aligned}$$

In our approximation, only the second term in (5.23) affects the level $2P^{1/2}$. The corresponding energy correction is calculated by the method from Subsection 3.2.4. It is equal to the second-order correction (3.27) times (α/π), i. e.,

$$\Delta\varepsilon(2P^{1/2}) = -\frac{m_e c^2\alpha^5}{48\pi}.$$

All the above results are summarized in Table 5.1. Finally, the full third- and fourth-order contribution to the Lamb shift

$$\begin{aligned}&\varepsilon(2S^{1/2}) - \varepsilon(2P^{1/2}) \\ &= \Delta\varepsilon(2S^{1/2}) - \Delta\varepsilon(2P^{1/2}) = \frac{m_e c^2\alpha^5}{6\pi}\left[-\frac{3}{40} - \ln(16.64\alpha^2)\right] = 3.91 \times 10^{-6}\,\text{eV} \\ &= 2\pi\hbar \times (945\,\text{MHz}) \qquad (5.31)\end{aligned}$$

appears to be in fairly good agreement with the experimental value (5.27). Taking this result into account, the relative positions of the lowest hydrogen levels are shown in Figure 3.1 (c).

So, in our approach the Lamb shift is a combination of two major physical effects: the radiative instability of the energy levels and the short-range "vertex" correction to the electron–proton potential. Both these effects are divergent in the infrared limit $\lambda \to 0$. However, these divergences almost exactly compensate each other, so that only a small gap (5.31) remains between the levels $2S^{1/2}$ and $2P^{1/2}$

Table 5.1: Corrections of the order α^5 to the energies of $2S^{1/2}$ and $2P^{1/2}$ hydrogen levels.

Perturbation	Origin of correction	$\Delta\varepsilon(2S^{1/2})$	$\Delta\varepsilon(2P^{1/2})$
(5.20)	"vacuum polarization"	$-\frac{m_e c^2\alpha^5}{30\pi}$	0
(5.21)	vertex integral	$-\frac{m_e c^2\alpha^5}{6\pi}\ln(\frac{\lambda}{m_e})$	0
(5.22)	spin–orbit interaction	0	$-\frac{m_e c^2\alpha^5}{48\pi}$
(5.8)	spontaneous emission	$\frac{m_e c^2\alpha^5}{6\pi}\ln(\frac{\lambda}{16.64\alpha^2 m_e})$	0
	Total	$-\frac{m_e c^2\alpha^5}{30\pi}[1 + 5\ln(16.64\alpha^2)]$	$-\frac{m_e c^2\alpha^5}{48\pi}$

6 Classical electrodynamics

All of physics is either impossible or trivial. It is impossible until you understand it and then it becomes trivial.
Ernest Rutherford

So far in this book, we were engaged in the quantum-mechanical description of electromagnetic phenomena. We were interested in scattering, decays and bound states of systems of charged particles, such as the hydrogen atom. We succeeded in reaching good agreement with experimental results in the fourth order of perturbation theory. We expect even better agreement in higher orders. But here we are going to consider a different class of electromagnetic phenomena, namely, the macroscopic motion of charges in the classical limit, i. e., classical electrodynamics.

Our goal is to show that the idea of charges interacting by means of direct instantaneous potentials is a worthy alternative to the classical theory based on the Maxwell equations, the Liénard–Wiechert potentials and the Lorentz force. This conclusion will give us additional arguments in favor of the corpuscular interpretation and RQD.

6.1 Maxwell's theory in a nutshell

6.1.1 Structure of fields and interactions

In its main features, the classical theory of electromagnetic phenomena was formulated in 1860s. It was based on a system of equations designed by Maxwell as a theoretical generalization of many experiments, carried out mostly by Faraday. We will call it the Maxwell theory, although Maxwell only laid its foundations. For example, in Maxwell's time, nothing was known about the electron and its point charge. Significant contributions to classical electrodynamics were made by Lorentz, Poynting, Hertz and many others.

Classical electrodynamics consists of several loosely connected parts. First, there are Maxwell's equations that say how charges (q) and their motion create fields (electric $\boldsymbol{E}$ and magnetic $\boldsymbol{B}$), and how these fields mutually affect each other. Next, there are definitions of the energy and momentum of these fields, which are supposed to be added to mechanical energies and momenta of the observed particles. Finally, the theory postulates the following formula for the Lorentz force:

$$\boldsymbol{F} = q\left(\boldsymbol{E} + \frac{1}{c}[\boldsymbol{v} \times \boldsymbol{B}]\right), \tag{6.1}$$

i. e., the force with which the fields act on charges.

The central idea of Maxwell's electrodynamics is that charged particles interact with each other not directly, but by means of (retarded) electric and magnetic

https://doi.org/10.1515/9783110493221-006

fields. Another important component is the statement that light is an electromagnetic wave, i. e., interconnected fields $\boldsymbol{E}$ and $\boldsymbol{B}$, which change sinusoidally in space and time.

For a long time it was believed that these ideas create a harmonious picture of electromagnetic phenomena. For example, if the retarded nature of the interaction is assumed, then the transferred momentum and energy should be temporarily stored in some form as long as they move from one charge to another. Then the laws of conservation of momentum and energy immediately imply the existence of *interaction carriers* with their own degrees of freedom, momenta and energies. In Maxwell's theory, these degrees of freedom are associated with electromagnetic fields.

> ...the interaction is a result of energy momentum exchanges between the particles through the field, which propagates energy and momentum and can transfer them to the particles by contact – F. Strocchi [243].

> The field concept came to dominate physics starting with the work of Faraday in the mid-nineteenth century. Its conceptual advantage over the earlier Newtonian program of physics, to formulate the fundamental laws in terms of forces among atomic particles, emerges when we take into account the circumstance, unknown to Newton (or, for that matter, Faraday) but fundamental in special relativity, that influences travel no farther than a finite limiting speed. For then the force on a given particle at a given time cannot be deduced from the positions of other particles at that time, but must be deduced in a complicated way from their previous positions. Faraday's intuition that the fundamental laws of electromagnetism could be expressed most simply in terms of fields filling space and time was of course brilliantly vindicated in Maxwell's mathematical theory – F. Wilczek [270].

6.1.2 Conservation laws

For the most part, Maxwell's theory is in good agreement with the experiments. However, on closer examination, numerous defects and problems in the classical theory become apparent. One important class of difficulties is associated with the nonconservation of total observables (energy, momentum, angular momentum) in systems of interacting charges and magnetic moments.

Maxwell's equations do not have built-in guarantees that the total observables will be conserved and that the energy–momentum is transformed as a 4-vector quantity. Therefore, it is not surprising that a large number of paradoxes are associated with applications of Maxwell's equations and the Lorentz force formula (6.1) [24, 197, 90, 3, 37, 39, 132, 133, 110, 231, 247, 119, 167, 136–138, 252, 164, 8, 25]. Proposed "solutions" of these paradoxes are usually unsatisfactory and require the introduction of various *ad hoc* constructions, such as "hidden momentum," "Poincaré stresses," alternative non-Lorentz forces, etc.

The simplest paradox in Maxwell's theory relates to an isolated system of two charges 1 and 2 that freely move in space without the influence of external forces. The

Lorentz magnetic forces acting on the charges in the system have the form[1]

$$\boldsymbol{f}_1^{\mathrm{magn}} = \frac{q_1 q_2}{4\pi c^2}\frac{[\boldsymbol{v}_1 \times [\boldsymbol{v}_2 \times \boldsymbol{r}]]}{r^3}, \tag{6.2}$$

$$\boldsymbol{f}_2^{\mathrm{magn}} = -\frac{q_1 q_2}{4\pi c^2}\frac{[\boldsymbol{v}_2 \times [\boldsymbol{v}_1 \times \boldsymbol{r}]]}{r^3}. \tag{6.3}$$

It is not difficult to see that the third Newton law ($\boldsymbol{f}_1^{\mathrm{magn}} = -\boldsymbol{f}_2^{\mathrm{magn}}$) is not satisfied in the general case [113].[2]

6.1.3 Energy and momentum of electromagnetic field

The usual explanation [130, 179, 225, 121] of this paradox is that in Maxwell's electrodynamics the two charges do not form a closed physical system. The charges are always surrounded by the electric $\boldsymbol{E}(\boldsymbol{r})$ and magnetic $\boldsymbol{B}(\boldsymbol{r})$ fields. These fields are assumed to have their own momentum and energy, expressed as integrals over the entire space,

$$\boldsymbol{P}^f = \frac{1}{c}\int d\boldsymbol{r}[\boldsymbol{E}(\boldsymbol{r}) \times \boldsymbol{B}(\boldsymbol{r})], \tag{6.4}$$

$$H^f = \frac{1}{2}\int d\boldsymbol{r}(E^2(\boldsymbol{r}) + B^2(\boldsymbol{r})). \tag{6.5}$$

Then the theory attempts to explain the paradox by postulating that only the sum of the momenta (particles + fields) = $(\boldsymbol{P}^p + \boldsymbol{P}^f)$ must be conserved. However, this explanation is not satisfactory as well.

First, there is no experimental evidence that the total momentum of the particles $\boldsymbol{P}^p$ is a nonconserved quantity. All arguments about the conservation of the sum $\boldsymbol{P}^p + \boldsymbol{P}^f$ are purely speculative. In addition, the integral of the "electromagnetic energy" (6.5) of a point charge (e. g., an electron) diverges.[3] To get around this difficulty, various "classical models" of the electron were proposed, the simplest of which is a charged sphere of a small finite radius. However, such models created new problems. One of them is the famous "4/3 paradox." It can be shown that the field's momentum associated with a finite-size electron does not form the expected 4-vector quantity together with the energy of the same field [197, 24, 38]. This gross violation of relativistic invariance can be "corrected" for if we introduce an additional factor of 4/3 in the formula

---

1 Here we omit Coulomb components of the forces, which, of course, obey the equality $\boldsymbol{f}_1^{\mathrm{Coulomb}} = -\boldsymbol{f}_2^{\mathrm{Coulomb}}$.

2 If we accept the traditional definition of force (6.16) $\boldsymbol{f} = d\boldsymbol{p}/dt$, then we get an even more serious problem – the nonconservation of the total momentum of the charges $\boldsymbol{P}^p = \boldsymbol{p}_1 + \boldsymbol{p}_2$.

3 See Chapter 28 in [78]. See also [86], where other difficulties are discussed related to the idea of the field's energy–momentum. An interesting critical review of Maxwell's electrodynamics and the Minkowski space–time picture can be found in Section 1 of [95].

for the field's momentum. To justify this *ad hoc* idea, the so-called *Poincaré stresses* are sometimes introduced.[4] All these arguments can hardly be called elegant. One often ends up with the conclusion that classical electrodynamics is unable to adequately describe point elementary particles and that for consistency one has to switch to quantum electrodynamics. But QED does not save the situation, because it faces its own set of problems, which we discussed earlier in this volume.

The problems described above are not present in RQD formulations of classical electrodynamics, since we do not recognize the concepts of the electromagnetic field, its momentum and energy, so we avoid the divergences and paradoxes associated with these quantities.

6.2 Interaction of classical charges in RQD

In our book, we question the universally accepted postulates of Maxwell. Our concern with the Maxwell theory is that its fundamental ingredients, such as electromagnetic fields, the Liénard–Wiechert potentials and the Lorentz force law, are not expressed in the language of Poincaré group representations and Hamiltonian dynamics,[5] which, in our opinion, is the most natural language of physics. For example, by using unitary representations of the Poincaré group one is guaranteed to observe conservation laws as well as correct transformations of observables between different frames. Unfortunately, in Maxwell's field theory, the fulfillment of these fundamental requirements is not at all obvious.

We claim that all results of classical electromagnetic theory can be described and explained with an equal (or even better) success in a Hamiltonian theory of directly interacting charged particles, where "electromagnetic fields" play absolutely no role.

6.2.1 Coulomb–Darwin–Breit Hamiltonian

In Chapter 3 we already derived the quantum Coulomb–Darwin–Breit Hamiltonian (3.5) for interacting charges. This Hamiltonian was obtained in the second perturbation order within the $(v/c)^2$ approximation of RQD.

The same Hamiltonian is applicable also in the classical limit $\hbar \to 0$. So, in this chapter we will be working in the classical approximation ignoring all quantum effects, not paying attention to the order of dynamical variables in their products and using Poisson brackets $[\ldots,\ldots]_P$ instead of quantum commutators $(-i/\hbar)[\ldots,\ldots]$. We

4 See Sections 16.4–16.6 in [118].

5 For some attempts to overcome this difficulty, see [141, 41].

will also use expansions in powers of (v/c) and keep only terms whose power is not higher than $(v/c)^2$.[6]

Recall that the Coulomb–Darwin–Breit Hamiltonian $H^d = H_0 + V^d$ for the system of two charges consists of the free part (3.4)

$$\begin{aligned} H_0 &= h_1 + h_2 \\ &= \sqrt{m_1^2c^4 + p_1^2c^2} + \sqrt{m_2^2c^4 + p_2^2c^2} \\ &\approx m_1c^2 + m_2c^2 + \frac{p_1^2}{2m_1} + \frac{p_2^2}{2m_2} - \frac{p_1^4}{8m_1^3c^2} - \frac{p_2^4}{8m_2^3c^2} \end{aligned} \tag{6.6}$$

and the potential energy (3.6)–(3.10)[7]

$$\begin{aligned} V^d \approx\ & \frac{q_1q_2}{4\pi r} - \frac{q_1q_2}{8\pi m_1m_2c^2r}\left((\boldsymbol{p}_1\cdot\boldsymbol{p}_2) + \frac{(\boldsymbol{p}_1\cdot\boldsymbol{r})(\boldsymbol{p}_2\cdot\boldsymbol{r})}{r^2}\right) \\ & - \frac{q_1q_2[\boldsymbol{r}\times\boldsymbol{p}_1]\cdot\boldsymbol{s}_1}{8\pi m_1^2c^2r^3} + \frac{q_1q_2[\boldsymbol{r}\times\boldsymbol{p}_2]\cdot\boldsymbol{s}_2}{8\pi m_2^2c^2r^3} + \frac{q_1q_2[\boldsymbol{r}\times\boldsymbol{p}_2]\cdot\boldsymbol{s}_1}{4\pi m_1m_2c^2r^3} \\ & - \frac{q_1q_2[\boldsymbol{r}\times\boldsymbol{p}_1]\cdot\boldsymbol{s}_2}{4\pi m_1m_2c^2r^3} + \frac{q_1q_2(\boldsymbol{s}_1\cdot\boldsymbol{s}_2)}{4\pi m_1m_2c^2r^3} - \frac{3q_1q_2(\boldsymbol{s}_1\cdot\boldsymbol{r})(\boldsymbol{s}_2\cdot\boldsymbol{r})}{4\pi m_1m_2c^2r^5}. \end{aligned} \tag{6.7}$$

In order to use this Hamiltonian in practical calculations, we will modify it a bit. First, we drop the rest energies of the two particles, because these constants have no effect on dynamics. Second, we notice that particle spins $\boldsymbol{s}_i$ are not easily accessible in classical experiments. It would be more convenient to replace them by magnetic moments $\boldsymbol{\mu}_i$, i. e.,[8]

$$\boldsymbol{\mu}_i = q_i\boldsymbol{s}_i/m_i.$$

Then the full Hamiltonian for two classical charges takes the form

$$H^d = \frac{p_1^2}{2m_1} + \frac{p_2^2}{2m_2} - \frac{p_1^4}{8m_1^3c^2} - \frac{p_2^4}{8m_2^3c^2} + \frac{q_1q_2}{4\pi r}$$

---

6 Of course, our approximation $(v/c)^2$ is only applicable to the cases of low velocities and accelerations, when one can neglect emission of electromagnetic radiation (photons) and radiation reaction. To describe these effects, we would need to add at least third-order potentials from Subsection 5.1.2. These studies are beyond the scope of our book.

7 Here q_1 and q_2 are charges of the two particles (they were chosen as $q_1 = -q_2 = -e$ in Subsection 3.1). We denote $\boldsymbol{r} \equiv \boldsymbol{r}_1 - \boldsymbol{r}_2$ throughout this chapter. We also drop contact interaction terms proportional to $\delta(\boldsymbol{r})$, because they have no effect on classical trajectories of particles. In Appendix E we verified that, together with a properly chosen boost generator $\boldsymbol{K}^d = \boldsymbol{K}_0 + \boldsymbol{Z}^d$, the Hamiltonian $H^d = H_0 + V^d$ satisfies all commutators of the Poincaré Lie algebra within the $(v/c)^2$ approximation.

8 See formula (5.25). In this approximation we assume that $g = 2$, i. e., ignore the “anomalous” contribution from Subsection 5.3.2.

$$
\begin{aligned}
&- \frac{q_1 q_2}{8\pi m_1 m_2 c^2 r}\left(\boldsymbol{p}_1 \cdot \boldsymbol{p}_2 + \frac{(\boldsymbol{r}\cdot\boldsymbol{p}_2)(\boldsymbol{r}\cdot\boldsymbol{p}_1)}{r^2}\right) \\
&- \frac{q_1[\boldsymbol{r}\times\boldsymbol{p}_1]\cdot\boldsymbol{\mu}_2}{4\pi m_1 c^2 r^3} + \frac{q_1[\boldsymbol{r}\times\boldsymbol{p}_2]\cdot\boldsymbol{\mu}_2}{8\pi m_2 c^2 r^3} - \frac{q_2[\boldsymbol{r}\times\boldsymbol{p}_1]\cdot\boldsymbol{\mu}_1}{8\pi m_1 c^2 r^3} \\
&+ \frac{q_2[\boldsymbol{r}\times\boldsymbol{p}_2]\cdot\boldsymbol{\mu}_1}{4\pi m_2 c^2 r^3} + \frac{(\boldsymbol{\mu}_1\cdot\boldsymbol{\mu}_2)}{4\pi c^2 r^3} - \frac{3(\boldsymbol{\mu}_1\cdot\boldsymbol{r})(\boldsymbol{\mu}_2\cdot\boldsymbol{r})}{4\pi c^2 r^5}.
\end{aligned} \tag{6.8}
$$

So, in the Coulomb–Darwin–Breit theory, point charges and magnetic moments interact with each other by means of instantaneous potentials. This approach uses neither "electromagnetic fields" nor their own energy–momentum. Our goal in this chapter is to demonstrate that the action-at-a-distance Hamiltonian (6.8) can be successfully applied for an accurate description of macroscopic electromagnetic phenomena.

6.2.2 Two charges

First, let us consider a system of two charged spinless particles. The Hamiltonian is obtained by discarding $\boldsymbol{\mu}$-dependent terms in (6.8), so we have

$$
\begin{aligned}
H^d &= \frac{p_1^2}{2m_1} + \frac{p_2^2}{2m_2} - \frac{p_1^4}{8m_1^3c^2} - \frac{p_2^4}{8m_2^3c^2} + \frac{q_1q_2}{4\pi r} \\
&\quad - \frac{q_1q_2}{8\pi m_1 m_2 c^2 r}\left((\boldsymbol{p}_1\cdot\boldsymbol{p}_2) + \frac{(\boldsymbol{p}_1\cdot\boldsymbol{r})(\boldsymbol{p}_2\cdot\boldsymbol{r})}{r^2}\right).
\end{aligned} \tag{6.9}
$$

This Hamiltonian determines the dynamics of charges through Hamilton's equations of motion (**1**-6.106)–(**1**-6.107) and the Poisson bracket (**1**-6.102). For example, the time derivative of the first particle's momentum is obtained from the first Hamilton equation,

$$
\begin{aligned}
\frac{d\boldsymbol{p}_1}{dt} &= [\boldsymbol{p}_1, H^d]_P = -\frac{\partial H^d}{\partial \boldsymbol{r}_1} \\
&= \frac{q_1q_2\boldsymbol{r}}{4\pi r^3} - \frac{q_1q_2(\boldsymbol{p}_1\cdot\boldsymbol{p}_2)\boldsymbol{r}}{8\pi m_1 m_2 c^2 r^3} + \frac{q_1q_2\boldsymbol{p}_1(\boldsymbol{p}_2\cdot\boldsymbol{r})}{8\pi m_1 m_2 c^2 r^3} + \frac{q_1q_2(\boldsymbol{p}_1\cdot\boldsymbol{r})\boldsymbol{p}_2}{8\pi m_1 m_2 c^2 r^3} \\
&\quad - \frac{3q_1q_2(\boldsymbol{p}_1\cdot\boldsymbol{r})(\boldsymbol{p}_2\cdot\boldsymbol{r})\boldsymbol{r}}{8\pi m_1 m_2 c^2 r^5}.
\end{aligned} \tag{6.10}
$$

Since the Hamiltonian (6.9) is symmetric with respect to the permutation of particles, the rate of change of the second particle's momentum is obtained by swapping indices $1 \leftrightarrow 2$ in (6.10), so we have

$$
\frac{d\boldsymbol{p}_2}{dt} = -\frac{d\boldsymbol{p}_1}{dt}. \tag{6.11}
$$

Velocities of particles 1 and 2 are obtained from the second Hamilton equation:

$$\boldsymbol{v}_1 \equiv \frac{d\boldsymbol{r}_1}{dt} = [\boldsymbol{r}_1, H^d]_P = \frac{\partial H^d}{\partial \boldsymbol{p}_1}$$
$$= \frac{\boldsymbol{p}_1}{m_1} - \frac{p_1^2\boldsymbol{p}_1}{2m_1^3c^2} - \frac{q_1q_2\boldsymbol{p}_2}{8\pi m_1m_2c^2r} - \frac{q_1q_2(\boldsymbol{p}_2\cdot\boldsymbol{r})\boldsymbol{r}}{8\pi m_1m_2c^2r^3}, \tag{6.12}$$

$$\boldsymbol{v}_2 \equiv \frac{d\boldsymbol{r}_2}{dt} = \frac{\boldsymbol{p}_2}{m_2} - \frac{p_2^2\boldsymbol{p}_2}{2m_2^3c^2} - \frac{q_1q_2\boldsymbol{p}_1}{8\pi m_1m_2c^2r} - \frac{q_1q_2(\boldsymbol{p}_1\cdot\boldsymbol{r})\boldsymbol{r}}{8\pi m_1m_2c^2r^3}. \tag{6.13}$$

From these results it is not difficult to calculate second derivatives of particle positions (= accelerations). We have[9]

$$\begin{aligned}
\frac{d^2\boldsymbol{r}_1}{dt^2} = \frac{d\boldsymbol{v}_1}{dt} &= \frac{\dot{\boldsymbol{p}}_1}{m_1} - \frac{p_1^2\dot{\boldsymbol{p}}_1}{2m_1^3c^2} - \frac{2(\boldsymbol{p}_1\cdot\dot{\boldsymbol{p}}_1)\boldsymbol{p}_1}{2m_1^3c^2} + \frac{q_1q_2\boldsymbol{p}_2(\boldsymbol{r}\cdot\dot{\boldsymbol{r}})}{8\pi m_1m_2c^2r^3}\\
&\quad - \frac{q_1q_2(\boldsymbol{p}_2\cdot\dot{\boldsymbol{r}})\boldsymbol{r}}{8\pi m_1m_2c^2r^3} + \frac{3q_1q_2(\boldsymbol{p}_2\cdot\boldsymbol{r})\boldsymbol{r}(\boldsymbol{r}\cdot\dot{\boldsymbol{r}})}{8\pi m_1m_2c^2r^5} - \frac{q_1q_2(\boldsymbol{p}_2\cdot\boldsymbol{r})\dot{\boldsymbol{r}}}{8\pi m_1m_2c^2r^3}\\
&\approx \frac{q_1q_2\boldsymbol{r}}{4\pi m_1r^3} - \frac{q_1q_2(\boldsymbol{p}_1\cdot\boldsymbol{p}_2)\boldsymbol{r}}{8\pi m_1^2m_2c^2r^3} + \frac{q_1q_2\boldsymbol{p}_1(\boldsymbol{p}_2\cdot\boldsymbol{r})}{8\pi m_1^2m_2c^2r^3} + \frac{q_1q_2(\boldsymbol{p}_1\cdot\boldsymbol{r})\boldsymbol{p}_2}{8\pi m_1^2m_2c^2r^3}\\
&\quad - \frac{3q_1q_2(\boldsymbol{p}_1\cdot\boldsymbol{r})(\boldsymbol{p}_2\cdot\boldsymbol{r})\boldsymbol{r}}{8\pi m_1^2m_2c^2r^5} - \frac{q_1q_2p_1^2\boldsymbol{r}}{8\pi m_1^3c^2r^3} - \frac{q_1q_2(\boldsymbol{p}_1\cdot\boldsymbol{r})\boldsymbol{p}_1}{4\pi m_1^3c^2r^3} + \frac{q_1q_2\boldsymbol{p}_2(\boldsymbol{r}\cdot\boldsymbol{p}_1)}{8\pi m_1^2m_2c^2r^3}\\
&\quad - \frac{q_1q_2\boldsymbol{p}_2(\boldsymbol{r}\cdot\boldsymbol{p}_2)}{8\pi m_1m_2^2c^2r^3} - \frac{q_1q_2(\boldsymbol{p}_2\cdot\boldsymbol{p}_1)\boldsymbol{r}}{8\pi m_1^2m_2c^2r^3} + \frac{q_1q_2p_2^2\boldsymbol{r}}{8\pi m_1m_2^2c^2r^3} + \frac{3q_1q_2(\boldsymbol{p}_2\cdot\boldsymbol{r})\boldsymbol{r}(\boldsymbol{r}\cdot\boldsymbol{p}_1)}{8\pi m_1^2m_2c^2r^5}\\
&\quad - \frac{3q_1q_2(\boldsymbol{p}_2\cdot\boldsymbol{r})\boldsymbol{r}(\boldsymbol{r}\cdot\boldsymbol{p}_2)}{8\pi m_1m_2^2c^2r^5} - \frac{q_1q_2(\boldsymbol{p}_2\cdot\boldsymbol{r})\boldsymbol{p}_1}{8\pi m_1^2m_2c^2r^3} + \frac{q_1q_2(\boldsymbol{p}_2\cdot\boldsymbol{r})\boldsymbol{p}_2}{8\pi m_1m_2^2c^2r^3}\\
&= \frac{q_1q_2\boldsymbol{r}}{4\pi m_1r^3} + \frac{q_1q_2(\boldsymbol{v}_1-\boldsymbol{v}_2)^2\boldsymbol{r}}{8\pi m_1c^2r^3} - \frac{q_1q_2v_1^2\boldsymbol{r}}{4\pi m_1c^2r^3}\\
&\quad + \frac{q_1q_2(\boldsymbol{v}_1\cdot\boldsymbol{r})(\boldsymbol{v}_2-\boldsymbol{v}_1)}{4\pi m_1c^2r^3} - \frac{3q_1q_2(\boldsymbol{v}_2\cdot\boldsymbol{r})^2\boldsymbol{r}}{8\pi m_1c^2r^5},
\end{aligned} \tag{6.14}$$

$$\begin{aligned}
\frac{d^2\boldsymbol{r}_2}{dt^2} &\approx -\frac{q_1q_2\boldsymbol{r}}{4\pi m_2r^3} - \frac{q_1q_2(\boldsymbol{v}_1-\boldsymbol{v}_2)^2\boldsymbol{r}}{8\pi m_2c^2r^3} + \frac{q_1q_2v_2^2\boldsymbol{r}}{4\pi m_2c^2r^3}\\
&\quad - \frac{q_1q_2(\boldsymbol{v}_2\cdot\boldsymbol{r})(\boldsymbol{v}_1-\boldsymbol{v}_2)}{4\pi m_2c^2r^3} + \frac{3q_1q_2(\boldsymbol{v}_1\cdot\boldsymbol{r})^2\boldsymbol{r}}{8\pi m_2c^2r^5}.
\end{aligned} \tag{6.15}$$

9 In this derivation we dropped small terms proportional to $q_1^2q_2^2$ (formally, they belong to the fourth perturbation order). Also, staying within the $(v/c)^2$ approximation, we set $\dot{\boldsymbol{r}} = d\boldsymbol{r}_1/dt - d\boldsymbol{r}_2/dt \equiv \boldsymbol{v}_1 - \boldsymbol{v}_2 \approx \boldsymbol{p}_1/m_1 - \boldsymbol{p}_2/m_2$ in the terms that already have the factor $(1/c)^2$. A dot above a symbol indicates the time derivative.

6.2.3 Definition of force

Two definitions of *force* are used in classical mechanics. In one definition, the force acting on a particle is identified with the time derivative of the particle's momentum, i. e.,

$$\boldsymbol{f}_i \equiv \frac{d\boldsymbol{p}_i}{dt}. \tag{6.16}$$

In the other definition [36], the force is the product of the particle's rest mass and its acceleration,[10]

$$\boldsymbol{f}_i \equiv m_i \frac{d^2\boldsymbol{r}_i}{dt^2}. \tag{6.17}$$

These two definitions coincide only for the simplest interaction potentials, which do not depend on the momenta (or velocities) of particles.[11] This is not true in the Coulomb–Darwin–Breit electrodynamics, where we are dealing with momentum-dependent potentials. So, momentum–velocity relationships (6.12)–(6.13) depend on the interaction, and we have to decide which definition of force we are going to use.

The more common definition (6.16) has the advantage that the Newton third law of motion (action = –reaction) in a two-particle system has a simple formulation,

$$\boldsymbol{f}_1 = -\boldsymbol{f}_2. \tag{6.18}$$

This is just a consequence of the momentum conservation law. Indeed, in the instant form of dynamics, the total momentum is free of interactions ($\boldsymbol{P}_0 = \boldsymbol{p}_1 + \boldsymbol{p}_2$), and from the Poisson bracket $[\boldsymbol{P}_0, H^d]_P = 0$ we obtain

$$\boldsymbol{f}_1 = \frac{d\boldsymbol{p}_1}{dt} = [\boldsymbol{p}_1, H^d]_P = [\boldsymbol{P}_0 - \boldsymbol{p}_2, H^d]_P = -[\boldsymbol{p}_2, H^d]_P = -\frac{d\boldsymbol{p}_2}{dt} = -\boldsymbol{f}_2. \tag{6.19}$$

Nevertheless, in our book we will use an alternative definition of force (6.17). Although this definition does not imply the existence of the *balance of forces* (6.18),[12] it is preferable for several reasons. First, the definition (6.17) is consistent with the usual expectation that the equilibrium state is reached when the forces (proportional to accelerations) vanish.[13] Second, the definition (6.17) is more convenient than (6.16), because in experiments particle velocities and accelerations are more readily available

10 This is equivalent to the Newton second law of motion.

11 In such cases, according to Hamilton's second equation of motion, we have $d\boldsymbol{r}_i/dt = \partial H^d/\partial \boldsymbol{p}_i = \boldsymbol{p}_i/m_i$ and $m_i d^2\boldsymbol{r}_i/dt^2 = d\boldsymbol{p}_i/dt$.

12 One can check that in our definition $\boldsymbol{f}_1 \neq -\boldsymbol{f}_2$ by comparing accelerations (6.14) and (6.15) multiplied by the corresponding masses.

13 This is Newton's first law.

than their momenta or $d\boldsymbol{p}_i/dt$. For example, by measuring the current in a wire, we are actually monitoring the number of charges that pass through the wire's cross section in unit time. This quantity is directly connected with velocities of the charges and is not related to their momenta.

6.2.4 Conservation laws in RQD

In contrast to Maxwell's theory, in our "fieldless" RQD, conservation laws and transformational properties of observables appear naturally, since they are direct consequences of the Poincaré group structure. For example, the Poisson bracket of any observable F with the complete Hamiltonian H^d determines the time dependence of this observable (**1**-6.105), i. e.,

$$\frac{dF(t)}{dt} = [F, H^d]_P.$$

Therefore, the conservation of the total energy H^d and the total momentum $\boldsymbol{P}_0$ of any isolated system of charges follows automatically from the fact that these quantities have zero Poisson brackets with H^d.[14] The boost transformation of any physical quantity F is obtained by solving equation (**1**-3.63), i. e.,

$$\frac{dF(\boldsymbol{\theta})}{d\boldsymbol{\theta}} = -c[F, \boldsymbol{K}^d]_P.$$

For the total 4-momentum $(H^d, c\boldsymbol{P}_0)$ of any system, the commutators (**1**-3.54)–(**1**-3.55) are valid independent of interactions acting there. So, the 4-vector transformation formulas (**1**-4.3)–(**1**-4.4) follow automatically, that is, in RQD, the conservation of total observables and their correct transformation laws are embedded in the formalism at the most fundamental level and cannot be violated even in principle.

6.2.5 Trouton–Noble paradox

Let us now discuss the conservation of the total angular momentum in the two competing theories.

From the Lie algebra of the Poincaré group it follows that in RQD the total angular momentum $\boldsymbol{J}_0$ of any isolated system is conserved, so

$$\frac{d\boldsymbol{J}_0}{dt} = [\boldsymbol{J}_0, H^d]_P = 0.$$

In other words, there can be no *torque*, which we define as the time derivative of $\boldsymbol{J}_0$.

14 For example, the conservation of $\boldsymbol{P}_0$ in RQD guarantees the resolution of the paradoxes 6 and 7 in [135] as well as the paradox from Subsection 6.1.2.

This result must be valid in any reference frame. For example, in a moving frame, we have the following dynamic variables[15]:

$$J_0(\boldsymbol{\theta}) = e^{-\frac{ic}{\hbar}\boldsymbol{K}^d\cdot\boldsymbol{\theta}} J_0 e^{\frac{ic}{\hbar}\boldsymbol{K}^d\cdot\boldsymbol{\theta}},$$
$$H^d(\boldsymbol{\theta}) = e^{-\frac{ic}{\hbar}\boldsymbol{K}^d\cdot\boldsymbol{\theta}} H^d e^{\frac{ic}{\hbar}\boldsymbol{K}^d\cdot\boldsymbol{\theta}}$$

and the equation of motion for the total angular momentum is[16]

$$\frac{d\boldsymbol{J}_0(\boldsymbol{\theta})}{dt'} = [\boldsymbol{J}_0(\boldsymbol{\theta}), H^d(\boldsymbol{\theta})]_P = [e^{-\frac{ic}{\hbar}\boldsymbol{K}^d\cdot\boldsymbol{\theta}} \boldsymbol{J}_0 e^{\frac{ic}{\hbar}\boldsymbol{K}^d\cdot\boldsymbol{\theta}}, e^{-\frac{ic}{\hbar}\boldsymbol{K}^d\cdot\boldsymbol{\theta}} H^d e^{\frac{ic}{\hbar}\boldsymbol{K}^d\cdot\boldsymbol{\theta}}]_P$$
$$= e^{-\frac{ic}{\hbar}\boldsymbol{K}^d\cdot\boldsymbol{\theta}} [\boldsymbol{J}_0, H^d]_P e^{\frac{ic}{\hbar}\boldsymbol{K}^d\cdot\boldsymbol{\theta}} = 0.$$

Maxwell's classical electrodynamics cannot make such a simple and unambiguous statement about the absence of the torque in all reference frames. This failure is in the center of the so-called Trouton–Noble paradox [251], which has haunted Maxwell's electrodynamics since the beginning of the 20th century.

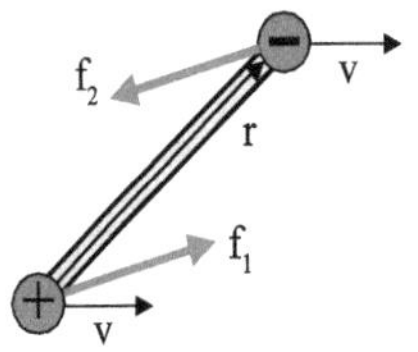

Figure 6.1: Trouton–Noble paradox in Maxwell's electrodynamics: two charges move with the same velocity $\boldsymbol{v}$; forces $\boldsymbol{f}_1$ and $\boldsymbol{f}_2$ create a nonzero torque, trying to rotate the vector $\boldsymbol{r}$ until it becomes perpendicular to the direction of motion.

Imagine two opposite charges,[17] connected by a rod and moving with equal velocities $\boldsymbol{v}$ in the configuration shown in Figure 6.1.[18] Maxwell's theory predicts the presence of a nonzero torque, trying to rotate the rod until it becomes perpendicular to the direction of motion [206]. Obviously, this torque vanishes in the comoving frame of reference, where the charges are at rest.[19] These results are paradoxical for two reasons. First, as we said in the beginning of this subsection, there should not be any torque due to the law of conservation of the total angular momentum. Second, the disappearance of the torque in the comoving frame contradicts the principle of relativity. Numerous attempts to explain this paradox from the standpoint of classical theory [179, 90, 231, 119, 24, 247, 122] do not seem convincing.

---

15 Here we use quantum notation with unitary transformation operators. It is easy to switch to the classical notation with Poisson brackets by using equation (**1**-E.14). The term $\boldsymbol{K}^d$ represents the full interaction-dependent operator of boost, e. g., in the $(v/c)^2$ approximation we can use (E.5)–(E.6).

16 Here t' is time measured by the moving observer's clock.

17 Or two parallel plates of a charged capacitor [251].

18 We assume that the angle between vectors $\boldsymbol{v}$ and $\boldsymbol{r}$ differs between 0° and 90°.

19 In this frame, there is only the Coulomb force, which is directed along the vector $\boldsymbol{r}$ and cannot create a torque.

6.2.6 Electromagnetic field or photons?

Maxwell's theory describes light in terms of electromagnetic waves $\boldsymbol{E}(t,\boldsymbol{r})$ and $\boldsymbol{B}(t,\boldsymbol{r})$, having momentum (6.4) and energy (6.5). This unification of electricity, magnetism and optics is usually celebrated as a significant achievement. Indeed, as we explained in Subsection **1**-1.1.2, the wave theory of light could easily explain such phenomena as diffraction and interference. However, this theory meets great difficulties when trying to describe the light of low intensity and the photoelectric effect.

In RQD, we claim that electromagnetic radiation is nothing but a cumulative flow of a huge number of massless particle–photons. We believe that the wave properties of light are manifestations of the quantum nature of individual photons. Coherent properties of light, for example in lasers, are adequately described in terms of wave functions of multiphoton states. Electromagnetic fields $\boldsymbol{E}$ and $\boldsymbol{B}$ as well as their energy–momentum play absolutely no role in our theory.

In our approach, the energy of light is the expectation value of the photon energy operator (**2**-1.33), calculated on the wave function of the appropriate multiphoton state. Likewise, the momentum of light is the expectation value of (**2**-1.35). From our point of view, the Huygens–Maxwell wave theory is, in essence, an attempt to approximate wave functions of multiphoton states by two surrogate functions $\boldsymbol{E}(t,\boldsymbol{r})$ and $\boldsymbol{B}(t,\boldsymbol{r})$ [80, 82, 50, 48, 49, 28]. From this we get a rather surprising conclusion [80] that observations of diffraction and interference made by Newton, Grimaldi and Young were, actually, the first experimental measurements of quantum effects.

6.3 Magnetic interactions

So, we have agreed that charges interact at a distance by direct instantaneous potentials without the mediation of electromagnetic fields. How can we verify the correctness of the Coulomb–Darwin–Breit potential? In principle, we could integrate equations (6.14)–(6.15) and obtain trajectories $\boldsymbol{r}_1(t)$, $\boldsymbol{r}_2(t)$ of the interacting charges. However, in experiments it is rather difficult to isolate two charged particles in an open space and trace their trajectories with an accuracy sufficiently high to test theoretical predictions. In many cases it is more convenient to study the behavior of electrons whose movements are confined within metallic conductors.

6.3.1 Charge + straight wire with current

Let us find the force exerted by a metal wire with current on a test charge located outside the conductor. The Hamiltonian of this system can be derived from the two-particle Hamiltonian (6.8). By the index 1 we indicate a test charge whose properties are denoted by q_1, $\boldsymbol{r}_1$ and $\boldsymbol{v}_1$. For simplicity, in this section we set the magnetic moment of this particle to zero

$$\boldsymbol{\mu}_1 = 0 \tag{6.20}$$

and discard all $\boldsymbol{\mu}_1$-dependent terms in (6.8). By the index 2, we label summarily all the charges in the wire. Then, to construct the complete Hamiltonian H_{c+w} of the system "test charge 1 + wire," we need to sum (6.8) over all charges 2.[20] The wire charges 2 are of two types: fixed positively charged ions and mobile negatively charged electrons. In most cases, the total ion charge exactly compensates the total charge of the electrons, so that the wire is electrically neutral. This means that the fifth (Coulomb) term in (6.8) does not contribute to H_{c+w}. We also assume that the wire is nonmagnetic, i. e., that magnetic moments $\boldsymbol{\mu}_2$ of electrons and ions in the wire are oriented randomly. Then all $\boldsymbol{\mu}_2$-dependent terms in (6.8) disappear after averaging over angles. If the wire moves as a single unit with velocity $\boldsymbol{w}$, then $\boldsymbol{w}$-dependent Darwin interactions (sixth term in equation (6.8)) of charge 1 with wire electrons and ions cancel out. Hence, the potential acting on charge 1 does not depend on the velocity of the wire, and without loss of generality we can assume that the heavy ions of the wire remain stationary and only the electrons move with velocity $\boldsymbol{v}_2$. At the moment, we are not interested in the influence of the test charge 1 on the dynamics of electrons 2 in the wire. Therefore, we ignore the kinetic energy of these electrons (second and fourth terms in (6.8)). After these approximations, our Hamiltonian takes the form

$$H_{c+w} = \frac{p_1^2}{2m_1} - \frac{p_1^4}{8m_1^3c^2} - \sum_i \frac{q_1 q_{2i}}{8\pi c^2 m_1 m_{2i} r_i}\left(\boldsymbol{p}_1 \cdot \boldsymbol{p}_{2i} + \frac{(\boldsymbol{r}_i \cdot \boldsymbol{p}_{2i})(\boldsymbol{r}_i \cdot \boldsymbol{p}_1)}{r_i^2}\right), \tag{6.21}$$

where index i runs over all electrons in the wire and $\boldsymbol{r}_i \equiv \boldsymbol{r}_1 - \boldsymbol{r}_{2i}$. These electrons participate in two types of motion: the *thermal* one and the *drift* one. Velocities of the thermal motion are very high, but their directions are randomly distributed. Since the magnetic part of our Hamiltonian (6.21) depends linearly on $\boldsymbol{v}_{2i} \approx \boldsymbol{p}_{2i}/m_{2i}$, the forces associated with the thermal motion cancel out after summation over i. Hence, $\boldsymbol{v}_{2i}$ should be interpreted as the drift velocity of the electrons along the wire. In a straight wire (as in Figure 6.2), this vector is the same for all electrons $\boldsymbol{v}_{2i} \equiv \boldsymbol{v}_2$. It is directed along the applied voltage, and its magnitude is rather small (less than 1 mm/s).

To find the force $\boldsymbol{f}_1$ felt by charge 1 in the vicinity of an infinite wire in the geometry shown in Figure 6.2, we solve the Hamilton equations by analogy with Subsection 6.2.2. For each pair of charges 1 and $2i$ we have

$$\frac{d\boldsymbol{p}_{1i}}{dt} = -\frac{q_1 q_{2i}(\boldsymbol{p}_1 \cdot \boldsymbol{p}_{2i})\boldsymbol{r}}{8\pi m_1 m_{2i} c^2 r^3} + \frac{q_1 q_{2i} \boldsymbol{p}_1(\boldsymbol{p}_{2i} \cdot \boldsymbol{r})}{8\pi m_1 m_{2i} c^2 r^3}$$

20 Here we are not interested in mutual interactions of charges inside the wire; see, however, Subsection 6.3.2.

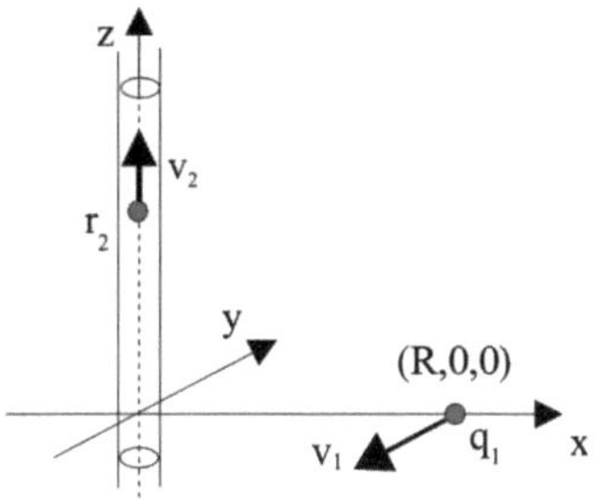

Figure 6.2: Test charge q_1 at the point $(R, 0, 0)$ and an infinite straight wire with current.

$$+\frac{q_1q_{2i}(\boldsymbol{p}_1\cdot\boldsymbol{r})\boldsymbol{p}_{2i}}{8\pi m_1 m_{2i}c^2r^3}-\frac{3q_1q_{2i}(\boldsymbol{p}_1\cdot\boldsymbol{r})(\boldsymbol{p}_{2i}\cdot\boldsymbol{r})\boldsymbol{r}}{8\pi m_1 m_{2i}c^2r^5},$$

$$\frac{d\boldsymbol{r}_{1i}}{dt}=\frac{\boldsymbol{p}_1}{m_1}-\frac{p_1^2\boldsymbol{p}_1}{2m_1^3c^2}-\frac{q_1q_{2i}\boldsymbol{p}_{2i}}{8\pi m_1 m_{2i}c^2r}-\frac{q_1q_{2i}(\boldsymbol{p}_{2i}\cdot\boldsymbol{r})\boldsymbol{r}}{8\pi m_1 m_{2i}c^2r^3},$$

$$\begin{aligned}\frac{d^2\boldsymbol{r}_{1i}}{dt^2}&=\frac{\dot{\boldsymbol{p}}_1}{m_1}-\frac{p_1^2\dot{\boldsymbol{p}}_1}{2m_1^3c^2}-\frac{2(\boldsymbol{p}_1\cdot\dot{\boldsymbol{p}}_1)\boldsymbol{p}_1}{2m_1^3c^2}+\frac{q_1q_{2i}\boldsymbol{p}_{2i}(\boldsymbol{r}\cdot\dot{\boldsymbol{r}})}{8\pi m_1 m_{2i}c^2r^3}\\
&\quad-\frac{q_1q_{2i}(\boldsymbol{p}_{2i}\cdot\dot{\boldsymbol{r}})\boldsymbol{r}}{8\pi m_1 m_{2i}c^2r^3}+\frac{3q_1q_{2i}(\boldsymbol{p}_{2i}\cdot\boldsymbol{r})\boldsymbol{r}(\boldsymbol{r}\cdot\dot{\boldsymbol{r}})}{8\pi m_1 m_{2i}c^2r^5}-\frac{q_1q_{2i}(\boldsymbol{p}_{2i}\cdot\boldsymbol{r})\dot{\boldsymbol{r}}}{8\pi m_1 m_{2i}c^2r^3}\\
&\approx-\frac{q_1q_{2i}(\boldsymbol{p}_1\cdot\boldsymbol{p}_{2i})\boldsymbol{r}}{8\pi m_1^2 m_{2i}c^2r^3}+\frac{q_1q_{2i}\boldsymbol{p}_1(\boldsymbol{p}_{2i}\cdot\boldsymbol{r})}{8\pi m_1^2 m_{2i}c^2r^3}+\frac{q_1q_{2i}(\boldsymbol{p}_1\cdot\boldsymbol{r})\boldsymbol{p}_{2i}}{8\pi m_1^2 m_{2i}c^2r^3}\\
&\quad-\frac{3q_1q_{2i}(\boldsymbol{p}_1\cdot\boldsymbol{r})(\boldsymbol{p}_{2i}\cdot\boldsymbol{r})\boldsymbol{r}}{8\pi m_1^2 m_{2i}c^2r^5}+\frac{q_1q_{2i}\boldsymbol{p}_{2i}(\boldsymbol{r}\cdot\boldsymbol{p}_1)}{8\pi m_1^2 m_{2i}c^2r^3}\\
&\quad-\frac{q_1q_{2i}\boldsymbol{p}_{2i}(\boldsymbol{r}\cdot\boldsymbol{p}_{2i})}{8\pi m_1 m_{2i}^2c^2r^3}-\frac{q_1q_{2i}(\boldsymbol{p}_{2i}\cdot\boldsymbol{p}_1)\boldsymbol{r}}{8\pi m_1^2 m_{2i}c^2r^3}+\frac{q_1q_{2i}p_{2i}^2\boldsymbol{r}}{8\pi m_1 m_{2i}^2c^2r^3}\\
&\quad+\frac{3q_1q_{2i}(\boldsymbol{p}_{2i}\cdot\boldsymbol{r})\boldsymbol{r}(\boldsymbol{r}\cdot\boldsymbol{p}_1)}{8\pi m_1^2 m_{2i}c^2r^5}-\frac{3q_1q_{2i}(\boldsymbol{p}_{2i}\cdot\boldsymbol{r})\boldsymbol{r}(\boldsymbol{r}\cdot\boldsymbol{p}_{2i})}{8\pi m_1 m_{2i}^2c^2r^5}\\
&\quad-\frac{q_1q_{2i}(\boldsymbol{p}_{2i}\cdot\boldsymbol{r})\boldsymbol{p}_1}{8\pi m_1^2 m_{2i}c^2r^3}+\frac{q_1q_{2i}(\boldsymbol{p}_{2i}\cdot\boldsymbol{r})\boldsymbol{p}_{2i}}{8\pi m_1 m_{2i}^2c^2r^3},\end{aligned}$$

so that after summation over all charges $2i$ and multiplication by m_1, the last expression yields the force acting on charge 1,

$$\boldsymbol{f}_1\approx q_1\sum_i q_{2i}\left(-\frac{(\boldsymbol{v}_1\cdot\boldsymbol{v}_{2i})\boldsymbol{r}_i}{4\pi c^2r_i^3}+\frac{v_{2i}^2\boldsymbol{r}_i}{8\pi c^2r_i^3}+\frac{(\boldsymbol{v}_1\cdot\boldsymbol{r}_i)\boldsymbol{v}_{2i}}{4\pi c^2r_i^3}-\frac{3(\boldsymbol{v}_{2i}\cdot\boldsymbol{r}_i)^2\boldsymbol{r}_i}{8\pi c^2r_i^5}\right).$$

Since the linear density (ρ_2) of conduction electrons in the metal is very high, we can proceed to the continuous limit, in which a small segment dr_{2z} of the wire acts on charge 1 by the following force[21]

$$d\boldsymbol{f}_1=q_1\rho_2dr_{2z}\left(\frac{[\boldsymbol{v}_1\times[\boldsymbol{v}_2\times\boldsymbol{r}]]}{4\pi c^2r^3}+\frac{v_2^2\boldsymbol{r}}{8\pi c^2r^3}-\frac{3(\boldsymbol{v}_2\cdot\boldsymbol{r})^2\boldsymbol{r}}{8\pi c^2r^5}\right)\tag{6.22}$$

21 Here we use formula (1-D.16) for the double vector product.

and the full force is obtained by integrating this expression along the full length of the wire.

First consider integrals of the second and third terms in (6.22). They express the "electric" part of the force, which is independent of the velocity $\boldsymbol{v}_1$. The y- and z-components of this integral vanish due to symmetry, and for the x-component we obtain

$$\frac{q_1\rho_2 v_2^2}{8\pi m_1 c^2}\int_{-\infty}^{\infty} dr_{2z}\left(\frac{R}{(R^2+r_{2z}^2)^{3/2}}-\frac{3r_{2z}^2 R}{(R^2+r_{2z}^2)^{5/2}}\right)=0.$$

This result means, in particular, that a neutral wire with current does not create a v_2^2-dependent "electric" force field in the surrounding space. An observation of such a field was (apparently erroneously) reported in the work [62]. Subsequent thorough studies [156, 224] did not confirm this report.

Thus, the total force acting on charge 1 is obtained by integrating the first term in (6.22) along the length of the wire. We have

$$\boldsymbol{F}_1=\frac{q_1\rho_2}{4\pi c^2}\int_{-\infty}^{\infty} dr_{2z}\frac{[\boldsymbol{v}_1\times[\boldsymbol{v}_2\times\boldsymbol{r}]]}{r^3}. \tag{6.23}$$

In this expression we easily recognize the *Biot–Savart law* (6.2) of the traditional Maxwell theory. This means that Maxwell's results related to magnetic properties of wires with current remain valid in our approach as well.

6.3.2 Longitudinal forces in wires

In Maxwell's electrodynamics, magnetic forces (6.2)–(6.3) are always perpendicular to a particle's velocity. Therefore, there can be no magnetic interaction between two electrons moving one behind another inside a straight wire, as in Figure 6.3. Indeed, if we substitute $\boldsymbol{v}_1=\boldsymbol{v}_2\equiv\boldsymbol{v}$ and $\boldsymbol{v}\cdot\boldsymbol{r}=vr$ into classical formulas (6.2)–(6.3), then we get $\boldsymbol{f}_1=\boldsymbol{f}_2=0$. However, making the same substitutions in the Coulomb–Darwin–Breit formulas (6.14)–(6.15), we obtain[22]

$$\boldsymbol{f}_1=-\frac{q^2v^2\boldsymbol{r}}{4\pi c^2r^3}-\frac{3q^2(\boldsymbol{v}\cdot\boldsymbol{r})^2\boldsymbol{r}}{8\pi c^2r^5}=-\frac{5q^2v^2(\boldsymbol{r}_1-\boldsymbol{r}_2)}{8\pi c^2r^3},$$

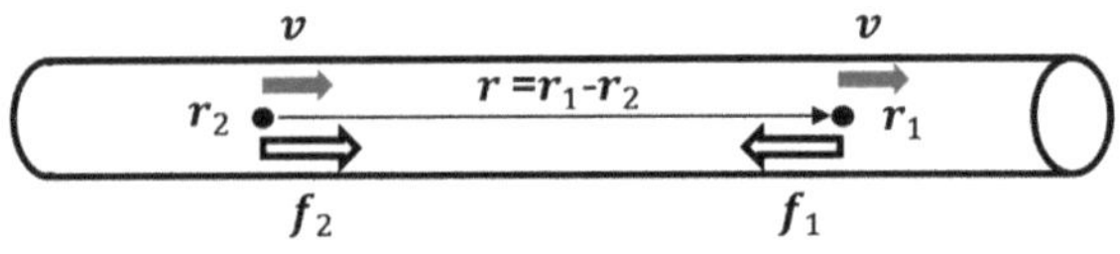

Figure 6.3: To the calculation of the longitudinal force in a wire with current.

22 Here we ignored the Coulomb component of the force, which is strongly screened in conductors.

$$f_2 = \frac{5q^2v^2(\boldsymbol{r}_1 - \boldsymbol{r}_2)}{8\pi c^2 r^3},$$

which indicates the presence of a (longitudinal) attractive force. Such a magnetic attraction of conduction electrons can contribute to superconductivity at low temperatures [71, 69, 68].

It is interesting that the existence of longitudinal forces in conductors has been discussed since Ampére proposed his interaction law in the beginning of the nineteenth century.[23] However, unlike our attraction, Ampère's formula predicted repulsion of electrons in the wire. Numerous experimental attempts to detect such a repulsion have not yielded convincing results. A recent study [100] announced a confirmation of the Ampère repulsion. However, this conclusion was questioned in [30]. Therefore, from the experimental point of view, the presence of longitudinal forces in conductors and their sign (i. e., attraction or repulsion) remains uncertain.

6.3.3 Charge + wire loop

Here we would like to calculate the interaction energy between a point charge and a circular wire loop of radius a with constant current in the geometry shown in Figure 6.4. As we saw in the preceding subsection, the movement of the wire as a whole does not affect its interaction with the charge. Therefore, we assume that the current loop is fixed at the origin. Then the potential energy of the interaction between charge 1 and a wire segment dl is given by the Darwin formula in (6.9),

$$V_{dl2-q1} \approx -\frac{q_1 \rho_2 dl}{8\pi m_1 c^2}\left(\frac{(\boldsymbol{p}_1 \cdot \boldsymbol{v}_2)}{r} + \frac{(\boldsymbol{p}_1 \cdot \boldsymbol{r})(\boldsymbol{v}_2 \cdot \boldsymbol{r})}{r^3}\right),$$

where ρ_2 is the linear charge density of the conduction electrons and $\boldsymbol{v}_2 \approx \boldsymbol{p}_2/m_2$ is their drift velocity. In the coordinate system shown in Figure 6.4, the wire element is

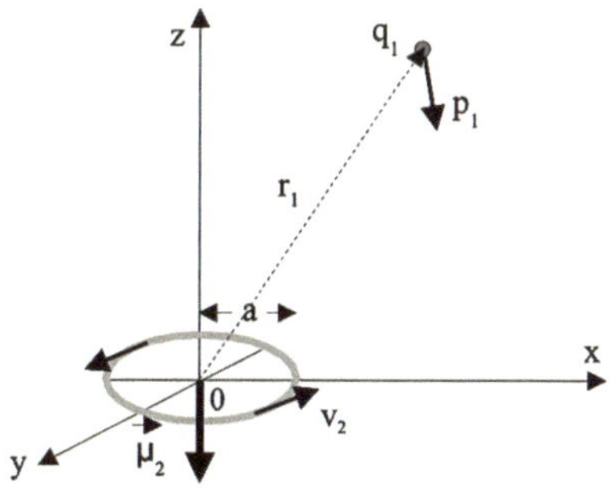

Figure 6.4: Interaction between a small current carrying coil and a charge q_1 located at an arbitrary point $\boldsymbol{r}_1 = (r_{1x}, r_{1y}, r_{1z})$ and having momentum $\boldsymbol{p}_1 = (p_{1x}, p_{1y}, p_{1z})$. The velocity of coil electrons is shown by the vector $\boldsymbol{v}_2$.

23 A good historical overview of the idea of longitudinal forces can be found in [123].

equal to $dl = ad\theta$ and

$$\begin{aligned}\boldsymbol{r}_2 &= (a\cos\theta, a\sin\theta, 0),\\ \boldsymbol{r} &= (r_{1x} - a\cos\theta, r_{1y} - a\sin\theta, r_{1z}),\\ \boldsymbol{v}_2 &= (-v_2\sin\theta, v_2\cos\theta, 0).\end{aligned}$$

For the infinitesimally small wire loop ($a \to 0$), we can leave only initial terms in Taylor expansions for r^{-1} and r^{-3}:

$$\begin{aligned}\frac{1}{r} &\equiv \frac{1}{|\boldsymbol{r}_1 - \boldsymbol{r}_2|} = \frac{1}{|\boldsymbol{r}_1 - \boldsymbol{r}_2|}\bigg|_{a=0} + a\left(\frac{d}{da}\frac{1}{|\boldsymbol{r}_1 - \boldsymbol{r}_2|}\right)\bigg|_{a=0} + O(a^2)\\ &\approx \frac{1}{r_1} + \frac{a(r_{1x}\cos\theta + r_{1y}\sin\theta)}{r_1^3},\end{aligned} \tag{6.24}$$

$$\frac{1}{r^3} \equiv \frac{1}{|\boldsymbol{r}_1 - \boldsymbol{r}_2|^3} \approx \frac{1}{r_1^3} + \frac{3a(r_{1x}\cos\theta + r_{1y}\sin\theta)}{r_1^5}. \tag{6.25}$$

The full interaction potential is obtained by integrating V_{dl2-q1} with respect to the angle θ from 0 to 2π. Then we have

$$\begin{aligned}V_{\text{loop2}-q1} \approx &-\frac{aq_1\rho_2}{8\pi m_1 c^2}\int_0^{2\pi} d\theta\bigg[(-v_2 p_{1x}\sin\theta + v_2 p_{1y}\cos\theta)\left(\frac{1}{r_1} + \frac{a(r_{1x}\cos\theta + r_{1y}\sin\theta)}{r_1^3}\right)\\ &+ (-v_2\sin\theta(r_{1x} - a\cos\theta) + v_2\cos\theta(r_{1y} - a\sin\theta))\\ &\times((\boldsymbol{p}_1\cdot\boldsymbol{r}_1) - p_{1x}a\cos\theta - p_{1y}a\sin\theta)\left(\frac{1}{r_1^3} + \frac{3a(r_{1x}\cos\theta + r_{1y}\sin\theta)}{r_1^5}\right)\bigg]\\ \approx &-\frac{aq_1\rho_2}{8\pi m_1 c^2}\int_0^{2\pi} d\theta\bigg[-\frac{v_2 p_{1x}\sin\theta}{r_1} - \frac{av_2 p_{1x}\sin\theta(r_{1x}\cos\theta + r_{1y}\sin\theta)}{r_1^3}\\ &+ \frac{v_2 p_{1y}\cos\theta}{r_1} + \frac{av_2 p_{1y}\cos\theta(r_{1x}\cos\theta + r_{1y}\sin\theta)}{r_1^3}\\ &+ (-v_2 r_{1x}(\boldsymbol{p}_1\cdot\boldsymbol{r}_1)\sin\theta + v_2 a(\boldsymbol{p}_1\cdot\boldsymbol{r}_1)\sin\theta\cos\theta\\ &+ v_2 r_{1y}(\boldsymbol{p}_1\cdot\boldsymbol{r}_1)\cos\theta - v_2 a(\boldsymbol{p}_1\cdot\boldsymbol{r}_1)\cos\theta\sin\theta\\ &+ v_2 r_{1x}p_{1x}a\sin\theta\cos\theta - v_2 a^2 p_{1x}\sin\theta\cos^2\theta\\ &- v_2 a r_{1y}p_{1x}\cos^2\theta + v_2 a^2 p_{1x}\sin\theta\cos^2\theta\\ &+ v_2 a r_{1x}p_{1y}\sin^2\theta - v_2 a^2 p_{1y}\cos\theta\sin^2\theta\\ &- v_2 a r_{1y}p_{1y}\cos\theta\sin\theta + v_2 a^2 p_{1y}\cos\theta\sin^2\theta)\left(\frac{1}{r_1^3} + \frac{3a(r_{1x}\cos\theta + r_{1y}\sin\theta)}{r_1^5}\right)\bigg].\end{aligned}$$

We can ignore small terms proportional to a^3. Moreover, we should only keep those terms whose angle dependence is expressed by squares $\sin^2\theta$ or $\cos^2\theta$ and who have a chance to survive under the angle integral. Then the formula simplifies to

$$V_{\text{loop2}-q1} \approx -\frac{aq_1\rho_2}{8\pi m_1 c^2}\int_0^{2\pi} d\theta$$

$$\times\left[-\frac{av_2 p_{1x} r_{1y}\sin^2\theta}{r_1^3} + \frac{av_2 p_{1y} r_{1x}\cos^2\theta}{r_1^3} - \frac{3av_2 r_{1x} r_{1y}(\boldsymbol{p}_1\cdot\boldsymbol{r}_1)\sin^2\theta}{r_1^5}\right.$$

$$\left.+\frac{3av_2 r_{1y} r_{1x}(\boldsymbol{p}_1\cdot\boldsymbol{r}_1)\cos^2\theta}{r_1^5} - \frac{v_2 a r_{1y} p_{1x}\cos^2\theta}{r_1^3} + \frac{v_2 a r_{1x} p_{1y}\sin^2\theta}{r_1^3}\right]$$

$$= -\frac{v_2 a^2 q_1 \rho_2}{4m_1 c^2 r_1^3}[r_{1x}p_{1y} - r_{1y}p_{1x}] = -\frac{q_1\mu_z\cdot[\boldsymbol{r}_1\times\boldsymbol{p}_1]_z}{4\pi m_1 c^2 r_1^3}.$$

Here, according to the usual definition,[24] we introduced the magnetic moment of the current loop, which is the vector $\boldsymbol{\mu}_2$ whose length is equal to the loop's area times the amperage $\mu_2 = \pi a^2\rho_2 v_2$ and whose direction is perpendicular to the loop's plane.[25] This formula can be generalized for arbitrary orientations of the loop:

$$V_{\text{loop2}-q1} \approx -\frac{q_1\boldsymbol{\mu}_2\cdot[\boldsymbol{r}\times\boldsymbol{p}_1]}{4\pi m_1 c^2 r^3} = -\frac{q_1[\boldsymbol{\mu}_2\times\boldsymbol{r}]\cdot\boldsymbol{p}_1}{4\pi m_1 c^2 r^3}. \tag{6.26}$$

So, the full Hamiltonian for the system "charge 1 + current loop 2" has the form

$$H = \frac{p_1^2}{2m_1} + \frac{p_2^2}{2m_2} - \frac{p_1^4}{8m_1^3c^2} - \frac{p_2^4}{8m_2^3c^2} - \frac{q_1[\boldsymbol{\mu}_2\times\boldsymbol{r}]\cdot\boldsymbol{p}_1}{4\pi m_1 c^2 r^3}. \tag{6.27}$$

Next we use this Hamiltonian to investigate the dynamics of charge 1 in the "loop + charge" system. The time derivative of the charge's momentum is obtained from the first Hamilton equation (**1**-6.106):

$$\frac{d\boldsymbol{p}_1}{dt} = -\frac{\partial H}{\partial \boldsymbol{r}_1} = \frac{q_1[\boldsymbol{p}_1\times\boldsymbol{\mu}_2]}{4\pi m_1 c^2 r^3} - \frac{3q_1([\boldsymbol{p}_1\times\boldsymbol{\mu}_2]\cdot\boldsymbol{r})\boldsymbol{r}}{4\pi m_1 c^2 r^5}.$$

The velocity is obtained from the second Hamilton equation (**1**-6.107):

$$\boldsymbol{v}_1 = \frac{\partial H}{\partial \boldsymbol{p}_1} = \frac{\boldsymbol{p}_1}{m_1} - \frac{p_1^2\boldsymbol{p}_1}{2m_1^3c^2} - \frac{q_1[\boldsymbol{\mu}_2\times\boldsymbol{r}]}{4\pi m_1 c^2 r^3}. \tag{6.28}$$

---

24 See equation (5.42) in [118].

25 The magnetic moment's orientation (up or down) is determined by the mnemonic "right-hand rule": place four fingers of your right hand in the direction of the current (this direction is opposite to the electron's movement, because $q_2 = -e$). Then the thumb will point in the direction of $\boldsymbol{\mu}_2$. In the geometry shown in Figure 6.4, $\boldsymbol{\mu}_2 = (0, 0, -\mu_2)$.

Acceleration is the time derivative of the velocity (6.28), so[26]

$$
\begin{aligned}
\boldsymbol{a}_1 &\equiv \frac{d\boldsymbol{v}_1}{dt} \\
&\approx \frac{\dot{\boldsymbol{p}}_1}{m_1} - \frac{q_1[\boldsymbol{\mu}_2 \times \dot{\boldsymbol{r}}]}{4\pi m_1 c^2 r^3} + \frac{3q_1[\boldsymbol{\mu}_2 \times \boldsymbol{r}](\boldsymbol{r} \cdot \dot{\boldsymbol{r}})}{4\pi m_1 c^2 r^5} \\
&= \frac{q_1[\boldsymbol{p}_1 \times \boldsymbol{\mu}_2]}{2\pi m_1^2 c^2 r^3} - \frac{3q_1([\boldsymbol{\mu}_2 \times \boldsymbol{r}] \cdot \boldsymbol{p}_1)\boldsymbol{r}}{4\pi m_1^2 c^2 r^5} + \frac{3q_1[\boldsymbol{\mu}_2 \times \boldsymbol{r}](\boldsymbol{r} \cdot \boldsymbol{p}_1)}{4\pi m_1^2 c^2 r^5} \\
&\quad + \frac{q_1[\boldsymbol{\mu}_2 \times \boldsymbol{p}_2]}{4\pi m_1 m_2 c^2 r^3} - \frac{3q_1[\boldsymbol{\mu}_2 \times \boldsymbol{r}](\boldsymbol{r} \cdot \boldsymbol{p}_2)}{4\pi m_1 m_2 c^2 r^5} \\
&= \frac{q_1[\boldsymbol{p}_1 \times \boldsymbol{\mu}_2]}{2\pi m_1^2 c^2 r^3} - \frac{3q_1[\boldsymbol{p}_1 \times [\boldsymbol{r} \times [\boldsymbol{\mu}_2 \times \boldsymbol{r}]]]}{4\pi m_1^2 c^2 r^5} - \left(\frac{d}{dt}\right)_2 \frac{q_1[\boldsymbol{\mu}_2 \times \boldsymbol{r}]}{4\pi m_1 c^2 r^3} \\
&= -\frac{q_1[\boldsymbol{p}_1 \times \boldsymbol{\mu}_2]}{4\pi m_1^2 c^2 r^3} + \frac{3q_1[\boldsymbol{p}_1 \times \boldsymbol{r}](\boldsymbol{\mu}_2 \cdot \boldsymbol{r})}{4\pi m_1^2 c^2 r^5} - \left(\frac{d}{dt}\right)_2 \frac{q_1[\boldsymbol{\mu}_2 \times \boldsymbol{r}]}{4\pi m_1 c^2 r^3}.
\end{aligned}
\tag{6.29}
$$

Our notation $(\frac{d}{dt})_2$ means that in the time derivative of the relative position $\boldsymbol{r}$ we take into account only loop's (2) contribution. For example

$$
\begin{aligned}
\left(\frac{d}{dt}\right)_2 \boldsymbol{r} &= -\boldsymbol{v}_2 \approx -\frac{\boldsymbol{p}_2}{m_2}, \\
\left(\frac{d}{dt}\right)_2 \frac{1}{r^3} &= \frac{3(\boldsymbol{r} \cdot \boldsymbol{v}_2)}{r^5}.
\end{aligned}
$$

6.3.4 Charge + spin's magnetic moment

Let us now consider a system consisting of a spinless charged particle 1 and a neutral particle 2 with a nonzero spin and magnetic moment.[27] The corresponding Hamiltonian is obtained from (6.8) by setting $\boldsymbol{\mu}_1 = 0$ and $q_2 = 0$, i. e.,

$$
H = \frac{p_1^2}{2m_1} + \frac{p_2^2}{2m_2} - \frac{p_1^4}{8m_1^3 c^2} + \frac{p_2^4}{8m_2^3 c^2} + \frac{q_1[\boldsymbol{\mu}_2 \times \boldsymbol{r}] \cdot \boldsymbol{p}_2}{8\pi m_2 c^2 r^3} - \frac{q_1[\boldsymbol{\mu}_2 \times \boldsymbol{r}] \cdot \boldsymbol{p}_1}{4\pi m_1 c^2 r^3}. \tag{6.30}
$$

Note that if particle 2 is at rest ($\boldsymbol{p}_2 = 0$), this expression coincides with the Hamiltonian (6.27) of the system "charge + magnetic moment of the current loop." However, in the case of a moving particle 2, the interaction energy has an additional term (the fifth term in (6.30)), which was absent in (6.27).

---

26 Here we notice that $\dot{\boldsymbol{p}}_1 \propto (v/c)^2$, so the time derivative of the second term on the right-hand side of (6.28) is proportional to $(v/c)^4$ and can be neglected. We also neglected the time derivative of the loop's magnetic moment. This can be done if we assume that the current in the loop is kept constant and if the loop has a large moment of inertia, which prevents the rotation of its axis. In this calculation we used vector identities (**1**-D.16) and (**1**-D.17).

27 An example of such a particle is the neutron.

As usual, we use Hamilton's equations of motion to calculate the time derivative of the momentum, the velocity and the acceleration of charge 1. We have

$$\begin{aligned}
\frac{d\boldsymbol{p}_1}{dt} &= [\boldsymbol{p}_1, H]_P = -\frac{\partial H}{\partial \boldsymbol{r}_1} \\
&= \frac{q_1[\boldsymbol{p}_1 \times \boldsymbol{\mu}_2]}{4\pi m_1 c^2 r^3} - \frac{3q_1([\boldsymbol{p}_1 \times \boldsymbol{\mu}_2] \cdot \boldsymbol{r})\boldsymbol{r}}{4\pi m_1 c^2 r^5} - \frac{q_1[\boldsymbol{p}_2 \times \boldsymbol{\mu}_2]}{8\pi m_2 c^2 r^3} + \frac{3q_1([\boldsymbol{p}_2 \times \boldsymbol{\mu}_2] \cdot \boldsymbol{r})\boldsymbol{r}}{8\pi m_2 c^2 r^5}, \\
\boldsymbol{v}_1 \equiv \frac{d\boldsymbol{r}_1}{dt} &= [\boldsymbol{r}_1, H]_P = \frac{\partial H}{\partial \boldsymbol{p}_1} = \frac{\boldsymbol{p}_1}{m_1} - \frac{p_1^2 \boldsymbol{p}_1}{2m_1^3 c} - \frac{q_1[\boldsymbol{\mu}_2 \times \boldsymbol{r}]}{4\pi m_1 c^2 r^3}, \\
\boldsymbol{a}_1 \equiv \frac{d\boldsymbol{v}_1}{dt} & \\
&\approx \frac{\dot{\boldsymbol{p}}_1}{m_1} - \frac{q_1[\boldsymbol{\mu}_2 \times \dot{\boldsymbol{r}}]}{4\pi m_1 c^2 r^3} + \frac{3q_1[\boldsymbol{\mu}_2 \times \boldsymbol{r}](\boldsymbol{r} \cdot \dot{\boldsymbol{r}})}{4\pi m_1 c^2 r^5} \\
&= \frac{q_1[\boldsymbol{p}_1 \times \boldsymbol{\mu}_2]}{2\pi m_1^2 c^2 r^3} - \frac{3q_1([\boldsymbol{p}_1 \times \boldsymbol{\mu}_2] \cdot \boldsymbol{r})\boldsymbol{r}}{4\pi m_1^2 c^2 r^5} + \frac{3q_1[\boldsymbol{\mu}_2 \times \boldsymbol{r}](\boldsymbol{r} \cdot \boldsymbol{p}_1)}{4\pi m_1^2 c^2 r^5} \\
&\quad - \frac{3q_1[\boldsymbol{v}_2 \times \boldsymbol{\mu}_2]}{8\pi m_1 c^2 r^3} + \frac{3q_1([\boldsymbol{\mu}_2 \times \boldsymbol{r}] \cdot \boldsymbol{v}_2)\boldsymbol{r}}{8\pi m_1 c^2 r^5} - \frac{3q_1[\boldsymbol{\mu}_2 \times \boldsymbol{r}](\boldsymbol{r} \cdot \boldsymbol{v}_2)}{4\pi m_1 c^2 r^5} \\
&= \frac{q_1[\boldsymbol{p}_1 \times \boldsymbol{\mu}_2]}{2\pi m_1^2 c^2 r^3} - \frac{3q_1[\boldsymbol{p}_1 \times [\boldsymbol{r} \times [\boldsymbol{\mu}_2 \times \boldsymbol{r}]]]}{4\pi m_1^2 c^2 r^5} \\
&\quad - \frac{3q_1[\boldsymbol{v}_2 \times \boldsymbol{\mu}_2]}{8\pi m_1 c^2 r^3} + \frac{3q_1([\boldsymbol{\mu}_2 \times \boldsymbol{r}] \cdot \boldsymbol{v}_2)\boldsymbol{r}}{8\pi m_1 c^2 r^5} - \frac{3q_1[\boldsymbol{\mu}_2 \times \boldsymbol{r}](\boldsymbol{r} \cdot \boldsymbol{v}_2)}{4\pi m_1 c^2 r^5} \\
&= -\frac{q_1[\boldsymbol{p}_1 \times \boldsymbol{\mu}_2]}{4\pi m_1^2 c^2 r^3} + \frac{3q_1[\boldsymbol{p}_1 \times \boldsymbol{r}](\boldsymbol{\mu}_2 \cdot \boldsymbol{r})}{4\pi m_1^2 c^2 r^5} - \left(\frac{d}{dt}\right)_2 \frac{q_1[\boldsymbol{\mu}_2 \times \boldsymbol{r}]}{4\pi m_1 c^2 r^3} \\
&\quad - \frac{d}{d\boldsymbol{r}_1} \frac{q_1([\boldsymbol{v}_2 \times \boldsymbol{\mu}_2] \cdot \boldsymbol{r})}{8\pi m_1 c^2 r^3}.
\end{aligned} \tag{6.31}$$

This means that acceleration of charge 1 in the field of the spin's magnetic moment is basically the same as in the field of the current loop (6.29). The only difference is in the presence of the additional last (gradient) term on the right-hand side of (6.31).

6.3.5 Two types of magnets

In practice there are two possible sources of the "magnetic field." First, it could be a piece of ferromagnetic material (a permanent magnet). The magnetization of such materials is composed of two components. One component comes from the orbital movement of electrons around the nuclei. The other component comes from electrons' spins.[28] The relative share of the "orbital" and "spin" magnetizations varies in different materials, but in most cases the spin part dominates [195].

28 Contributions from heavy nuclear spins are usually much smaller, because in the formula for the magnetic moment (5.25), the particle's mass is in the denominator. Of course, for nonzero macroscopic

The other type of magnets are electromagnets, solenoids, magnetic coils. They have only the orbital part of magnetization due to electrons moving along the wire.

As we explained in the preceding subsection, the "orbital" and "spin" magnetic moments have different interactions with external charges if the magnets are in motion. This difference vanishes for magnetic moments at rest. Their interaction energy with a moving charge 1 is

$$V^{\text{magn}} = -\frac{q_1[\boldsymbol{\mu}_2 \times \boldsymbol{r}] \cdot \boldsymbol{p}_1}{4\pi m_1 c^2 r^3}. \tag{6.32}$$

In this case, the force acting on the charge is[29]

$$\boldsymbol{f}_1 = m_1 \boldsymbol{a}_1 = -\frac{q_1[\boldsymbol{p}_1 \times \boldsymbol{\mu}_2]}{4\pi m_1 c^2 r^3} + \frac{3q_1[\boldsymbol{p}_1 \times \boldsymbol{r}](\boldsymbol{\mu}_2 \cdot \boldsymbol{r})}{4\pi m_1 c^2 r^5} \approx \frac{q_1}{c}[\boldsymbol{v}_1 \times \boldsymbol{b}_1]. \tag{6.33}$$

This coincides with the magnetic part of the standard *Lorentz force* (6.1), if another standard expression,

$$\boldsymbol{b}_1 = -\frac{\boldsymbol{\mu}_2}{4\pi c r^3} + \frac{3(\boldsymbol{\mu}_2 \cdot \boldsymbol{r})\boldsymbol{r}}{4\pi c r^5}, \tag{6.34}$$

is accepted for the "magnetic field," created by the magnetic moment $\boldsymbol{\mu}_2$ at the point $\boldsymbol{r}_1$.[30]

On the other hand, if we assume that

$$\boldsymbol{B}(\boldsymbol{r}_2) = \frac{q_1[\boldsymbol{v}_1 \times (\boldsymbol{r}_2 - \boldsymbol{r}_1)]}{4\pi c r^3} \tag{6.35}$$

is the "magnetic field," created by the moving charge 1 at $\boldsymbol{r}_2$, then the interaction energy (6.32) takes another textbook form $V^{\text{magn}} = -(\boldsymbol{\mu}_2 \cdot \boldsymbol{B})/c$.

Despite these encouraging coincidences, there is an important difference between our formulas and the traditional Maxwell theory. In our approach, there are no fields (neither electric nor magnetic) that exist independently at each point of space. We recognize only direct interparticle forces. This is why we have placed the phrase "magnetic field" in quotation marks.

6.3.6 Thin long magnet/solenoid

The full "magnetic field" of macroscopic permanent magnets or solenoids is obtained by summation/integration of contributions like (6.34) over the magnet's volume. For

magnetization, it is necessary that magnetic moments of different atoms point predominantly in one direction.

29 See the first two terms in equation (6.29) or the first two terms in equation (6.31).

30 See equation (5.56) in [118].

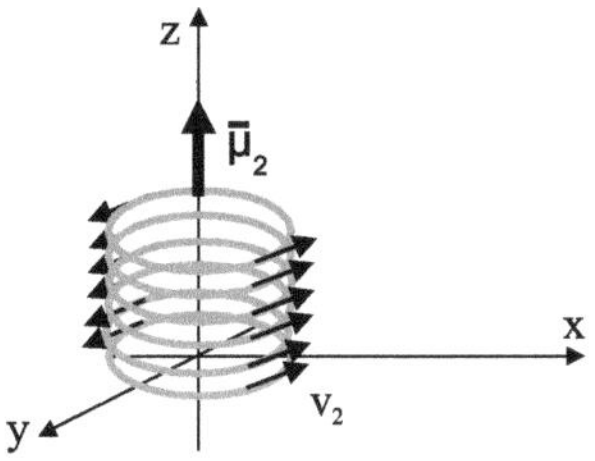

Figure 6.5: A thin solenoid can be represented as many current loops stacked one above the other. The arrows indicate the direction of the current (against the motion of electrons). The magnetization vector $\overline{\boldsymbol{\mu}}_2$ is directed along the axis of the solenoid.

example, an infinite straight thin solenoid can be regarded as a set of small current loops stacked on top of each other, as in Figure 6.5. In this case, individual loops have coordinates $\boldsymbol{r}_2 = (0, 0, z)$, and the observation point $\boldsymbol{r}_1 = (x_1, y_1, 0)$ can be chosen in the plane $z_1 = 0$ without loss of generality. The full "magnetic field" is obtained by integrating (6.34) over the length of the stack[31]:

$$\boldsymbol{B}_{\text{long}}(\boldsymbol{r}_1) = -\int_{-\infty}^{\infty} dz \left(\frac{(0, 0, \overline{\mu}_2)}{4\pi c(x_1^2 + y_1^2 + z^2)^{3/2}} + \frac{3\overline{\mu}_2 z(x_1, y_1, -z)}{4\pi c(x_1^2 + y_1^2 + z^2)^{5/2}} \right) = 0. \tag{6.36}$$

This result implies that an infinite thin solenoid does not act on surrounding charges. However, the vanishing force does not mean that the interaction energy vanishes as well. This energy can be found by integrating (6.32) along the solenoid's length and noticing that the mixed product $-[\boldsymbol{v}_1 \times \overline{\boldsymbol{\mu}}_2] \cdot \boldsymbol{r}_1 = \overline{\mu}_2(v_{1x}y_1 - v_{1y}x_1)$ does not depend on z. Denoting $r \equiv (x_1^2 + y_1^2)^{1/2}$ the charge–solenoid distance, we obtain

$$V_{\text{long}} \approx -\int_{-\infty}^{\infty} dz \frac{q_1[\boldsymbol{v}_1 \times \overline{\boldsymbol{\mu}}_2] \cdot \boldsymbol{r}_1}{4\pi c^2(x_1^2 + y_1^2 + z^2)^{3/2}} = -\frac{q_1[\boldsymbol{v}_1 \times \overline{\boldsymbol{\mu}}_2] \cdot \boldsymbol{r}_1}{2\pi c^2 r^2}. \tag{6.37}$$

This is a curious example of a nonzero interaction potential, which, nevertheless, does not accelerate charges moving in it. Experimental manifestations of such potentials will be discussed in Section 7.2.

6.3.7 Cylindrical magnet/solenoid of arbitrary cross section

In the previous subsection we considered a thin solenoid. However, real electromagnets have nonzero cross sections. It is not hard to generalize our theory to this case. Consider a hollow solenoid, whose cross section is shown in Figure 6.6. This wire coil can be represented as a superposition of small current loops. Thus, any cylindrical solenoid of macroscopic dimensions can be represented as a bundle of thin parallel solenoids joined together.

31 Here $\overline{\mu}_2$ is the magnetization per unit length and $\boldsymbol{r} \equiv \boldsymbol{r}_1 - \boldsymbol{r}_2 = (x_1, y_1, -z)$.

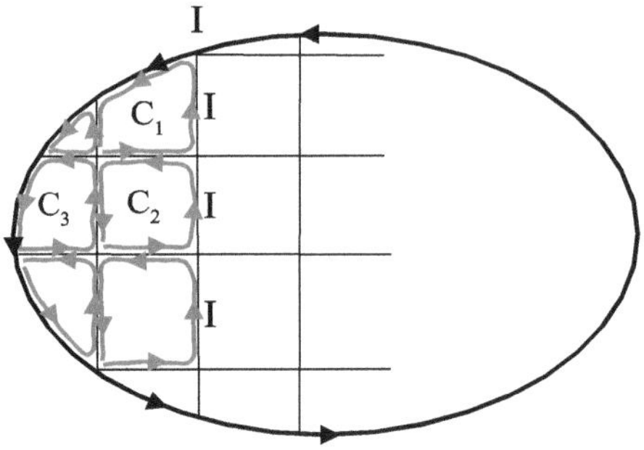

Figure 6.6: Wire coil (black bold line) with current I can be represented as a union of small loops $C_1, C_2, C_3, \ldots$ (grey lines) touching each other. All loops have the same current I, so that all currents inside the coil cancel each other out, and only the real peripheral current remains.

If the observation point $\boldsymbol{r}_1$ is outside the solenoid's volume, then equality (6.36) holds for all thin components in the bundle. Therefore, the "magnetic field" around such "thick" solenoid is zero. The same analysis is applicable to infinitely long cylindrical permanent magnets of arbitrary cross section. Such magnets do not exert any force on outside charges. This agrees with Maxwell's theory; see, for example, Problem 5.2 (a) in [118].

6.3.8 Cullwick paradox

By "looping" a solenoid one can create a toroidal electromagnet as shown in Figure 6.7. Its interaction with moving charges is interesting because Maxwell's classical electrodynamics encounters serious problems when analyzing the total momentum in this system. These difficulties are called the "Cullwick paradox" [42, 1, 166].

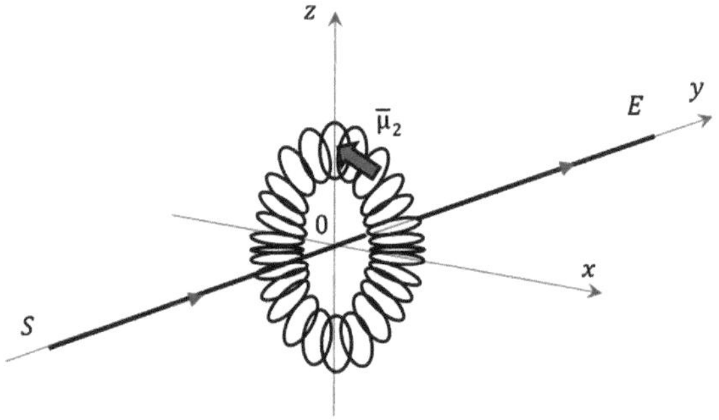

Figure 6.7: Toroidal solenoid with current and a charge moving along the y axis. The term $\overline{\mu}_2$ represents the linear magnetization density of the solenoid.

According to Maxwell's theory, there is no "magnetic field" outside the toroidal magnet, so no force is acting on outside charges.[32] However, a moving charge creates its own "magnetic field" (6.35), which acts on the torus with a nonzero force. Is this

32 This result is valid in the RQD approach as well.

not a violation of Newton's third law? According to McDonald [166], the balance of forces can be restored if one takes into account the hypothetical "momentum of the electromagnetic field" (6.4). Somewhat absurdly, the field's momentum appears to be nonzero, even when both the magnet and the charge are at rest. It is argued that the paradox can be resolved by assuming the existence of the so-called *hidden momentum* [225, 36] in the magnet. However, these explanations do not look satisfactory, and we are going to suggest another version of the events based on the Coulomb–Darwin–Breit theory.

First we have to derive the Hamiltonian describing the dynamics of the system "charge 1 + toroidal magnet 2." We introduce a Cartesian coordinate system as shown in Figure 6.7. We denote the radius of the torus a and its linear magnetization density $\overline{\mu}_2$. We suppose that charge 1 moves directly along the symmetry axis of the torus with momentum $\boldsymbol{p}_1 = (0, p_{1y}, 0)$, and we assume that both the charge and the magnet can move freely along the y axis. The component y of the magnet's velocity is denoted by $V_{2y} \approx P_{2y}/M_2$, where M_2 is the total mass of the magnet and $\boldsymbol{P}_2 = (0, P_{2y}, 0)$ is its momentum. Then the potential interaction energy V_{torus} between the charge and the toroidal magnet is obtained by integrating the potential energy density (6.30) along the torus circumference.[33] Then we find

$$V_{\text{torus}} = \int_{\text{torus}} dl \left(-\frac{q_1 [\overline{\boldsymbol{\mu}}_2 \times \boldsymbol{r}] \cdot \boldsymbol{p}_1}{4\pi m_1 c^2 r^3} + \frac{q_1 [\overline{\boldsymbol{\mu}}_2 \times \boldsymbol{r}] \cdot \boldsymbol{p}_2}{8\pi m_2 c^2 r^3} \right). \tag{6.38}$$

In the geometry of Figure 6.7,[34]

$$\begin{aligned}
\boldsymbol{r}_2 &= (a\cos\theta, 0, a\sin\theta), \\
\boldsymbol{r} = \boldsymbol{r}_1 - \boldsymbol{r}_2 &= (-a\cos\theta, r_{1y} - R_{2y}, -a\sin\theta), \\
\overline{\boldsymbol{\mu}}_2 &= (-\overline{\mu}_2 \sin\theta, 0, \overline{\mu}_2\cos\theta), \\
[\overline{\boldsymbol{\mu}}_2 \times \boldsymbol{r}]_y &= -a\overline{\mu}_2 \sin^2\theta - a\overline{\mu}_2\cos^2\theta = -a\overline{\mu}_2, \\
[\overline{\boldsymbol{\mu}}_2 \times \boldsymbol{r}] \cdot \boldsymbol{p}_1 &= -a\overline{\mu}_2 p_{1y}, \\
[\overline{\boldsymbol{\mu}}_2 \times \boldsymbol{r}] \cdot \boldsymbol{P}_2 &\approx -a\overline{\mu}_2 P_{2y},
\end{aligned} \tag{6.39}$$

$$\begin{aligned}
V_{\text{torus}} &= \int_{\text{torus}} dl \left(-\frac{q_1 [\overline{\boldsymbol{\mu}}_2 \times \boldsymbol{r}] \cdot \boldsymbol{p}_1}{4\pi m_1 c^2 r^3} + \frac{q_1 [\overline{\boldsymbol{\mu}}_2 \times \boldsymbol{r}] \cdot \boldsymbol{p}_2}{8\pi m_2 c^2 r^3} \right) \\
&= -\int_0^{2\pi} d\theta \left(-\frac{q_1 a^2 \overline{\mu}_2 p_{1y}}{4\pi m_1 c^2 (a^2\cos^2\theta + (r_{1y} - R_{2y})^2 + a^2\sin^2\theta)^{3/2}} \right.
\end{aligned}$$

33 This formula is written for a permanent magnet. For a toroidal electromagnet, the second term in the integrand will be absent.

34 We are integrating with respect to the angle θ, which is related to the line element by the formula $dl \equiv a d\theta$. Here $\boldsymbol{R}_2$ is the position of the magnet's center of mass.

$$+\frac{q_1 a^2 \overline{\mu}_2 P_{2y}}{8\pi M_2 c^2 (a^2\cos^2\theta + (r_{1y}-R_{2y})^2 + a^2\sin^2\theta)^{3/2}}\Bigg)$$
$$= \frac{q_1 a^2 \overline{\mu}_2 p_{1y}}{2m_1 c^2 (a^2 + (r_{1y}-R_{2y})^2)^{3/2}} - \frac{q_1 a^2 \overline{\mu}_2 P_{2y}}{4M_2 c^2 (a^2 + (r_{1y}-R_{2y})^2)^{3/2}} \tag{6.40}$$

and the full Hamiltonian takes the form

$$H = \frac{p_{1y}^2}{2m_1} + \frac{P_{2y}^2}{2M_2} - \frac{p_{1y}^4}{8m_1^3 c^2} - \frac{P_{2y}^4}{8M_2^3 c^2} + \frac{q_1 a^2 \overline{\mu}_2 p_{1y}}{2m_1 c^2 (a^2 + (r_{1y}-R_{2y})^2)^{3/2}} - \frac{q_1 a^2 \overline{\mu}_2 P_{2y}}{4M_2 c^2 (a^2 + (r_{1y}-R_{2y})^2)^{3/2}}.$$

The first Hamilton equation of motion gives the following result:

$$\frac{dp_{1y}}{dt} = -\frac{\partial V_{\text{torus}}}{\partial r_{1y}} = \frac{3q_1 a^2 \overline{\mu}_2 p_{1y}(r_{1y}-R_{2y})}{2m_1 c^2 (a^2 + (r_{1y}-R_{2y})^2)^{5/2}} - \frac{3q_1 a^2 \overline{\mu}_2 P_{2y}(r_{1y}-R_{2y})}{4M_2 c^2 (a^2 + (r_{1y}-R_{2y})^2)^{5/2}},$$
$$\frac{dP_{2y}}{dt} = -\frac{dp_{1y}}{dt}.$$

In contrast to Maxwell's theory, the rate of change of the momentum $\boldsymbol{p}_1$ is nonzero, and the momentum conservation law

$$\frac{d}{dt}(p_{1y} + P_{2y}) = 0 \tag{6.41}$$

holds without the need to involve the hypothetical "field" momentum and/or "hidden" momentum. The acceleration of charge 1 is calculated as follows:

$$\frac{dr_{1y}}{dt} = \frac{p_{1y}}{m_1} - \frac{p_{1y}^3}{2m_1^3 c} + \frac{\partial V_{\text{tor}}}{\partial p_{1y}} \approx \frac{p_{1y}}{m_1} - \frac{p_{1y}^3}{2m_1^3 c} + \frac{q_1 a^2 \overline{\mu}_2}{2m_1 c^2 (a^2 + (r_{1y}-R_{2y})^2)^{3/2}},$$
$$\frac{d^2 r_{1y}}{dt^2} \approx \frac{\dot{p}_{1y}}{m_1} - \frac{3q_1 a^2 \overline{\mu}_2 (r_{1y}-R_{2y})(v_{1y}-V_{2y})}{2m_1 c^2 (a^2 + (r_{1y}-R_{2y})^2)^{5/2}} = \frac{3q_1 a^2 \overline{\mu}_2 V_{2y}(r_{1y}-R_{2y})}{4m_1 c^2 (a^2 + (r_{1y}-R_{2y})^2)^{5/2}}.$$

When the magnet is at rest ($V_{2y} = 0$), this acceleration vanishes. Thus, just as in the Maxwell theory, there is no force acting on the charge. The force (acceleration) acting on the magnet is found as follows

$$\frac{dR_{2y}}{dt} = \frac{P_{2y}}{M_2} - \frac{P_{2y}^3}{2M_2^3 c} + \frac{\partial V_{\text{tor}}}{\partial P_{2y}}$$

$$
\begin{aligned}
&= \frac{P_{2y}}{M_2} - \frac{P_{2y}^3}{2M_2^3 c} - \frac{q_1 a^2 \overline{\mu}_2}{4M_2 c^2 (a^2 + (r_{1y} - R_{2y})^2)^{3/2}}, \\
\frac{d^2 R_{2y}}{dt^2} &= \frac{\dot{P}_{2y}}{M_2} + \frac{3q_1 a^2 \overline{\mu}_2 (r_{1y} - R_{2y})(v_{1y} - V_{2y})}{4M_2 c^2 (a^2 + (r_{1y} - R_{2y})^2)^{5/2}} \\
&\approx -\frac{3q_1 a^2 \overline{\mu}_2 v_{1y} (r_{1y} - R_{2y})}{4M_2 c^2 (a^2 + (r_{1y} - R_{2y})^2)^{5/2}}.
\end{aligned}
$$

This acceleration does not vanish even when $V_{2y} = 0$. This is an example of the situation described in Subsection 6.2.3: The forces are not balanced; nevertheless, the momentum conservation law (6.41) is not violated. This is our solution of Cullwick's paradox.

7 Experimental support of RQD

An intelligence that would know at a certain moment all the forces existing in nature and the situations of the bodies that compose nature and if it would be powerful enough to analyze all these data, would be able to grasp in one formula the movements of the biggest bodies of the Universe as well as of the lightest atom.
Pierre-Simon Laplace

Today, all experiments in electrodynamics are interpreted through the lens of the Maxwell field-based approach. For example, the phenomenon of electromagnetic induction is explained by one of Maxwell's equations connecting the derivatives of electric and magnetic fields,

$$\left[\frac{\partial}{\partial \boldsymbol{r}} \times \boldsymbol{E}\right] = -\frac{1}{c}\frac{d\boldsymbol{B}}{dt}, \tag{7.1}$$

and the Aharonov–Bohm effect [2] is described by the action of the vector potential $\boldsymbol{\mathcal{A}}$ on the wave function of a charged particle.

However, in our RQD, we abandoned the fields $\boldsymbol{E}, \boldsymbol{B}$ as well as the electromagnetic potentials $\mathcal{A}_0$, $\boldsymbol{\mathcal{A}}$. Therefore, in Sections 7.1 and 7.2 we are going to explain how the mentioned experiments can be understood in our fieldless theory.

One important feature of RQD is the prediction of instantaneous propagation of electric and magnetic interactions between charges. In Section 7.3 we will focus on experiments demonstrating superluminal propagation of signals in the near-field region of the emitter. In Section 7.4 we will be interested in the Frascati experiment, which is the most unambiguous confirmation of the RQD-predicted faster-than-light propagation of the Coulomb field.

7.1 Electromagnetic induction

We already know that a magnet at rest does not act on resting charges. Or, as is said, it "does not create an electric field." One of Faraday's outstanding discoveries was the realization that the magnet must be moving in order to create an electric field. The presence of the electric field around a moving magnet is called the *electromagnetic induction*. Here we will consider this effect from the point of view of RQD.

7.1.1 Moving magnets

Consider the force acting on charge 1 at rest ($\boldsymbol{p}_1 = 0$) from the moving point-like magnet $\boldsymbol{\mu}_2$. In the traditional Maxwell theory, a permanent solid-state magnet and a coil electromagnet create indistinguishable fields and forces. However, in our approach

https://doi.org/10.1515/9783110493221-007

this is not so. If the magnetic moment $\boldsymbol{\mu}_2$ is created by a spinning particle, then the force acting on the resting charge 1 is given by equation (6.31)

$$\boldsymbol{f}_1^{\text{spin}} = -\frac{d}{d\boldsymbol{r}_1}\frac{q_1[\boldsymbol{v}_2 \times \boldsymbol{\mu}_2] \cdot \boldsymbol{r}}{8\pi c^2 r^3} - \left(\frac{d}{dt}\right)_2 \frac{q_1[\boldsymbol{\mu}_2 \times \boldsymbol{r}]}{4\pi c^2 r^3}, \tag{7.2}$$

but if the magnetic moment $\boldsymbol{\mu}_2$ is created by a current loop, we have to use equation (6.29)

$$\boldsymbol{f}_1^{\text{coil}} = -\left(\frac{d}{dt}\right)_2 \frac{q_1[\boldsymbol{\mu}_2 \times \boldsymbol{r}]}{4\pi c^2 r^3}. \tag{7.3}$$

In mechanics, a force is called *conservative* if it can be represented as a gradient of a scalar function.[1] The distinguishing feature of conservative forces is that their integrals along any closed contour are equal to zero. Therefore, the action of conservative forces on electrons in metals cannot be detected by measuring a current in a closed circuit. In other words, conservative interactions do not create an electromotive force.

For this reason, experimental studies of the first (conservative) contribution in (7.2) are rather difficult, and we will not discuss them in our book. Here we consider only the nonconservative component of force (7.3), which is common for both types of magnets and which is easily detected by the current induced in a closed circuit L, as shown in Figure 7.1.

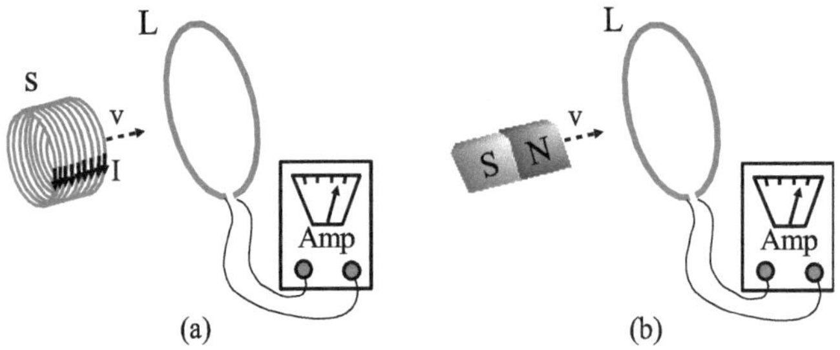

Figure 7.1: Electromagnetic induction. The current in the wire coil L can be induced by a moving (a) electromagnet or (b) permanent magnet.

In the case of macroscopic magnets, equation (7.3) should be integrated over the magnet's volume $\mathcal{V}$, i. e., the total nonconservative force acting on the resting charge q_1 is[2]

$$\boldsymbol{F}_1^{\text{noncons}} = -\left(\frac{d}{dt}\right)_2 \int_{\mathcal{V}} \frac{q_1[\overline{\boldsymbol{\mu}}_2 \times \boldsymbol{r}]}{4\pi c^2 r^3} d\boldsymbol{r}_2. \tag{7.4}$$

1 For example, as in the first term on the right-hand side of (7.2).

2 Here $\overline{\boldsymbol{\mu}}_2$ is the bulk magnetization density.

Let us check that this RQD prediction is in qualitative agreement with the Maxwell electrodynamics. By the symbol $\boldsymbol{E}_1$ we denote the force with which the moving magnet acts on a unit charge at rest. We have[3]

$$\boldsymbol{E}_1 \equiv \boldsymbol{F}_1^{\text{noncons}}/q_1.$$

If we take the *curl* of this vector function, we get[4]

$$\begin{aligned}\left[\frac{\partial}{\partial \boldsymbol{r}_1} \times \boldsymbol{E}_1\right] &= -\frac{1}{4\pi c^2}\left(\frac{d}{dt}\right)_2 \int_{\mathcal{V}} \left[\frac{\partial}{\partial \boldsymbol{r}_1} \times \frac{[\overline{\boldsymbol{\mu}}_2 \times \boldsymbol{r}]}{r^3}\right] d\boldsymbol{r}_2 \\ &= -\frac{1}{c}\left(\frac{d}{dt}\right)_2 \int_{\mathcal{V}} \left(\frac{1}{4\pi c}\right)\left(-\frac{\overline{\boldsymbol{\mu}}_2}{r^3} + \frac{3(\overline{\boldsymbol{\mu}}_2 \cdot \boldsymbol{r})\boldsymbol{r}}{r^5}\right) d\boldsymbol{r}_2. \end{aligned} \tag{7.5}$$

In the integrand we recognize the "magnetic field" (6.34), created at the point $\boldsymbol{r}_1$ by the magnet's unit volume. Thus, the integral expresses the full "magnetic field" $\boldsymbol{B}_1$, and (7.5) formally assumes the form of the Maxwell–Faraday induction law (7.1)

$$\left[\frac{\partial}{\partial \boldsymbol{r}_1} \times \boldsymbol{E}_1\right] = -\frac{1}{c}\left(\frac{d}{dt}\right)_2 \boldsymbol{B}_1.$$

However, it is important to emphasize that in our approach the causes of electromagnetic induction are fundamentally different from those proclaimed in Maxwell's theory. In the traditional interpretation, electromagnetic induction is the result of the interdependence of alternating electric and magnetic fields. In our approach, electromagnetic induction is a consequence of the characteristic interaction potential between magnetic moments and charges.

7.1.2 Homopolar generator

One interesting manifestation of electromagnetic induction is the *homopolar generator*[5] shown in Figure 7.2. This device consists of a conducting disk C and a cylindrical magnet M, each affixed to their own shafts in such a way that they can independently rotate about the same vertical axis. Magnetization vectors $\overline{\boldsymbol{\mu}}_2$ of each small volume element of the magnet are directed along the same axis. The shaft AB is conducting. Points A and C are connected to sliding contacts (shown by arrows), and the circuit is closed through an ammeter.

3 In Maxwell's electrodynamics this is the definition of the *electric field* $\boldsymbol{E}_1$.

4 Here we used the known formula $[\nabla \times [\boldsymbol{A} \times \boldsymbol{B}]] = \boldsymbol{A}(\nabla \cdot \boldsymbol{B}) - \boldsymbol{B}(\nabla \cdot \boldsymbol{A}) + (\boldsymbol{B} \cdot \nabla)\boldsymbol{A} - (\boldsymbol{A} \cdot \nabla)\boldsymbol{B}$, where $\nabla \equiv (\partial/\partial \boldsymbol{r})$.

5 Also known as Faraday's disk.

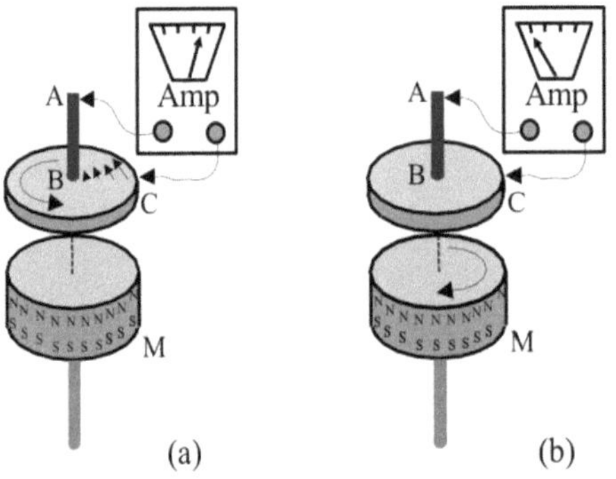

Figure 7.2: Homopolar generator. (a) Conducting disk C rotates. (b) Magnet M rotates.

This device can work in two modes. In the first mode (see Figure 7.2 (a)), the magnet is stationary, while the conducting disk C rotates about its axis. In this case, the ammeter registers the flow of direct current through the circuit. This effect has a simple explanation. Generally speaking, the force acting on the electrons in the circuit can be obtained by integrating formula (6.33) over the magnet's volume $\mathcal{V}$, so we have

$$\boldsymbol{F}_1(\boldsymbol{r}_1, \boldsymbol{v}_1) = \int_{\mathcal{V}} \frac{q_1}{c} [\boldsymbol{v}_1 \times \overline{\boldsymbol{b}}_1(\boldsymbol{r}_1, \boldsymbol{r}_2)] d\boldsymbol{r}_2. \tag{7.6}$$

Here

$$\overline{\boldsymbol{b}}_1(\boldsymbol{r}_1, \boldsymbol{r}_2) = \frac{1}{4\pi c} \left(\frac{-\overline{\boldsymbol{\mu}}_2}{r^3} + \frac{3(\overline{\boldsymbol{\mu}}_2 \cdot \boldsymbol{r})\boldsymbol{r}}{r^5} \right)$$

denotes the "magnetic field" created at the point $\boldsymbol{r}_1$ by the magnet's unit volume in the vicinity of $\boldsymbol{r}_2$.

The full electromotive force is obtained by integrating (7.6) along the closed contour $A \to B \to C \to$ ammeter $\to A$. The velocity $\boldsymbol{v}_1$ is nonzero only in the segment $B \to C$,[6] where the force $\boldsymbol{F}_1$ is directed radially along the segment $B \to C$, so it makes a nonzero contribution to the integral, and the theory correctly predicts a nonvanishing current through the ammeter.

In the second mode of operation (see Figure 7.2 (b)), the disk C is fixed, while the magnet M rotates. Careful experiments [47, 248] showed that in this case there is no current in the circuit. At first glance, this result looks surprising. Does the principle of relativity not tell us that the physical result (the current in the circuit) can depend only on the relative motion of the magnet and the disk? The answer is: "No. The principle of relativity can only be applied to inertial motions. It is easy to make a mistake when blindly extending this principle to rotational motions."

6 Velocities of the electrons in the rotating disk are shown by small arrows in Figure 7.2 (a). The "magnetic-field" vector is directed vertically in this region of space.

To find out the electromotive force in the case of a rotating magnet, we need to integrate again the force vector acting on (now resting) electrons along the closed circuit $A \to B \to C \to$ ammeter $\to A$. As we have already explained, the conservative part of this force[7] does not contribute to this integral. To calculate the nonconservative force, we take into account that, due to the cylindrical symmetry of the magnet, when it rotates about the axis, the volume integral in equation (7.4) does not depend on time. Therefore $\boldsymbol{F}_1^{\text{noncons}} = 0$, which agrees with the observed absence of the current.

7.2 Aharonov–Bohm effect

In our approach to classical electrodynamics, we deny the presence of electromagnetic fields $\boldsymbol{E}$, $\boldsymbol{B}$ as well as electromagnetic potentials $\mathcal{A}_0$ and $\boldsymbol{\mathcal{A}}$. In the traditional Maxwell theory, these potentials are also considered unobservable. However, there is a class of experiments in which – it is claimed – the reality of electromagnetic potentials is manifested even in those regions of space where $\boldsymbol{E} = \boldsymbol{B} = 0$. The most famous representative of this class is the Aharonov–Bohm experiment [2]. In this section we are going to demonstrate that the Aharonov–Bohm effect can be successfully explained without involving the notions of electromagnetic potentials. Our explanation will be based on the idea of quantum particles moving under the action of direct Coulomb–Darwin–Breit forces.

7.2.1 Aharonov–Bohm effect with linear magnet

Let us consider an idealized version of the Aharonov–Bohm experiment, shown in Figure 7.3. An infinitely long solenoid or ferromagnetic rod with a linear magnetization density $\overline{\mu}_2$ is erected vertically at the origin ($x = y = 0$). This magnet is bombarded by wave packets of electrons 1. At the point S (start), the wave packets are divided into two parts,[8] which continue to move on both sides of the magnet. The two packets meet again at the point E (end), where their interference is measured. Two alternative trajectories of the wave packets $S \to (-R) \to E$ and $S \to R \to E$ are shown by solid lines in Figure 7.3.

We already know (from Subsections 6.3.6–6.3.7) that outside the solenoid, no electric or magnetic forces act on charge 1. However, it was found experimentally that the interference at the point E depends on the solenoid's magnetization $\overline{\mu}_2$ [31, 250, 178]. Aharonov and Bohm [2] predicted this effect and explained that outside the solenoid there is a nonzero vector potential $\boldsymbol{\mathcal{A}}$, which affects phases of wave packets of charged

7 Originating from the first term on the right-hand side of (7.2).

8 Using some kind of beamsplitter or simply two holes, as in the interference experiment.

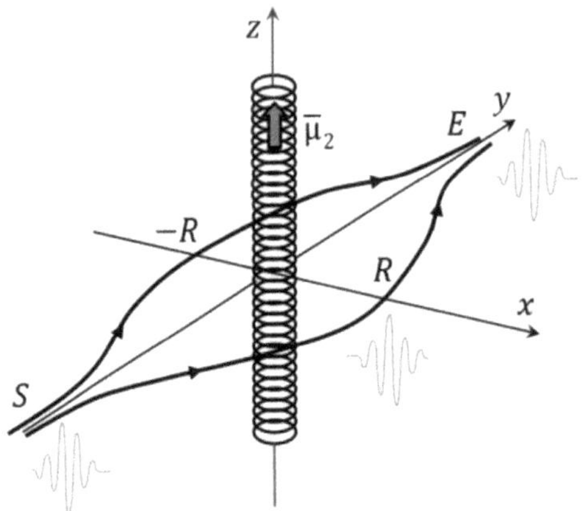

Figure 7.3: The Aharonov–Bohm experiment. An infinite solenoid with linear magnetization density $\overline{\mu}_2$ is placed at the origin. Electron wave packets bypass it from the left and from the right and then interfere at point *E*.

particles. There were also alternative attempts to explain this effect [230, 268, 186, 109]. For example, a number of works [201, 16, 17, 19, 18] entertained the idea that there *is* a nonzero force acting on the electrons. If this force acted differently on the electrons moving along the two paths, then a difference in the arrival time of the two wave packets would be created and the phase shift at point *E* may be observed. However, this idea appears to be in conflict with recent measurements [26], which found no relative delay.

Here we propose our own explanation, based on the Hamiltonian electrodynamics of Coulomb, Darwin and Breit [238].

To simplify calculations, we assume that the distance $|SE|$ is large enough, so that the two alternative paths $S \to (-R) \to E$ and $S \to R \to E$ can be regarded as parallel to the *y*-axis. We also know from (6.36) that both wave packets move without acceleration and arrive at the point *E* simultaneously, regardless of the magnetization $\overline{\boldsymbol{\mu}}_2 = (0, 0, \overline{\mu}_2)$. So, their trajectories can be written as

$$\begin{aligned} \boldsymbol{v}_1 &= (0, v_{1y}, 0), \\ \boldsymbol{r}_1(t) &= (\pm R, v_{1y}t, 0). \end{aligned} \tag{7.7}$$

To estimate the magnet's effect on the interference, we should turn to the quasiclassical dynamics from Subsection **1**-6.6.6. We found there that the phase factor of the wave packet changes with time as $\exp(\frac{i}{\hbar}\phi(t))$, where the *action integral* is given by the formula[9]

$$\phi(t) \equiv \int_{-\infty}^{t} \left(\frac{m_e v_{1y}^2(t')}{2} - V_{\text{long}}(t') \right) dt' \tag{7.8}$$

9 In this case, particle 1 is the electron, so we put $m_1 = m_e$ and $q_1 = -e$.

and $V_{long}(t)$ is the time dependence of the potential (6.37) acting on the electron. In the geometry shown in Figure 7.3, the phase factors accumulated along the two alternative paths $S \to (-R) \to E$ and $S \to R \to E$ are different and the interference of the "left" and "right" wave packets at the point E depends on this difference, so we have

$$\Delta\phi = \phi_{right}(\infty) - \phi_{left}(\infty).$$

The kinetic energy term in (7.8) does not contribute to this difference, because electron velocities are constant and equal for the two paths. However, the electron's potential energies on two sides of the magnet are different. For all points on the "right" path the numerator of (6.37) is equal to $e\overline{\mu}_2 v_{1y} R$, while for the "left" path this numerator is $-e\overline{\mu}_2 v_{1y} R$. Therefore, the full accumulated phase difference

$$\Delta\phi = \int_{-\infty}^{\infty} \frac{e\overline{\mu}_2 R v_{1y}}{\pi c^2 (R^2 + v_1^2 t^2)} dt = \frac{e\overline{\mu}_2}{c^2} \tag{7.9}$$

depends neither on the electron's velocity (energy) nor on the value of the *impact parameter R*. This difference depends only on the magnetization $\overline{\mu}_2$, as in the original Aharonov–Bohm formula.

7.2.2 Aharonov–Bohm effect with toroidal magnet

Now consider the Aharonov–Bohm effect with the toroidal magnet described in Subsection 6.3.8. Toroidal magnets were used in the experiments of Tonomura and coauthors [250, 178]. These experiments are still regarded as the most reliable confirmations of the Aharonov and Bohm prediction.

As in the preceding subsection, we switch to the semiclassical approximation in which electron 1 is described by a localized wave packet whose center moves according to the laws of classical mechanics (see Subsection 6.3.8). In other words, the packets move without acceleration along the paths $S \to 0 \to E$ and $S \to R \to E$ in Figure 7.4. We have

$$\boldsymbol{r}_1^{SOE}(t) = (0, v_{1y}t, 0), \tag{7.10}$$

$$\boldsymbol{r}_1^{SRE}(t) = (R, v_{1y}t, 0). \tag{7.11}$$

Now turn to the packets' phases expressed by the action integral (**1**-6.111). We have to integrate the potential (6.38) along the trajectories (7.10) and (7.11).[10] In the former

10 The solenoid is assumed to be fixed ($\boldsymbol{P}_2 = \boldsymbol{0}, \boldsymbol{R}_2 = \boldsymbol{0}$), so the second term in (6.38) is omitted.

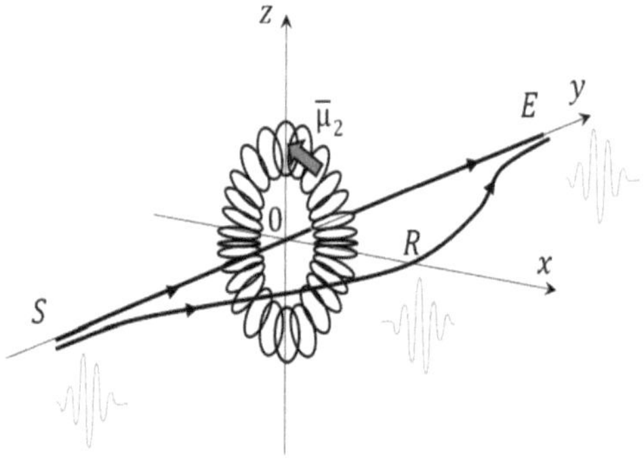

Figure 7.4: Toroidal electromagnet and a moving charge. The charge's path $S - 0 - E$ passes through the center of the torus; the path $S - R - E$ is outside the torus.

case, we obtain the action integral for the passage of the electron through the center of the toroidal magnet, i. e.,

$$\phi^{SOE} \approx -\int_{-\infty}^{\infty} dt \frac{ea^2\bar{\mu}_2 v_{1y}}{2c^2(a^2 + v_{1y}^2 t^2)^{3/2}} = -\frac{e\bar{\mu}_2}{c^2}. \tag{7.12}$$

For the latter case, we need to evaluate the charge's potential energy along the path $S \to R \to E$, where

$$\begin{aligned} \boldsymbol{r}_1 &= (r_{1x}, r_{1y}, r_{1z}), \\ \boldsymbol{r} &= (r_{1x} - a\cos\theta, r_{1y}, r_{1z} - a\sin\theta). \end{aligned}$$

From equation (6.39) for $\bar{\boldsymbol{\mu}}_2$ we get the following components in (6.38):

$$\begin{aligned} [\bar{\boldsymbol{\mu}}_2 \times \boldsymbol{r}] &= \bar{\mu}_2(-r_{1y}\cos\theta, r_{1z}\sin\theta + r_{1x}\cos\theta - a, -r_{1y}\sin\theta), \\ [\bar{\boldsymbol{\mu}}_2 \times \boldsymbol{r}] \cdot \boldsymbol{p}_1 &= \bar{\mu}_2(-p_{1x}r_{1y}\cos\theta + p_{1y}r_{1z}\sin\theta + p_{1y}r_{1x}\cos\theta - p_{1y}a - p_{1z}r_{1y}\sin\theta) \\ &= -\frac{1}{a}(\boldsymbol{N}_2 \cdot \boldsymbol{p}_1) + \bar{\mu}_2[\boldsymbol{r}_1 \times \boldsymbol{p}_1]_z\cos\theta - \bar{\mu}_2[\boldsymbol{r}_1 \times \boldsymbol{p}_1]_x\sin\theta. \end{aligned}$$

Here we characterized magnetic properties of the toroidal magnet by the vector $\boldsymbol{N}_2 = (0, \bar{\mu}_2 a^2, 0)$ perpendicular to the torus' plane and having the length of $\bar{\mu}_2 a^2$. We assume that the size of the torus is small and that along the entire particle trajectory we can approximate $r \gg a$, $\boldsymbol{r}_1 \approx \boldsymbol{r}$. Then, using approximation (6.25) and integrating along the torus' circumference, we obtain

$$\begin{aligned} V_{\text{torus}} &= \int_0^{2\pi} d\theta \frac{ea[\bar{\boldsymbol{\mu}}_2 \times \boldsymbol{r}] \cdot \boldsymbol{p}_1}{4\pi m_e c^2 r^3} \\ &\approx \frac{e}{4\pi m_e c^2} \int_0^{2\pi} d\theta(-(\boldsymbol{N}_2 \cdot \boldsymbol{p}_1) + \bar{\mu}_2 a[\boldsymbol{r} \times \boldsymbol{p}_1]_z\cos\theta - \bar{\mu}_2 a[\boldsymbol{r} \times \boldsymbol{p}_1]_x\sin\theta) \\ &\quad \times \left(\frac{1}{r^3} + \frac{3a(r_x\cos\theta + r_z\sin\theta)}{r^5}\right) \end{aligned}$$

$$
\begin{aligned}
&= -\frac{e(\boldsymbol{N}_2 \cdot \boldsymbol{p}_1)}{2m_e c^2 r^3} + \frac{3e\bar{\mu}_2 a^2}{4\pi m_e c^2 r^5} \int_0^{2\pi} d\theta([\boldsymbol{r} \times \boldsymbol{p}_1]_z r_x \cos^2\theta - [\boldsymbol{r} \times \boldsymbol{p}_1]_x r_z \sin^2\theta) \\
&= -\frac{e(\boldsymbol{N}_2 \cdot \boldsymbol{p}_1)}{2m_e c^2 r^3} + \frac{3e\bar{\mu}_2 a^2}{4m_e c^2 r^5}([\boldsymbol{r} \times \boldsymbol{p}_1]_z r_x - [\boldsymbol{r} \times \boldsymbol{p}_1]_x r_z) \\
&= -\frac{e(\boldsymbol{N}_2 \cdot \boldsymbol{p}_1)}{2m_e c^2 r^3} - \frac{3e}{4m_e c^2 r^5}([[\boldsymbol{p}_1 \times \boldsymbol{r}] \times \boldsymbol{r}] \cdot \boldsymbol{N}_2) \\
&= -\frac{e(\boldsymbol{N}_2 \cdot \boldsymbol{p}_1)}{2m_e c^2 r^3} + \frac{3e}{4m_e c^2 r^5}((\boldsymbol{p}_1 \cdot \boldsymbol{N}_2) r^2 - (\boldsymbol{r} \cdot \boldsymbol{N}_2)(\boldsymbol{p}_1 \cdot \boldsymbol{r})) \\
&= \frac{e(\boldsymbol{N}_2 \cdot \boldsymbol{p}_1)}{4m_e c^2 r^3} - \frac{3e(\boldsymbol{r} \cdot \boldsymbol{N}_2)(\boldsymbol{p}_1 \cdot \boldsymbol{r})}{4m_e c^2 r^5}.
\end{aligned} \tag{7.13}
$$

The time dependence $V_{\text{torus}}(t)$ of this potential energy is obtained by substituting (7.11) into (7.13). The corresponding action integral vanishes:

$$
\begin{aligned}
\phi^{SRE} &= -\int_{-\infty}^{\infty} V_{\text{torus}}(t)dt \\
&\approx -\int_{-\infty}^{\infty} dt\left(\frac{eN_2 v_{1y}}{4c^2(R^2 + v_{1y}^2 t^2)^{3/2}} - \frac{3eN_2 v_{1y}^3 t^2}{4c^2(R^2 + v_{1y}^2 t^2)^{5/2}}\right) = 0.
\end{aligned}
$$

Comparing this result with (7.12), we see that the phase difference between the two paths (inside the torus and outside it) does not depend on the magnet's radius a, the electron's velocity v_{1y} or the impact parameter R. We have

$$
\Delta\phi = \phi^{SRE} - \phi^{SOE} = \frac{e\bar{\mu}_2}{c^2}.
$$

This is the same result as in the case of the infinite linear solenoid (7.9). It is in full agreement with Tonomura's experiments [250, 178].

7.3 Experimental studies of bound fields

7.3.1 Three types of force fields

Summarizing our discussions of electromagnetic interactions in Chapter 6, it will be convenient to introduce the notion of *force fields*.[11] We distinguish three types of force fields corresponding to three ways by which one group of charges can interact with another group at a distance. The main characteristics of such fields are listed in Table 7.1.

First, we should mention the *electric* force field, whose most important representative is the Coulomb potential (3.6). The characteristic feature of the electric force is

11 We call them "force" fields, to stress that there are no independent electromagnetic fields in RQD.

Table 7.1: Three electromagnetic force fields in RQD.

Force field	Acts on charges at rest?	Propagation speed	Long range?	Carrier
electric	yes	∞	no	n/a
magnetic	no	∞	no	n/a
radiation	yes	c	yes	photon

that it acts even on charges at rest. Second is the *magnetic* force field, such as the Darwin potential (3.8), which acts only on moving charges. Both electric and magnetic fields propagate instantly.[12] In other words, they are tightly bound to the charges that generate them, and, as a rule, their range is very limited. Therefore, we will often combine them into the class of *bound* force fields.

The third type is the *radiation* force field. Unlike bound fields, the radiation field is not described by any analytical potential. The radiative transfer of energy–momentum between systems of charges is the result of emission and absorption of real particles – photons. As we showed in Subsection 5.1.2, accelerated charge[13] emits a large number of photons. These particles propagate in space at the speed of light. Some of the photons can reach another charge (in the receiving antenna) and interact with it either by being absorbed or by Compton-like scattering (see Subsection **2**-3.2.5). In any case, the receiver gets a part of the energy lost by the emitter. Unlike instantaneous direct Coulomb and magnetic interactions, the energy–momentum transfer by means of real photons occurs at the speed of light. The radiation force field is responsible for TV and radio signal propagating in vacuum, as well as for the light that comes to us from distant stars and galaxies. Hence, this is a long-range force field.

Thus, in RQD, the total force field created by a group of moving charges can be conditionally divided into instant (bound) and retarded (radiative) components, i. e.,[14]

$$\boldsymbol{e}^{\mathrm{RQD}} = \boldsymbol{e}^{\mathrm{inst}}_{\mathrm{bound}} + \boldsymbol{e}^{\mathrm{ret}}_{\mathrm{radiation}}.$$

The infinite propagation speed of $\boldsymbol{e}^{\mathrm{inst}}_{\mathrm{bound}}$ is the most controversial prediction of RQD.[15]

---

12 In RQD, we do not have interaction carriers or intermediate fields with their additional degrees of freedom. So, there is no place where the transferred momentum can be temporarily stored during the interaction. Therefore, when particle 1 loses a part of its momentum, particle 2 receives exactly this part without delay. Otherwise, the momentum conservation law would be violated. Hence in our purely corpuscular theory, the instantaneous nature of the interactions is not an approximation, but a necessity. For more details, see Section 8.4.

13 For example, in a radiating antenna.

14 Similar ideas about the separation of the electromagnetic field into a short-range instantaneous and a long-range retarded part can be found in [33, 79, 81, 134, 139]. For brevity, we consider only the "electric" parts of (both bound and radiative) force fields, dropping their magnetic parts.

15 A number of investigations of quantum field models [217, 218, 127, 126] have concluded that the propagation speed of electromagnetic interactions coincides with the speed of light. However, these

On the other hand, Maxwell's electrodynamics and special relativity forbid interactions that propagate faster than light. Both the bound and radiative components of Maxwell's field are considered to be retarded, i. e.,

$$\boldsymbol{E}^{\text{Maxwell}} = \boldsymbol{E}^{\text{ret}}_{\text{bound}} + \boldsymbol{E}^{\text{ret}}_{\text{radiation}}.$$

We will not pay much attention to the retarded radiation fields $\boldsymbol{e}^{\text{ret}}_{\text{radiation}}$ and $\boldsymbol{E}^{\text{ret}}_{\text{radiation}}$, since their properties are very similar in the two theories. Instead, we will focus on the more significant and interesting difference between the bound fields $\boldsymbol{e}^{\text{inst}}_{\text{bound}}$ and $\boldsymbol{E}^{\text{ret}}_{\text{bound}}$.

7.3.2 Force fields emitted by antennas

In experimental studies of the electromagnetic field propagation in the microwave and radio bands, two antennas are usually used: the *emitter* and the *receiver*. The signal arriving at the receiver is usually a combined effect of both bound and radiation force fields created by the emitting antenna.

It is well known that the radiation and bound force fields depend in different ways on the distance to the emitter. In RQD, the radiation signal is formed by freely propagating photons. Around the spherically symmetric emitter, the photon energy density decreases with distance, like r^{-2}. On the other hand, the energy density is proportional to the square of the "electric-field" vector, which leads us to the estimate $e^{\text{ret}}_{\text{radiation}} \propto r^{-1}$.

Usually, the bound fields are damped like $e^{\text{inst}}_{\text{bound}} \propto r^{-2}$ or faster [140]. Therefore, at large distances from the emitter, the bound field is lost against the background of the much stronger radiation.

In this section, we will be interested in experimental methods capable of testing our hypothesis about the instantaneous propagation of $\boldsymbol{e}^{\text{inst}}_{\text{bound}}$. Such measurements are very nontrivial. The bound (Coulomb) electric fields are easily observed in static or almost static situations, when accelerations of the charges are small. However, as we are interested in the *speed* of propagation of the force field, we have to perturb the equilibrium state of the charges and give them substantial accelerations. This will inevitably lead to the appearance of the radiation contribution $\boldsymbol{e}^{\text{ret}}_{\text{radiation}}$, which would mask the desired effect. A successful experiment of this kind should somehow minimize the effect of radiation, so as to allow us to study dynamic properties of the bound field $\boldsymbol{e}^{\text{inst}}_{\text{bound}}$ in its pure form. Here we will discuss several works that have tried to solve this difficult problem. Descriptions of other interesting experiments can be found in the reviews [194, 271].

---

models confirmed only the indisputable finite rate of propagation of radiation force fields (carried by real photons) and did not say anything about the bound fields of charges.

7.3.3 Studies in the near field

A remarkable study of the electromagnetic field propagation was undertaken by Kholmetskii and coauthors [140, 139, 169].[16] They used the classical scheme of the Hertz experiment with two antennas. The radiating (or emitting) antenna (E) created short electromagnetic pulses with a carrier frequency of $\approx$ 125 MHz. The receiving antenna (R) was placed at different distances r from E, and the received signals were recorded as functions of time. The authors demonstrated that at large distances (r up to 3 m) the dominant signal in R was created by the radiation field $\boldsymbol{e}^{\text{ret}}_{\text{radiation}}$. In the near-field zone ($r < 50$ cm), a mixture of the bound and radiation fields was observed with the predominance of a short-range bound field. Assuming that the radiation field propagates at the constant speed c in the entire range of distances, the authors were able to subtract this contribution from the total signal and thus obtained the pure time dependencies $\boldsymbol{e}^{\text{inst}}_{\text{bound}}(t)$ for all r. From their analysis, the authors were able to estimate the propagation speed of this bound component. In the near-field zone, this value turned out to be much greater than the speed of light, and possibly even infinite.[17] This result is in agreement with the RQD idea about the instantaneous propagation of the bound field.

7.3.4 Microwave horn antennas

Probably the first convincing experimental observation of the superluminal propagation of bound electromagnetic force fields was made by Giakos and Ishii in 1991 [94, 93]. They studied the transmission of microwave pulses between two horn antennas arranged as shown in Figure 7.5. In the first variant of the experiment, the emitter (E) and the receiver ($R1$) were located directly opposite each other, separated by the

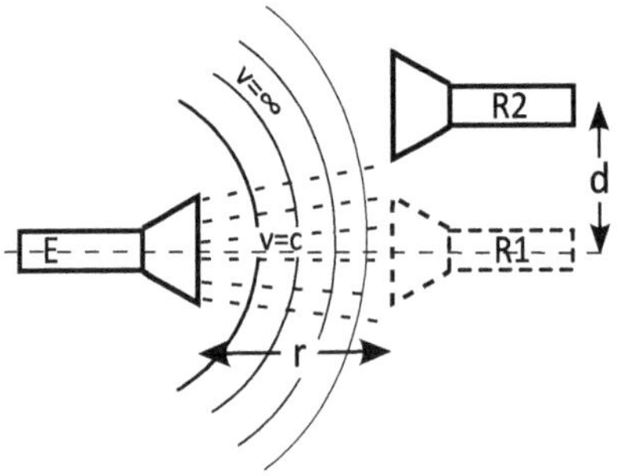

Figure 7.5: Schematics of the experiment with microwave horn antennas and its interpretation.

16 See also [177].
17 See also [263].

distance of $r = 71.5$ cm. The signal traveled this distance in 2.378 ns, which roughly corresponded to the speed of light, 3.01×10^8 m/s, as expected. In the second variant, the receiver ($R2$) was shifted away from the emitter's axis.[18] Up to shifts of $d = 34$ cm, the signal propagation time remained practically constant, despite a significant increase in the distance E–$R2$. In such situations, signal propagation rates of up to 3.32×10^8 m/s were recorded, i. e., exceeding the speed of light by 10 %.

These results were later confirmed in the experiments of Ranfagni and coworkers [192, 191, 171, 193] over a wider range of radiation frequencies and antenna separations.

These observations are not difficult to explain from the point of view of RQD: a radiating horn antenna generates both bound and radiation fields. Due to the specific shape of the emitter E, the photon flux[19] is concentrated near the antenna axis. At the same time, the bound field[20] is more diffuse, short-range and spreads instantly. When the receiver was located directly on the axis of the emitter, the microwave photons dominated in the signal and the apparent propagation speed was close to the normal photon speed c. When the receiver was displaced from the axis, the contribution of the photons in the signal decreased, a more important role was played by the instantaneous bound force field and the effective signal speed increased. The radiation field still had a significant contribution, so the effective signal speed exceeded c by only a few percent. If it were possible to completely suppress the radiation part of the force field, the experimenters would see an infinite propagation speed of the signal.

7.4 Relativistic electron bunches

In Subsection 7.4.5 we are going to discuss the experiment of Professor Pizzella and his colleagues from the Frascati laboratory at the Italian Institute of Nuclear Physics. To date, this experiment is the most serious challenge to Maxwell's electrodynamics and the best evidence of the validity of RQD. To discuss this experiment, we will have to return briefly to theory and discuss the electric fields of high-energy electron beams.

In Subsection 6.2.2, we calculated forces (6.14)–(6.15) acting between two moving charges. These formulas were approximate, because they took into account only terms of the order $(v/c)^2$ and lower. Moreover, we assumed that accelerations of the charges are negligible, ignored bremsstrahlung interactions and neglected the possibility of the photon emission.

In Subsections 7.4.1–7.4.4, we try to fill these gaps and discuss (albeit qualitatively) RQD-predicted force fields around charges moving with high velocities and acceler-

---

18 The authors also tried to tilt the receiving antenna towards the emitter.

19 Which is the radiation field $\boldsymbol{e}^{\text{ret}}_{\text{radiation}}$ shown by broken lines in Figure 7.5.

20 That is, $\boldsymbol{e}^{\text{inst}}_{\text{bound}}$, shown by concentric arcs in Figure 7.5.

ations. We will also see that these predictions diverge from the standard Maxwell theory.

7.4.1 Fast moving charge (RQD)

Let us now derive the RQD interaction potential between two charges beyond the $(v/c)^2$ approximation. We will be interested in a specific geometry in which the charge q_1 moves with a large constant velocity $v_1 \approx c$ and momentum $p_1 \gg m_1 c$ along the z-axis, while charge 2 rests at the distance x from the line of motion, as shown in Figure 7.6. We choose Cartesian axes in such a way that the point $z_1 = 0$ on the line of motion corresponds to the closest approach of the two charges. Suppose that the test charge 2 is very small ($q_2 \ll q_1$), so that its presence has no apparent effect on the rectilinear motion of particle 1. Moreover, we assume the mass m_2 to be infinitely large, so that in the course of our thought experiment this particle does not move ($v_2 = 0$).[21] Our goal is to calculate the force $\boldsymbol{f}_2$ acting on charge 2. More precisely, we are interested in the ratio

$$\boldsymbol{e} \equiv \boldsymbol{f}_2/q_2, \tag{7.14}$$

which goes by the name "electric field" in Maxwell's electrodynamics.

We begin by calculating the interaction energy between charges 1 and 2. Near the energy surface, we can use the relation $V_2^d = (\Phi_2^n)^{\text{phys}}$ and obtain the required interac-

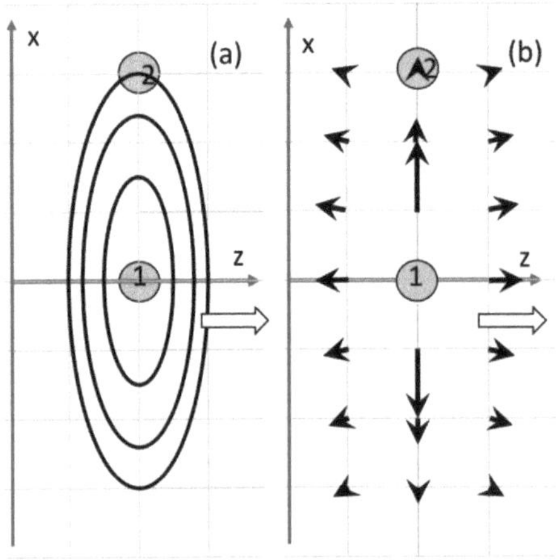

Figure 7.6: Force field around electric charge 1, which moved uniformly and rectilinearly for a long time. (a) Equipotential lines (7.18). (b) Vectors of the electric (force) field (7.19)–(7.21).

21 By these approximations we are trying to model the actual Frascati experiment from Subsection 7.4.5: charge 1 represents a relativistic beam leaving the accelerator's pipe and charge 2 is a simplified model of the electric field sensor.

tion operator from equation (2-3.28). We have[22]

$$V_2^d = (\Phi_2^n)^{\text{phys}}$$
$$\approx -\frac{q_1 q_2 \hbar^2 c^2}{(2\pi\hbar)^3} \int \frac{d\boldsymbol{k} d\boldsymbol{p}_2 d\boldsymbol{p}_1 m_1 c^2}{\sqrt{\omega_{\boldsymbol{p}_1}\omega_{\boldsymbol{p}_1+\boldsymbol{k}}}} \frac{\mathcal{U}^\mu(\boldsymbol{p}_1+\boldsymbol{k};\boldsymbol{p}_1)\mathcal{W}_\mu(\boldsymbol{p}_2-\boldsymbol{k};\boldsymbol{p}_2)}{(\omega_{\boldsymbol{p}_1}-\omega_{\boldsymbol{p}_1+\boldsymbol{k}})^2 - c^2k^2} d^\dagger_{\boldsymbol{p}_2-\boldsymbol{k}} a^\dagger_{\boldsymbol{p}_1+\boldsymbol{k}} d_{\boldsymbol{p}_2} a_{\boldsymbol{p}_1}$$
$$\equiv \int d\boldsymbol{k} d\boldsymbol{p}_2 d\boldsymbol{p}_1 \mathfrak{v}_2^d(\boldsymbol{p}_1+\boldsymbol{k},\boldsymbol{p}_1,\boldsymbol{k}) d^\dagger_{\boldsymbol{p}_2-\boldsymbol{k}} a^\dagger_{\boldsymbol{p}_1+\boldsymbol{k}} d_{\boldsymbol{p}_2} a_{\boldsymbol{p}_1}. \tag{7.15}$$

According to Subsection 2-1.2.7, the position–space representation of this potential is obtained by the Fourier transform of the coefficient function,

$$\mathfrak{v}_2^d(\boldsymbol{p}_1+\boldsymbol{k},\boldsymbol{p}_1,\boldsymbol{k}) \approx -\frac{q_1 q_2 \hbar^2 c^2}{(2\pi\hbar)^3} \frac{m_1 c^2}{\sqrt{\omega_{\boldsymbol{p}_1}\omega_{\boldsymbol{p}_1+\boldsymbol{k}}}} \frac{\mathcal{U}^0(\boldsymbol{p}_1+\boldsymbol{k},\boldsymbol{p}_1)}{(\omega_{\boldsymbol{p}_1}-\omega_{\boldsymbol{p}_1+\boldsymbol{k}})^2 - c^2k^2},$$

with respect to $\boldsymbol{k}$. We are interested only in the long-range behavior of this potential, which results from integration in the region of low frequencies $k = 0$. So, we will assume that $k \ll p_1$ and $\omega_{\boldsymbol{p}} \approx cp$. Then

$$\frac{m_1 c^2}{\sqrt{\omega_{\boldsymbol{p}_1}\omega_{\boldsymbol{p}_1+\boldsymbol{k}}}} \approx \frac{m_1 c}{p_1},$$
$$\mathcal{U}^0(\boldsymbol{p}_1+\boldsymbol{k};\boldsymbol{p}_1) = \Bigg(\sqrt{\omega_{\boldsymbol{p}_1+\boldsymbol{k}}+m_1c^2}\sqrt{\omega_{\boldsymbol{p}_1}+m_1c^2}$$
$$+\sqrt{\omega_{\boldsymbol{p}_1+\boldsymbol{k}}-m_1c^2}\sqrt{\omega_{\boldsymbol{p}_1}-m_1c^2}\frac{(\boldsymbol{p}_1+\boldsymbol{k})\cdot\boldsymbol{p}_1}{|\boldsymbol{p}_1+\boldsymbol{k}|p_1}\Bigg)\frac{1}{2m_1c^2} \approx \frac{p_1}{m_1 c},$$
$$\mathfrak{v}_2^d(\boldsymbol{p}_1+\boldsymbol{k},\boldsymbol{p}_1,\boldsymbol{k}) \approx -\frac{q_1 q_2 \hbar^2}{(2\pi\hbar)^3} \frac{c^2}{(\omega_{\boldsymbol{p}_1}-\omega_{\boldsymbol{p}_1+\boldsymbol{k}})^2 - c^2k_x^2 - c^2k_y^2 - c^2k_z^2}. \tag{7.16}$$

The nonnegative expression $\Omega(k_x,k_y,k_z) \equiv (\omega_{\boldsymbol{p}_1}-\omega_{\boldsymbol{p}_1+\boldsymbol{k}})^2$ in the denominator is a function vanishing at $k_x = k_y = k_z = 0$ and having a zero derivative there; indeed[23]

$$\left.\frac{\partial\Omega}{\partial k_y}\right|_{\boldsymbol{k}=0} = -2(\omega_{\boldsymbol{p}_1}-\omega_{\boldsymbol{p}_1+\boldsymbol{k}})\left.\frac{c^2k_y}{\omega_{\boldsymbol{p}_1+\boldsymbol{k}}}\right|_{\boldsymbol{k}=0} = 0,$$
$$\left.\frac{\partial\Omega}{\partial k_z}\right|_{\boldsymbol{k}=0} = -2(\omega_{\boldsymbol{p}_1}-\omega_{\boldsymbol{p}_1+\boldsymbol{k}})\left.\frac{c^2(p_{1z}+k_z)}{\omega_{\boldsymbol{p}_1+\boldsymbol{k}}}\right|_{\boldsymbol{k}=0} = 0.$$

---

22 Here we conditionally identified charges 1 and 2 with the electron and proton, respectively. As usual, the coefficient function of the interaction operator V_2^d was obtained from the S-matrix dividing it by $(-2\pi i)$. We ignored spins of the particles and used formulas (**2**-B.81), (**2**-B.83) that are valid in the limit $m_2 \to \infty$.

23 We took into account that $p_{1x} = p_{1y} = 0$.

For the second derivatives we get

$$\left.\frac{\partial^2\Omega}{\partial k_z^2}\right|_{\boldsymbol{k}=0} = -2\frac{c^2(p_{1z}+k_z)}{\omega_{\boldsymbol{p}_1+\boldsymbol{k}}}\frac{\partial}{\partial k_z}(\omega_{\boldsymbol{p}_1} - \omega_{\boldsymbol{p}_1+\boldsymbol{k}})\right|_{\boldsymbol{k}=0}$$

$$= 2\frac{c^2(p_{1z}+k_z)}{\omega_{\boldsymbol{p}_1+\boldsymbol{k}}}\cdot\left.\frac{c^2(p_{1z}+k_z)}{\omega_{\boldsymbol{p}_1+\boldsymbol{k}}}\right|_{\boldsymbol{k}=0} = \frac{2c^4p_{1z}^2}{\omega_{\boldsymbol{p}_1}^2} = 2v_{1z}^2,$$

$$\left.\frac{\partial^2\Omega}{\partial k_y^2}\right|_{\boldsymbol{k}=0} = \left.\frac{\partial^2\Omega}{\partial k_y\partial k_z}\right|_{\boldsymbol{k}=0} = \left.\frac{\partial^2\Omega}{\partial k_x\partial k_y}\right|_{\boldsymbol{k}=0} = 0.$$

Therefore, the Taylor expansion near $\boldsymbol{k} = 0$ has the form $\Omega(k_x, k_y, k_z) \approx k_z^2 v_1^2$. Substituting this into (7.16), we get

$$\mathfrak{v}_2^d(\boldsymbol{p}_1+\boldsymbol{k}, \boldsymbol{p}_1, \boldsymbol{k}) \approx \frac{q_1q_2\hbar^2}{(2\pi\hbar)^3}\cdot\frac{1}{k_x^2+k_y^2+k_z^2(1-v_1^2/c^2)}$$

and the desired position-dependent potential is obtained from (**2**-A.3):

$$\mathfrak{v}_2^d(\boldsymbol{p}_1, \boldsymbol{r}) = \int d\boldsymbol{k} w_2^d(\boldsymbol{p}_1+\boldsymbol{k}, \boldsymbol{p}_1, \boldsymbol{k})e^{\frac{i}{\hbar}\boldsymbol{k}\cdot\boldsymbol{r}} \approx \frac{q_1q_2\hbar^2}{(2\pi\hbar)^3}\int d\boldsymbol{k}\frac{e^{\frac{i}{\hbar}\boldsymbol{k}\cdot\boldsymbol{r}}}{k_x^2+k_y^2+k_z^2/\gamma^2}$$

$$= \frac{q_1q_2\gamma}{4\pi\sqrt{x^2+y^2+\gamma^2z^2}}, \tag{7.17}$$

where we defined $\gamma \equiv 1/\sqrt{1-v_1^2/c^2} \gg 1$ and $\boldsymbol{r} \equiv \boldsymbol{r}_2 - \boldsymbol{r}_1 \equiv (x, y, z)$.

So, within our approximations, the full Hamiltonian of the two-charge system is

$$H \approx m_2c^2 + cp_1 + \frac{q_1q_2\gamma}{4\pi\sqrt{x^2+y^2+\gamma^2z^2}}. \tag{7.18}$$

The force acting on particle 2 is

$$\boldsymbol{f}_2 \equiv m_2\frac{d^2\boldsymbol{r}_2}{dt^2} = \frac{d\boldsymbol{p}_2}{dt} = -\frac{\partial H}{\partial \boldsymbol{r}_2}$$

and the "electric field" (7.14) at the moment of the closest approach[24] is obtained by taking the gradient of the potential (7.17), i. e.,

$$e_x^{(\gamma)}(x, y, z) = -\frac{1}{q_2}\cdot\frac{\partial\mathfrak{v}_2^d}{\partial x_2} = \frac{q_1\gamma x}{4\pi(x^2+y^2+\gamma^2z^2)^{3/2}}, \tag{7.19}$$

$$e_y^{(\gamma)}(x, y, z) = -\frac{1}{q_2}\cdot\frac{\partial\mathfrak{v}_2^d}{\partial y_2} = \frac{q_1\gamma y}{4\pi(x^2+y^2+\gamma^2z^2)^{3/2}}, \tag{7.20}$$

24 When $\boldsymbol{r}_1 = (0, 0, 0)$, as shown in Figure 7.6.

$$e_z^{(\gamma)}(x,y,z) = -\frac{1}{q_2} \cdot \frac{\partial \mathfrak{v}_2^d}{\partial z_2} = \frac{q_1 \gamma^3 z}{4\pi(x^2 + y^2 + \gamma^2 z^2)^{3/2}}. \tag{7.21}$$

It is interesting to compare this result with the spherically symmetric field created by charge 1 at rest ($\gamma = 1$),

$$\boldsymbol{e}^{(\gamma=1)}(x,y,z) = \frac{q_1 \boldsymbol{r}}{4\pi(x^2 + y^2 + z^2)^{3/2}}. \tag{7.22}$$

The motion of charge 1 has two effects on its "electric field." First, the equipotential surfaces are squeezed by the factor of γ in the direction of motion. Instead of spheres, they take the shape of oblate ellipsoids, as in Figure 7.6 (a). Second, the peak value of the field increases with increasing speed. This means that the "electric field" of the fast moving charge is concentrated in a narrow disk perpendicular to the direction of motion, as shown in Figure 7.6 (b).

7.4.2 Fast moving charge (Maxwell's theory)

Let us now see how the same problem is solved in Maxwell's electrodynamics.

This theory does not recognize the difference between bound and radiation fields of the charge. Both fields are described by the same vectors $\boldsymbol{E}$ and $\boldsymbol{B}$ that can be obtained from Liénard–Wiechert formulas.[25] The idea is that the electric (and magnetic) fields are not rigidly attached to the moving charge, but propagate radially at a speed equal to the speed of light.[26] Thus, the field configuration around the charge is not determined by its instantaneous position, but depends in a complex way on the charge's previous trajectory.

In the Liénard–Wiechert method, the following formulas are obtained for the electric field around the uniformly moving charge 1 in the configuration shown in Figure 7.6[27]:

$$E_x(x,y,z) = \frac{q_1 \gamma x}{4\pi(x^2 + y^2 + \gamma^2 z^2)^{3/2}}, \tag{7.23}$$

$$E_y(x,y,z) = \frac{q_1 \gamma y}{4\pi(x^2 + y^2 + \gamma^2 z^2)^{3/2}}, \tag{7.24}$$

$$E_z(x,y,z) = \frac{q_1 \gamma z}{4\pi(x^2 + y^2 + \gamma^2 z^2)^{3/2}}. \tag{7.25}$$

25 See [114, 27] and Section 14.1 in [118].

26 Note that such propagation delay has not yet been verified experimentally. In addition, the idea of the retarded propagation of the Coulomb field leads to a number of paradoxes [142, 65, 66, 132–134, 40, 102, 232], which do not have satisfactory resolution within the Maxwell theory.

27 For detailed derivation, see, for example, Sections 11.10 and 14.1 in [118].

The transverse components E_x and E_y coincide with our results (7.19)–(7.20), but the longitudinal component (7.25) is γ^2 times smaller than ours (7.21).

So, despite some quantitative difference, both theories predict a similar disk-shaped form of the force field surrounding the fast moving charge.

7.4.3 Charge leaving accelerator (Maxwell's theory)

In Subsection 7.4.5 we will be interested in the "electric field" of a charge leaving the accelerator pipe. This situation is somewhat more complicated than the case of a uniformly moving charge considered above. Let us first find out how this situation is described in traditional electrodynamics. If necessary, the predicted field dynamics can be obtained by Liénard–Wiechert formulas [29]. However, we will be satisfied with only a qualitative picture obtained from the analysis of Figure 7.7.

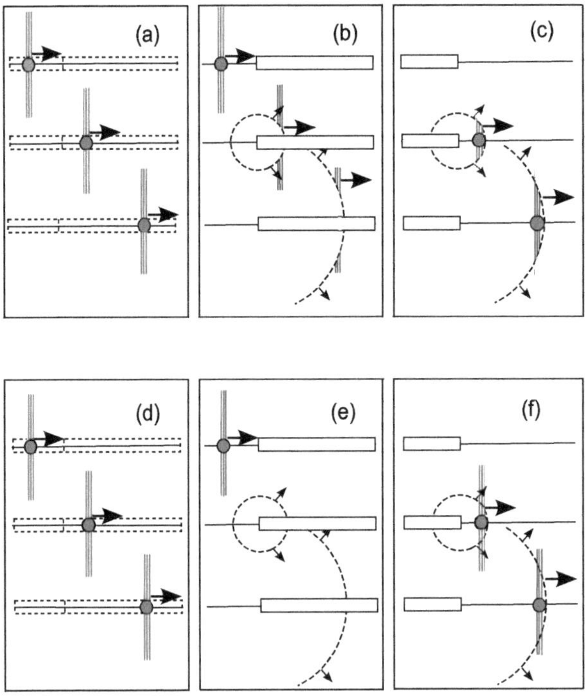

Figure 7.7: Electric force field dynamics in the cases of retarded (Liénard–Wiechert) propagation (a), (b), (c) and instantaneous (RQD) propagation (d), (e), (f). Explanations in the text.

In Figure 7.7 (a) we showed three frames from the "life" of a free charge moving uniformly at high speed. As we have already discussed, its electric field has the form of a disk perpendicular to the direction of motion. If the charge moved inside an infinite metal pipe (shown by a dashed line), then its field would be screened and the total electric field outside the pipe would practically vanish. So for an external observer the charge in the pipe "disappears."

Figure 7.7 (b) shows what happens when the charge goes from empty space to a semiinfinite pipe, shown by an elongated rectangle on the right side of the frame. The entry of a charge into the pipe creates the so-called *transition radiation*,[28] which is a spherical electromagnetic wave, shown by a dashed circle in the second and third frames. The points outside this sphere cannot "know" about the disappearance of the charge inside the pipe, because information about this event moves only with the speed of light and has not reached these points yet. This means that a part of the field disk continues its movement, as if there were no pipe. In the second and third frames one can see how, over time, the spherical electromagnetic wave "eats out" the central part of the field disk. The radius d of the "eaten" part is proportional to the distance L traversed by the disk from the pipe's entrance. To derive this formula, we take into account that the electron beam (and the electric field disk attached to it) has traveled the distance L in time $t = L/v$, where $v \approx c$ is the beam's speed. During this time, the spherical electromagnetic wave will cover a slightly larger distance $L' = tc = L(c/v) > L$. Hence we get

$$d = \sqrt{(L')^2 - L^2} \approx L/\gamma. \tag{7.26}$$

By analyzing our drawings, it is not difficult to understand that the field dynamics of the beam *leaving* the pipe can be obtained by "subtracting" the field in Figure 7.7 (b) from that in Figure 7.7 (a). This subtraction is done in Figure 7.7 (c). The accelerator, from which the beam emerges, is shown by the elongated rectangle in the left part of this figure.

In the first frame in Figure 7.7 (c) (before the beam has left the pipe), the electric field is zero everywhere. The disk-shaped Coulomb field begins to grow, immediately after the beam leaves the pipe (the second and third frames) [174, 173, 29]. The size of the disk is given by the same formula (7.26) as in the reverse process (see also Figure 7.8). After leaving the accelerator pipe, the total field disk grows gradually and reaches the steady-state shape (7.23)–(7.25) only in the long time limit.

The sum of the fields shown in Figures 7.7 (b) and 7.7 (c) is approximately equal to the field of a free uniformly moving charge as in Figure 7.7 (a).

7.4.4 Charge leaving accelerator (RQD)

Now we would like to consider the situations described above from the point of view of RQD, where bound force fields propagate instantaneously. Of course, both the incom-

28 Transition radiation is emitted when a charged particle crosses a boundary between two media with different electrical properties. In our case, one medium is the empty space, the other medium is the inside of the metal pipe.

ing charge in Figure 7.7 (e) and the departing charge in Figure 7.7 (f) create spherical fluxes of transition radiation photons depicted by dashed circles. In this prediction, RQD differs from Maxwell's theory only in that RQD does not recognize the description of the photon flux in terms of electric and magnetic fields.

The force field of a freely moving charge in Figure 7.7 (d) is qualitatively similar to the traditional prediction in Figure 7.7 (a). However, RQD results for the fields of charges entering and leaving the metal pipe differ significantly from the Maxwell–Liénard–Wiechert theory. When entering the pipe, the disk of the Coulomb field disappears not gradually, but instantaneously in the entire space, as shown in Figure 7.7 (e). Similarly, the Coulomb field of a charge leaving the pipe (Figure 7.7 (f)) emerges instantly, fully formed throughout the entire space. As before, the consistency of these qualitative drawings is confirmed by checking the sum 7.7 (e) + 7.7 (f) = 7.7 (d).

The differences between the predictions of the two theories are so obvious that they can be observed in a relatively simple experimental setup, which we will discuss in the next subsection.

7.4.5 Frascati experiment

Unfortunately, the experiments that we discussed in Section 7.3 could not confirm or deny the superluminal propagation of bound force fields with 100 % certainty. The problem was that in those experiments, instantaneous bound fields were almost always present in a mixture with retarded radiation fields. A clear separation of these two effects proved to be very difficult. In this subsection we will discuss a unique experimental situation [51, 53, 52], in which the bound fields of moving charges can be seen in their pure form and the superluminal speed of such fields can be established beyond doubt.

The idea of the Frascati experiment[29] [51] was to measure the disk-shaped electric field of relativistic electron beams exiting the accelerator pipe. The setup is shown schematically in Figure 7.8. A beam of electrons with energies of 500 MeV[30] left the accelerator pipe (P) and flew through the experimental hall, where electric field sensors were positioned along the beam axis. The sensors covered a rather large volume: the longitudinal shifts of the sensors from the exit flange of the pipe were $L = 92 \div 552.5$ cm and their transverse distances from the beam axis varied within $d = 3 \div 55$ cm. Time-dependent signals from the sensors were recorded with resolution not worse than 50 ps. So, it was possible to measure accurately the distribution of the "electric field" around the beam, both in space and in time. Did this field look more like the prediction of Maxwell's theory (Figure 7.7 (c)) or like the prediction of RQD (Figure 7.7 (f))?

---

29 The author is grateful to Professor G. Pizzella for information on the progress of the experiments and for illuminating discussions.

30 Which corresponds to the parameter $\gamma \approx 10^3$.

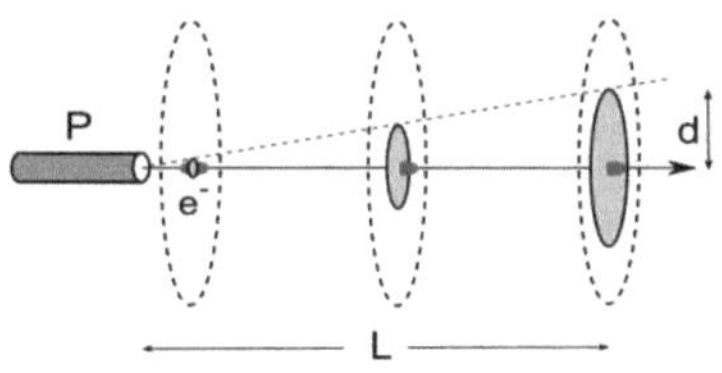

Figure 7.8: Schematics of the Frascati experiment. Dark solid disks show the evolution of the electric field according to the Maxwell–Liénard–Wiechert theory (7.26). Dashed ovals are predictions of RQD.

As follows from Figures 7.7 (c) and (f), we can expect two types of contributions to sensor signals. First is the disk-shaped bound field of the moving charge. Second is the radiation field of the transition radiation. Pizzella and coworkers conducted a series of calibration measurements, which confirmed that the intensity of the transition radiation is much less than the signal from the Coulomb field disk. So in our arguments we will neglect the former effect: $e^{\text{inst}}_{\text{bound}} \gg e^{\text{ret}}_{\text{radiation}}$.

The Liénard–Wiechert theory states that within the experimental hall the disk of the electric field should be only at an early stage of its formation (as in Figure 7.7 (c)). Even at a distance of $L = 552.5\,\text{cm}$ from the accelerator pipe, the field disk is fully formed only at transverse distances $d < L/\gamma = 0.6\,\text{cm}$ from the beam's axis. Therefore, according to the classical theory, the sensors should not have registered any signal at all.

On the other hand, RQD predicts that the force field disk (7.19)–(7.21) is completely formed immediately after the beam left the accelerator pipe. After that, the field does not change its shape, but only moves along with the beam in space. Therefore, the amplitudes of signals recorded by the sensors at different distances from the pipe ($L = 92, 172, 329.5$ and $552.5\,\text{cm}$) should not depend on L. This is exactly what was observed in the experiment.

Thus, the Frascati experiment established the instantaneous formation of the Coulomb field of the beam immediately after its emergence from the pipe. This conclusion is in direct contradiction with the prohibition of superluminal signals in special relativity (see Assertion A.3 in Appendix A.4). In Section 8.4 we will explain this paradox and demand the abolition of the special-relativistic ban.

8 Particles and relativity

How often have I said to you that when you have eliminated the impossible, whatever remains, however improbable, *must be the truth?*
Sherlock Holmes

In previous chapters, we formulated a "dressed" version of quantum electrodynamics, which we called *relativistic quantum dynamics* or RQD. An important property of RQD is that this theory reproduces accurately the S-matrix of the standard renormalized QED. Therefore, RQD is able to describe numerous experiments in elementary particle physics (such as scattering cross sections, energies of bound states and their lifetimes) no worse than QED. However, RQD is fundamentally different from QED. The main actors in RQD are particles (not fields). These particles interact with each other by instantaneous potentials. Such statements are usually regarded as controversial and physically impossible [243, 106, 264, 270, 111]. In this chapter we will try to sort out these conflicting claims.

For example, the orthodoxy advocates argue that the existence of localized states of particles is incompatible with the principles of relativity and causality. We will analyze these objections in Section 8.1 and show that there is no cause for concern: the Newton–Wigner position operator and strictly localized states do not contradict any of the fundamental principles. In particular, we will consider in detail the famous causality "paradox" associated with the superluminal spreading of localized wave packets.

In Sections 8.2 and 8.3 we will define the concept of a localized physical *event* and derive transformations of space–time coordinates of such events between different inertial frames of reference. We note that boost transformations must be dynamical, i. e., depending on interactions. This implies that the true relationships between space–time coordinates of events in moving frames should differ from familiar Lorentz formulas. This casts doubt on the foundations of special relativity.

In Section 8.5, we will conclude that Minkowski's idea of combining the three-dimensional space with one-dimensional time in one 4D space–time continuum cannot be rigorous. In the presence of interactions, physical laws do not have to obey the requirements of manifest covariance. We will also try to dispel the common misconception about the incompatibility between instantaneous action-at-a-distance and causality. In Section 8.4 we will see that instantaneous force fields in RQD do not violate the principles of relativity and causality.

Section 8.6 is devoted to somewhat more speculative reasoning on the role of quantum fields and their interpretations. In the same section, we will reflect on possible directions for further development of RQD.

https://doi.org/10.1515/9783110493221-008

8.1 Localizability of particles

In Section **1**-4.3, we saw that in relativistic quantum mechanics, the position observable is represented by the Newton–Wigner operator. However, this idea is often considered controversial. At least three arguments are usually given, which supposedly "explain" why there can be no position operator and localized states of particles:

(1) Ideal localization of particles is impossible, because it requires infinite energy (due to the Heisenberg uncertainty relation) and inevitably leads to the creation of new particles (due to Einstein's formula $E = mc^2$) [14]:

> The argument is always that, to localize the electric charge on a particle with an accuracy better than the Compton wavelength of the electron, so much energy should be put in, that electron–positron pairs would be formed. This would make the concept of position meaningless – Th. W. Ruijgrok [203].

(2) Newton–Wigner localization is unacceptable, because it is relative, i. e., different moving observers disagree on whether the particle is localized or not.

(3) Strictly localized wave packets spread out with superluminal velocities, which contradicts the principle of causality [108]:

> The 'elementary particles' of particle physics are generally understood as pointlike objects, which would seem to imply the existence of position operators for such particles. However, if we add the requirement that such operators are covariant (so that, for instance, a particle localized at the origin in one Lorentz frame remains so localized in another), or the requirement that the wave-functions of the particles do not spread out faster than light, then it can be shown that no such position operator exists. (See Halvorson and Clifton (2001) [106] and references therein, for details.) – D. Wallace [264].

In this section, we are going to show that localized states of relativistic particles have a well-defined and noncontroversial meaning, in spite of the arguments given above.

Quite often one encounters the opinion that in relativistic quantum physics there is no point to discuss such things as observables (positions and momenta) of particles, their wave functions and also their time evolutions in the interacting regime. For example, in the Introduction to Volume IV of the influential Landau and Lifshitz "Course of Theoretical Physics" [14], we read

> The foregoing discussion suggests that the theory will not consider the time dependence of particle interaction processes. It will show that in these processes there are no characteristics precisely definable (even within the usual limitations of quantum mechanics); the description of such a process as occurring in the course of time is therefore just as unreal as the classical paths are in non-relativistic quantum mechanics. The only observable quantities are the properties (momenta, polarizations) of free particles: the initial particles which come into interaction, and the final particles which result from the process (L. D. Landau, R. E. Peierls, 1930).

Contrary to this quote, we believe that relativistic QFT should not differ fundamentally from quantum mechanics. All laws of quantum mechanics – including those relating to measurements of particle positions and momenta – should remain valid in QFT, and even more so in its dressed version, RQD. The interacting time evolution of states and observables should be also reachable both by experiment and by theory.

8.1.1 Measurements of position

First we consider the idea that precise measurements of position can change the number of particles in the system.

There is no doubt that due to the Heisenberg uncertainty relation (**1**-6.95), strictly localized one-particle states do not have well-defined momentum and energy. The energy uncertainty ΔE can indeed exceed the threshold for the formation of particle–antiparticle pairs. However, the large value of ΔE does not automatically mean the uncertainty of the number of particles, and strict localization does not automatically mean pair production. The number of particles in the localized state would indeed be indeterminate if the particle number operator did not commute with position operators of individual particles. However, the latter is not true. It is not difficult to show (see Subsection **2**-1.1.2) that Newton–Wigner positions of particles commute with particle number operators in the Fock space. This conclusion is valid for both interacting and noninteracting systems, because in our approach the structure of the Fock space and the definitions of single-particle observables do not depend on the interaction.

8.1.2 Localized states in moving frame

Does the noninvariance of Newton–Wigner localization undermine the idea of point particles?

Let us consider the following example of a position–space wave function of a massive spinless particle localized at the origin with zero expected momentum (**1**-6.92):

$$\psi_{\mathbf{0},\mathbf{0}}(\boldsymbol{r}) = Ne^{-r^2/d^2}. \tag{8.1}$$

The corresponding momentum–space wave function is (**1**-6.93)

$$\psi_{\mathbf{0},\mathbf{0}}(\boldsymbol{p}) = \frac{Nd^3}{(2\hbar)^{3/2}} e^{-p^2d^2/(4\hbar^2)}. \tag{8.2}$$

Let us now find the wave function of this state from the point of view of a moving observer O'. Applying the boost transformation (**1**-5.33) to (8.2)

$$e^{-\frac{ic}{\hbar}\hat{K}_x\theta}\psi_{\mathbf{0},\mathbf{0}}(\boldsymbol{p}) = \frac{Nd^3}{(2\hbar)^{3/2}}\sqrt{\frac{\omega_{\boldsymbol{p}}\cosh\theta - cp_x\sinh\theta}{\omega_{\boldsymbol{p}}}}$$
$$\times \exp\left(-\frac{[(p_x\cosh\theta - \omega_{\boldsymbol{p}}\sinh\theta/c)^2 + p_y^2 + p_z^2]d^2}{4\hbar^2}\right)$$

and going back to the position representation by equation (**1**-5.49), we obtain

$$e^{-\frac{ic}{\hbar}\hat{K}_x\theta}\psi_{\mathbf{0},0}(\boldsymbol{r}) = \frac{Nd^3\pi^{3/2}}{(2\pi\hbar)^3}\int d\boldsymbol{p}\sqrt{\cosh\theta - (cp_x/\omega_{\boldsymbol{p}})\sinh\theta}e^{\frac{i}{\hbar}\boldsymbol{p}\cdot\boldsymbol{r}}$$
$$\times\exp\left(-\frac{[(p_x\cosh\theta - \omega_{\boldsymbol{p}}\sinh\theta/c)^2 + p_y^2 + p_z^2]d^2}{4\hbar^2}\right). \tag{8.3}$$

We are not going to calculate this integral explicitly, but one property of the function (8.3) should be clear without calculation: for nonzero θ, this function does not vanish for all (no matter how large) values of $\boldsymbol{r}$, even in the limit of perfect initial localization $d \to 0$.[1] In other words, the moving observer O' has a nonzero probability to detect the particle at any point in space. This means that the concept of localization is relative: different observers may disagree on whether a state is localized or not.

The noninvariant nature of localization is unusual from the point of view of classical physics or nonrelativistic quantum mechanics. Although this property has not been observed in experiments, it does not contradict any postulate of quantum theory and there is no reason to doubt the possibility of particle localization in each specific frame of reference.

8.1.3 Spreading of localized states

Here we are going to discuss the widely held view that the superluminal spreading of localized wave packets violates the principle of causality [106, 264, 163, 108, 23].

In the previous subsection we found out how the localized state (8.1) looks from the point of view of the moving observer. Now we would like to define the wave function of the state (8.1) from the point of view of the observer shifted in time. As before, we first go to the momentum representation (8.2), apply the time translation operator (**1**-6.96)

$$\psi_{\mathbf{0},0}(t,\boldsymbol{p}) = e^{-\frac{i}{\hbar}\hat{H}t}\psi(0,\boldsymbol{p}) = \frac{Nd^3}{(2\hbar)^{3/2}}e^{-p^2d^2/(4\hbar^2)}e^{-\frac{it}{\hbar}\sqrt{m^2c^4+p^2c^2}}$$

and then use the Fourier transform (**1**-5.49) to return to the position representation, now at $t > 0$. So we have

$$\psi_{\mathbf{0},0}(t,\boldsymbol{r}) = \frac{1}{(2\pi\hbar)^{3/2}}\int d\boldsymbol{p}\psi(t,\boldsymbol{p})e^{\frac{i}{\hbar}\boldsymbol{p}\cdot\boldsymbol{r}}$$
$$= \frac{Nd^3\pi^{3/2}}{(2\pi\hbar)^3}\int d\boldsymbol{p}e^{-p^2d^2/(4\hbar^2)}e^{-\frac{it}{\hbar}\sqrt{m^2c^4+p^2c^2}}e^{\frac{i}{\hbar}\boldsymbol{p}\cdot\boldsymbol{r}}. \tag{8.4}$$

1 This property follows from the nonanalyticity of the square root in the integrand [243]. .

Again, we are not interested in the exact value of this integral. The most important result is that even for $d \to 0$, the wave function (8.4) does not disappear outside the "light cone," i. e., at distances greater than ct from the initial location $\boldsymbol{r} = 0$.[2] Although the probability density outside the light cone is very small, there is a nonzero chance that the particle propagates faster than the speed of light.

Note that the superluminal propagation of the wave function does not at all mean that the expectation value of the particle's velocity exceeds c. As we found in Subsection **1**-5.1.3, eigenvalues of the velocity operator are strictly limited by c. So, the faster-than-light spreading of wave packets is a purely quantum effect associated with the noncommutativity of the position $\boldsymbol{R}$ and velocity $\boldsymbol{V}$ operators.

8.1.4 Superluminal spreading and causality

The superluminal spreading of localized wave packets is observed in relativistic quantum theory under very general assumptions [108]. This effect is usually regarded as a sign of serious theoretical problems [106, 264, 163, 23, 260], because the superluminal propagation of any signals is strictly prohibited in special relativity (see Appendix A.4). This contradiction is often presented as a proof of the impossibility of the corpuscular interpretation in QFT. But this interpretation is in the center of our approach. Therefore, we have to provide some explanations.

Let us first examine the reasons why the superluminal propagation of wave functions is traditionally considered unacceptable. The usual argument is that such an effect could be used to make a device that violates the causality principle, as explained in Appendix A.4. Imagine two observers, Alice and Bob, with Bob moving away from Alice at high speed $v = c \tanh \theta$. Let us now assume that Alice and Bob can send each other superluminal signals using quantum wave packets. Namely, we will assume that both Alice and Bob have signaling devices, which are small impermeable boxes containing quantum particles. Until the time $t = 0$, both boxes are tightly closed, so that wave functions of the particles are confined inside. The walls of Alice's box for $t < 0$ are shown by two bold vertical parallel lines on the space–time diagram in Figure 8.1. At the time $t = 0$ (point A in Figure 8.1), Alice opens her box and thus sends a signal $A \to B$ to Bob. The spreading wave function of the released particles for $t > 0$ is shown schematically in Figure 8.1 by wide arrows parallel to the x-axis. Due to the superluminal (moreover – instantaneous) propagation of this wave function, there is a

---

2 This conclusion is supported by the same nonanalyticity argument as in the footnote on page 150; see also [261] and Section 2.1 in [183].

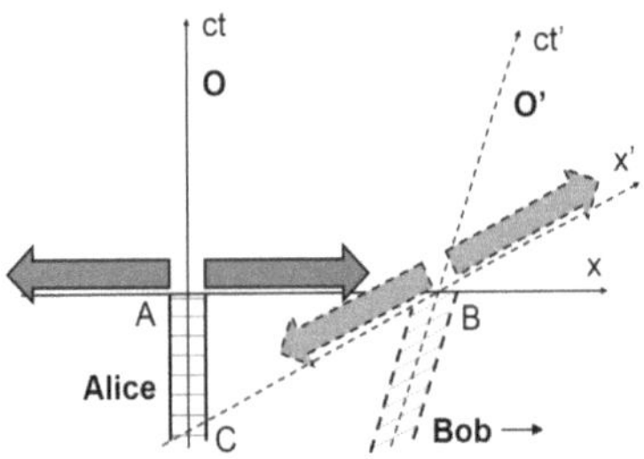

Figure 8.1: Space–time diagram illustrating the imaginary causality paradox, created by the superluminal propagation of a localized wave packet. Observers O and O' have coordinate systems with axes (ct, x) and (ct', x'), respectively. Both observers send each other superluminal signals by opening containers with localized particles. Detailed explanations are in the text.

nonvanishing probability of detecting particles in Bob's vicinity (point B in Figure 8.1) at the time $t \approx 0$.

Bob has a similar box with particles. Until now, he kept his box closed, but after receiving a signal from Alice, he immediately opens his box. It is clear that the resultant wave packet $\psi(t, \mathbf{r})$ spreads instantly in Bob's own frame of reference. The question is, what kind of spreading will Alice see?

If we literally understand the Lorentz formulas from Appendix A.1, then we could try to transform the wave function $\psi(t, \mathbf{r})$ into Alice's reference frame by simply applying the (reverse) transformation (A.2)–(A.5) to the arguments of ψ [182]. Usually, such a transformation,

$$\psi(t, x, y, z) \to \psi(t \cosh\theta + (x/c)\sinh\theta, x\cosh\theta + ct\sinh\theta, y, z) = \psi(\tilde{\theta}_x \tilde{x}), \qquad (8.5)$$

is depicted by a pseudo-rotation on the space–time diagram. The result is schematically shown in Figure 8.1 with dashed wide arrows. It predicts a nonzero probability of finding the particles released by Bob at the point C of Alice's world line at $t < 0$. This means that the response signal $B \to C$, dispatched by Bob, reaches Alice *earlier* than the original signal $A \to B$ was sent by her. This is a clear violation of the principle of causality [96].

In fact, the time evolution of the wave packet (8.5) released by Bob looks utterly absurd from the point of view of Alice. The particles do not seem to emanate from point B at all. Alice sees that the wave function (8.5) was out of Bob's box long before this box was opened. As can be seen from Figure 8.1, this wave function actually approaches Alice from the opposite side (from the side of negative values of x) and moves in Bob's direction. So Alice would not even agree that the signal was sent from Bob to her!

Where is the error? The traditional theory claims that the source of the problem is in the superluminal spreading of wave functions. Then the solution is to prohibit such spreading, and if particle wave functions refuse to respect this ban, then we should prohibit wave functions and particles themselves. Such a drastic conclusion contradicts the entire theory developed in our book. Is there another solution to this paradox?

8.1.5 Transformations of quantum fields

It is obvious that the key link in the above paradox is the moving-frame transformation (8.5) of the wave function. However, there are serious reasons to doubt that this formula is applicable, even approximately. Where could such a formula come from? Most likely it was borrowed from quantum field transformations (**2**-3.1). But can these transformations be applied to wave functions of particles? Let us discuss this issue in more detail.

Consider the simple scalar quantum field[3]

$$\begin{aligned}\psi(t,\boldsymbol{x}) &= \psi^-(t,\boldsymbol{x}) + \psi^+(t,\boldsymbol{x})\\ &= \int \frac{d\boldsymbol{p}}{\sqrt{2\omega_{\boldsymbol{p}}}(2\pi\hbar)^{3/2}} e^{\frac{i}{\hbar}(\boldsymbol{p}\cdot\boldsymbol{x}-\omega_{\boldsymbol{p}}t)}\alpha_{\boldsymbol{p}} + \int \frac{d\boldsymbol{p}}{\sqrt{2\omega_{\boldsymbol{p}}}(2\pi\hbar)^{3/2}} e^{-\frac{i}{\hbar}(\boldsymbol{p}\cdot\boldsymbol{x}-\omega_{\boldsymbol{p}}t)}\alpha^{\dagger}_{\boldsymbol{p}}.\end{aligned}$$

For simplicity, we limit ourselves to the creation part $\psi^+(t,\boldsymbol{x})$ of the field operator. Let us check that this operator is Lorentz-covariant. For example, by introducing the 4-vectors of energy–momentum $\tilde{p} = (\omega_{\boldsymbol{p}}, \boldsymbol{p}c)$ and time–position $\tilde{x} = (t, \boldsymbol{x}/c)$, we get[4]

$$\begin{aligned}e^{-\frac{ic}{\hbar}\boldsymbol{K}_0\cdot\boldsymbol{\theta}}\psi^+(\tilde{x})e^{\frac{ic}{\hbar}\boldsymbol{K}_0\cdot\boldsymbol{\theta}} &= \int \frac{d\boldsymbol{p}}{\sqrt{2\omega_{\boldsymbol{p}}}(2\pi\hbar)^{3/2}} e^{-\frac{i}{\hbar}\tilde{p}\cdot\tilde{x}} e^{-\frac{ic}{\hbar}\boldsymbol{K}_0\cdot\boldsymbol{\theta}}\alpha^{\dagger}_{\boldsymbol{p}} e^{\frac{ic}{\hbar}\boldsymbol{K}_0\cdot\boldsymbol{\theta}}\\ &= \int \frac{d\boldsymbol{p}}{\sqrt{2\omega_{\boldsymbol{p}}}(2\pi\hbar)^{3/2}} e^{-\frac{i}{\hbar}\tilde{p}\cdot\tilde{x}} \sqrt{\frac{\omega_{\boldsymbol{\theta p}}}{\omega_{\boldsymbol{p}}}}\alpha^{\dagger}_{\boldsymbol{\theta p}}\\ &= \int \frac{d(\boldsymbol{\theta p})}{\omega_{\boldsymbol{\theta p}}\sqrt{2}(2\pi\hbar)^{3/2}} e^{-\frac{i}{\hbar}\tilde{p}\cdot\tilde{x}} \sqrt{\omega_{\boldsymbol{\theta p}}}\alpha^{\dagger}_{\boldsymbol{\theta p}}\\ &= \int \frac{d\boldsymbol{q}}{\sqrt{2\omega_{\boldsymbol{q}}}(2\pi\hbar)^{3/2}} e^{-\frac{i}{\hbar}\tilde{\boldsymbol{\theta}}^{-1}\tilde{q}\cdot\tilde{x}}\alpha^{\dagger}_{\boldsymbol{q}}\\ &= \int \frac{d\boldsymbol{q}}{\sqrt{2\omega_{\boldsymbol{q}}}(2\pi\hbar)^{3/2}} e^{-\frac{i}{\hbar}\tilde{q}\cdot\tilde{\boldsymbol{\theta}}\tilde{x}}\alpha^{\dagger}_{\boldsymbol{q}}\\ &= \psi^+(\tilde{\boldsymbol{\theta}}\tilde{x}),\end{aligned} \tag{8.6}$$

i. e., a formula similar to (8.5). However, it would be incorrect to apply this transformation to the particle wave function. The fact is that the component $\psi^+(0,\boldsymbol{x})$ cannot be interpreted as an operator creating a particle in location $\boldsymbol{x}$ at time 0.

Correct expressions for operators creating and annihilating a particle at the Newton–Wigner position $\boldsymbol{x}$ are obtained from particle operators in the momentum

---

3 See formulas (5.2.3)–(5.2.4) in [266].

4 Here we use formulas (**2**-1.41), (**1**-J.6) and (**1**-5.31). For a similar derivation for the Dirac field, see Appendix **2**-B.8.

representation using the Fourier transform from Subsection **1**-5.3.2, i. e.,[5]

$$\alpha_{\boldsymbol{x}} = \int \frac{d\boldsymbol{p}}{(2\pi\hbar)^{3/2}} e^{\frac{i}{\hbar}(\boldsymbol{p}\cdot\boldsymbol{x})} \alpha_{\boldsymbol{p}}, \tag{8.7}$$

$$\alpha_{\boldsymbol{x}}^{\dagger} = \int \frac{d\boldsymbol{p}}{(2\pi\hbar)^{3/2}} e^{-\frac{i}{\hbar}(\boldsymbol{p}\cdot\boldsymbol{x})} \alpha_{\boldsymbol{p}}^{\dagger}. \tag{8.8}$$

Then the operator

$$\alpha_{\boldsymbol{x}}^{\dagger}(t) \equiv e^{\frac{i}{\hbar}H_0 t} \alpha_{\boldsymbol{x}}^{\dagger} e^{-\frac{i}{\hbar}H_0 t} = \int \frac{d\boldsymbol{p}}{(2\pi\hbar)^{3/2}} e^{-\frac{i}{\hbar}(\boldsymbol{p}\cdot\boldsymbol{x}-\omega_{\boldsymbol{p}} t)} \alpha_{\boldsymbol{p}}^{\dagger}$$

describes the creation of a particle with a spread-out wave packet at time t. Note that this expression differs from $\psi^{+}(t,\boldsymbol{x})$ only by the absence of the factor $\sqrt{2\omega_{\boldsymbol{p}}}$ in the denominator. This difference is the reason why the true position–space particle creation operator (8.8) (as well as the particle's wave function) is *not* transformed by the Lorentz formulas (8.5)–(8.6).

From this we conclude that:

(1) Localized wave packets do spread out faster than c.
(2) One can use these wave packets to send superluminal signals.
(3) However, one cannot prove the violation of causality (like in Subsection 8.1.4), because the covariant transformation law (8.5) does not apply to wave functions.

8.2 Inertial transformations without interaction

One important goal of theoretical physics is the derivation of transformations of observables between different inertial frames of reference. In Chapter **1**-4 and in Subsection **1**-6.2.3 we discussed inertial transformations of *total* observables in a multiparticle system and found that these transformations have a universal form that does not depend on the composition of the system or on the forces acting in it. Here we will be interested in inertial transformations of observables of *individual particles* in a multiparticle system. In this section we will consider noninteracting particles for which usual Lorentz transformations are applicable in the classical limit. In Section 8.3, we will turn to interacting particles and arrive at the controversial conclusion that formulas of special relativity are only approximate in the presence of interaction.

8.2.1 Events and observables

One of the most fundamental ideas in physics is the concept of *event*. Generally speaking, an event is a physical process or phenomenon occurring in a small volume of

5 The physical meaning of the product $\alpha_{\boldsymbol{x}}^{\dagger}\alpha_{\boldsymbol{x}}$ is the spatial density of the particles [261].

space in a short period of time. Thus, each event can be characterized by four numbers: its time t and three Cartesian coordinates $\boldsymbol{r} = (x, y, z)$. These numbers will be called *space–time coordinates* $(t, \boldsymbol{r})$ of the event or simply *4-coordinates*. It is important to emphasize that the four numbers $(t, \boldsymbol{r})$ are not just labels of points in an empty space–time. In order for the event to have an objective meaning which could be agreed upon by different observers at time t there should be some material particles present at the position $\boldsymbol{r}$. The simplest example of such an event is the intersection of classical trajectories (= collision) of two particles. The time t of this event is the reading of the laboratory clock at the collision moment, and $\boldsymbol{r}$ is the (coinciding) position of the colliding particles.

Strictly speaking, in quantum mechanics, the concept of an event is not always well-defined, because the localization of particles is not absolute. If the observer at rest sees a localized event (or localized particles that make up this event), then moving observers may disagree that the particles are localized (see Subsection 8.1.2) or that the event has happened at all. To avoid these complications, we will work in the classical limit ($\hbar \to 0$), where states are described by quasiclassical wave packets with unambiguous trajectories and negligibly small spreading.

Of course, different observers will assign different quadruples of numbers $(t, \boldsymbol{r})$ to the same event. In particular, in this section we would like to derive the relations connecting 4-coordinates of the event $(t, \boldsymbol{r})$, measured in the frame at rest O, and coordinates $(t', \boldsymbol{r}')$ of the same event in the moving frame O'. According to our definitions, finding the correspondence $\boldsymbol{r} \to \boldsymbol{r}'$ is just a simple exercise in applying boosts to the Newton–Wigner position operator (see Subsection **1**-4.3.10). Following this idea, we can derive analogs of the Lorentz transformations (A.2)–(A.5) for our events without artificial special-relativistic assumptions. Consequently, we will be able to test the foundations of the special theory of relativity. This is the plan of the present section.

8.2.2 Two noninteracting particles

Consider a system of two massive spinless particles in the Hilbert space $\mathcal{H} = \mathcal{H}_1 \otimes \mathcal{H}_2$, where single-particle observables (position, momentum, velocity, angular momentum, spin, energy, etc.) will be denoted by lower case letters. We have

$$\boldsymbol{r}_1, \boldsymbol{p}_1, \boldsymbol{v}_1, \boldsymbol{j}_1, \boldsymbol{s}_1, h_1, \ldots, \tag{8.9}$$

$$\boldsymbol{r}_2, \boldsymbol{p}_2, \boldsymbol{v}_2, \boldsymbol{j}_2, \boldsymbol{s}_2, h_2, \ldots. \tag{8.10}$$

Transformations of these observables between two frames of reference O and O' are found by general rules formulated in Subsection **1**-3.2.6. Namely, suppose that the observers O and O' are connected by an inertial transformation, which is generated in $\mathcal{H}$ by the Hermitian operator F. If $\mathfrak{g}$ is a single-particle observable (an Hermitian operator from the lists (8.9)–(8.10)) in the reference frame O, then the same observable in the reference frame O' is obtained by equations like (**1**-3.59)–(**1**-3.61), i. e.,

$$\mathfrak{g}(b) = e^{-\frac{i}{\hbar}Fb}\mathfrak{g}e^{\frac{i}{\hbar}Fb} = \mathfrak{g} - \frac{ib}{\hbar}[F,\mathfrak{g}] - \frac{b^2}{2!\hbar^2}[F,[F,\mathfrak{g}]] + O(b^2/\hbar^2), \tag{8.11}$$

where b is the transformation parameter. In the classical approximation, we will replace quantum commutators in (8.11) by Poisson brackets (**1**-6.101) and obtain

$$\mathfrak{g}(b) \approx \mathfrak{g} + b[F,\mathfrak{g}]_P + \frac{b^2}{2!}[F,[F,\mathfrak{g}]_P]_P + O(b^2). \tag{8.12}$$

To perform calculations using this formula, we need to have two basic things. First, we need to know the Poisson brackets between all single-particle observables (8.9)–(8.10). This problem was solved in Chapters **1**-4 and **1**-5 and in Section **1**-6.1. The brackets of same-particle observables are

$$[r_i,r_j]_P = [p_i,p_j]_P = [r_i,s_j]_P = [p_i,s_j]_P = 0, \tag{8.13}$$
$$[r_i,p_j]_P = \delta_{ij}, \tag{8.14}$$
$$[s_i,s_j]_P = \sum_{k=1}^{3}\epsilon_{ijk}s_k, \tag{8.15}$$
$$[\boldsymbol{p},h]_P = [\boldsymbol{s},h]_P = 0, \tag{8.16}$$
$$[\boldsymbol{r},h]_P = \frac{\boldsymbol{p}c^2}{h}, \tag{8.17}$$

where $i,j,k = 1,2,3$. Observables of different particles have vanishing Poisson brackets.

Second, we should know expressions for the Poincaré generators F in terms of single-particle quantities (8.9)–(8.10). This is equivalent to knowing the complete dynamical description of our two-particle system. In this section we are interested in a noninteracting system, whose generators are simply sums of single-particle generators, i. e.,

$$H_0 = h_1 + h_2, \tag{8.18}$$
$$\boldsymbol{P}_0 = \boldsymbol{p}_1 + \boldsymbol{p}_2, \tag{8.19}$$
$$\boldsymbol{J}_0 = \boldsymbol{j}_1 + \boldsymbol{j}_2, \tag{8.20}$$
$$\boldsymbol{K}_0 = \boldsymbol{k}_1 + \boldsymbol{k}_2. \tag{8.21}$$

8.2.3 Boosts of trajectories

Trajectory of particle 1 in the frame O can be obtained from formula (**1**-4.36), i. e.,

$$\boldsymbol{r}_1(t) = e^{\frac{i}{\hbar}H_0 t}\boldsymbol{r}_1 e^{-\frac{i}{\hbar}H_0 t} = e^{\frac{i}{\hbar}(h_1+h_2)t}\boldsymbol{r}_1 e^{-\frac{i}{\hbar}(h_1+h_2)t} = e^{\frac{i}{\hbar}h_1 t}\boldsymbol{r}_1 e^{-\frac{i}{\hbar}h_1 t}$$

$$\approx \boldsymbol{r}_1 - t[h_1, \boldsymbol{r}_1]_P + \frac{t^2}{2!}[h_1, [h_1, \boldsymbol{r}_1]_P]_P + O(t^3) = \boldsymbol{r}_1 + \boldsymbol{v}_1 t. \tag{8.22}$$

Applying a boost transformation to (8.22) and taking into account (**1**-4.7)–(**1**-4.9) and (**1**-4.55)–(**1**-4.58), we obtain the particle trajectory in the frame O' moving with the speed $v = c \tanh\theta$ along the x-axis. We have[6]

$$r_{1x}(\theta, t') = \beta\left(\frac{r_{1x}}{\cosh\theta} + (v_{1x} - v)t'\right), \tag{8.23}$$

$$r_{1y}(\theta, t') = \beta\left(r_{1y} + \frac{j_{1z}v}{h_1} + \frac{v_{1y}t'}{\cosh\theta}\right) = r_{1y} + \beta\left(\frac{r_{1x}v_{1y}v}{c^2} + \frac{v_{1y}t'}{\cosh\theta}\right), \tag{8.24}$$

$$r_{1z}(\theta, t') = \beta\left(r_{1z} - \frac{j_{1y}v}{h_1} + \frac{v_{1z}t'}{\cosh\theta}\right) = r_{1z} + \beta\left(\frac{r_{1x}v_{1z}v}{c^2} + \frac{v_{1z}t'}{\cosh\theta}\right), \tag{8.25}$$

where we denote $\beta \equiv (1 - v_{1x}vc^{-2})^{-1}$. Similar relations are valid for particle 2.

An important property of these results is that inertial transformations of observables of the two particles are completely independent of each other. This is not at all surprising, since we assumed that the two particles do not interact.

8.2.4 Lorentz transformations

Now let us consider a localized event associated with the intersection of two particle trajectories. Suppose that from the point of view of O this event had space–time coordinates (ct, x, y, z). This means that

$$x \equiv r_{1x}(t) = r_{2x}(t),$$
$$y \equiv r_{1y}(t) = r_{2y}(t),$$
$$z \equiv r_{1z}(t) = r_{2z}(t).$$

Obviously, this intersection should be seen also by the moving observer O'. What are the space–time coordinates from the point of view of O'? The answer is given by the following theorem.

Theorem 8.1 (Lorentz transformation). *Space–time coordinates of events defined as intersection of trajectories of noninteracting particles transform by Lorentz formulas* (A.2)–(A.5).

Proof. Let us first verify that Lorentz formulas (A.2)–(A.5) are suitable for transforming the trajectory of particle 1 between the two observers. For simplicity, we consider the

6 If we set $t' = 0$, then these formulas coincide with (23)–(24) in [170]. Setting also $\boldsymbol{v}_1 = 0$, we obtain the usual Lorentz formula of *length contraction*: $r_{1x}(\theta, 0) = r_{1x}/(\cosh\theta)$; $r_{1y}(\theta, 0) = r_{1y}$; $r_{1z}(\theta, 0) = r_{1z}$. Compare with (A.6).

case when this particle moves along the x-axis,

$$r_{1y}(t) = r_{1z}(t) = v_{1y} = v_{1z} = 0, \tag{8.26}$$

and we omit y- and z-components in our proof.

First we calculate the transformed position $r_{1x}(\theta, t')$ by the Lorentz–Einstein (LE) formula (A.3), taking into account that in the rest frame $r_{1x}(0,t) = r_{1x} + v_{1x}t$. We have

$$r_{1x}^{\mathrm{LE}}(\theta, t') = r_{1x}(0,t)\cosh\theta - ct\sinh\theta = (r_{1x} + v_{1x}t)\cosh\theta - ct\sinh\theta. \tag{8.27}$$

We want to compare this expression with the Wigner–Dirac (WD) formula (8.23), where the time parameter is taken from LE (A.2). We have

$$t' = t\cosh\theta - \frac{r_{1x}(t)}{c}\sinh\theta,$$

i. e.,

$$r_{1x}^{\mathrm{WD}}(\theta, t') = \beta\left(\frac{r_{1x}}{\cosh\theta} + (v_{1x} - v)\left(t\cosh\theta - \frac{r_{1x}(t)}{c}\sinh\theta\right)\right). \tag{8.28}$$

To show the identity of equations (8.27) and (8.28), we will verify that their difference vanishes. Indeed,

$$\begin{aligned}
&r_{1x}^{\mathrm{WD}}(\theta, t') - r_{1x}^{\mathrm{LE}}(\theta, t') \\
&\quad = \frac{\beta r_{1x}}{\cosh\theta} + (v_{1x} - v)\beta(t\cosh\theta - c^{-1}(r_{1x} + v_{1x}t)\sinh\theta) - (r_{1x} + v_{1x}t)\cosh\theta + ct\sinh\theta \\
&\quad = \frac{\beta}{\cosh\theta}[r_{1x} + v_{1x}t\cosh^2\theta - vt\cosh^2\theta - (v_{1x}r_{1x}/c)\sinh\theta\cosh\theta \\
&\quad\quad + (vr_{1x}/c)\sinh\theta\cosh\theta - (v_{1x}^2/c)t\sinh\theta\cosh\theta + (vv_{1x}/c)t\sinh\theta\cosh\theta \\
&\quad\quad - r_{1x}\cosh^2\theta + r_{1x}(v_{1x}v/c^2)\cosh^2\theta - v_{1x}t\cosh^2\theta + (v_{1x}^2v/c^2)t\cosh^2\theta \\
&\quad\quad + ct\sinh\theta\cosh\theta - (v_{1x}v/c)t\sinh\theta\cosh\theta] \\
&\quad = \frac{\beta}{\cosh\theta}[r_{1x} - vt\cosh^2\theta - (v_{1x}r_{1x}/c)\sinh\theta\cosh\theta + (vr_{1x}/c)\sinh\theta\cosh\theta \\
&\quad\quad - (v_{1x}^2/c)t\sinh\theta\cosh\theta - r_{1x}\cosh^2\theta + r_{1x}(v_{1x}v/c^2)\cosh^2\theta + (v_{1x}^2v/c^2)t\cosh^2\theta \\
&\quad\quad + ct\sinh\theta\cosh\theta] \\
&\quad = \frac{\beta}{\cosh\theta}[r_{1x} - ct\sinh\theta\cosh\theta - (v_{1x}r_{1x}/c)\sinh\theta\cosh\theta + r_{1x}\sinh^2\theta \\
&\quad\quad - (v_{1x}^2/c)t\sinh\theta\cosh\theta - r_{1x}\cosh^2\theta + (r_{1x}v_{1x}/c)\sinh\theta\cosh\theta \\
&\quad\quad + (v_{1x}^2/c)t\sinh\theta\cosh\theta + ct\sinh\theta\cosh\theta] \\
&\quad = 0.
\end{aligned}$$

This means that the boost-transformed trajectory (8.23) of particle 1 agrees with Lorentz formulas (A.2) and (A.3). The same conclusion is valid for particle 2. This implies that times and positions of intersections of the two trajectories also transform by Lorentz (A.2)–(A.5).

The same result can be obtained in the general case, i. e., when restrictions (8.26) are lifted. □

Thus, we have proved that 4-coordinates of classical events with noninteracting particles are transformed by Lorentz formulas.

8.3 Inertial transformations with interaction

This time we will assume that the two particles 1 and 2 interact with each other. In the quantum-mechanical case, we say that the unitary representation U_g of the Poincaré group in $\mathscr{H}$ differs from the noninteracting representation U_g^0. This means that generators of inertial transformations are no longer expressed by simple sums of single-particle generators, as in (8.18)–(8.21). Generators can be written in the form (**1**-6.14)–(**1**-6.17), so we have

$$H = h_1 + h_2 + V(\boldsymbol{r}_1, \boldsymbol{p}_1, \boldsymbol{r}_2, \boldsymbol{p}_2), \tag{8.29}$$

$$\boldsymbol{P} = \boldsymbol{p}_1 + \boldsymbol{p}_2 + \boldsymbol{U}(\boldsymbol{r}_1, \boldsymbol{p}_1, \boldsymbol{r}_2, \boldsymbol{p}_2), \tag{8.30}$$

$$\boldsymbol{J} = \boldsymbol{j}_1 + \boldsymbol{j}_2 + \boldsymbol{Y}(\boldsymbol{r}_1, \boldsymbol{p}_1, \boldsymbol{r}_2, \boldsymbol{p}_2), \tag{8.31}$$

$$\boldsymbol{K} = \boldsymbol{k}_1 + \boldsymbol{k}_2 + \boldsymbol{Z}(\boldsymbol{r}_1, \boldsymbol{p}_1, \boldsymbol{r}_2, \boldsymbol{p}_2), \tag{8.32}$$

where V, $\boldsymbol{U}$, $\boldsymbol{Y}$ and $\boldsymbol{Z}$ are interaction operators which are functions of one-particle observables. One goal of this section is to obtain more information about the operators V, $\boldsymbol{U}$, $\boldsymbol{Y}$ and $\boldsymbol{Z}$. In particular, we will try to decide which form of relativistic dynamics is prevalent in nature.

8.3.1 Time translations

The most obvious manifestation of interaction is the difference between temporal evolutions of the interacting and free systems. We estimate the strength of interaction between particles by how much their trajectories deviate from straight lines (8.22). Therefore, in any relativistic form of dynamics, the Hamiltonian – the generator of time translations – must contain a nontrivial potential energy term V. Then the time evolution of the position of particle 1 is

$$\begin{aligned}
\boldsymbol{r}_1(t) &= e^{\frac{i}{\hbar}Ht}\boldsymbol{r}_1 e^{-\frac{i}{\hbar}Ht} = e^{\frac{i}{\hbar}(h_1+h_2+V)t}\boldsymbol{r}_1 e^{-\frac{i}{\hbar}(h_1+h_2+V)t} \\
&= \boldsymbol{r}_1 - t[(h_1 + V), \boldsymbol{r}_1]_P + \frac{t^2}{2}[(h_1 + h_2 + V), [(h_1 + V), \boldsymbol{r}_1]_P]_P + O(t^3) \\
&= \boldsymbol{r}_1 + \boldsymbol{v}_1 t - t[V, \boldsymbol{r}_1]_P - \frac{t^2}{2}[V, \boldsymbol{v}_1]_P + \frac{t^2}{2}[(h_1 + h_2), [V, \boldsymbol{r}_1]_P]_P
\end{aligned}$$

$$+\frac{t^2}{2}[V,[V,\boldsymbol{r}_1]_P]_P + O(t^3). \tag{8.33}$$

In the simplest case, when interaction V commutes with particle positions and in the nonrelativistic approximation $\boldsymbol{v}_1 \approx \boldsymbol{p}_1/m_1$, this formula simplifies to

$$\boldsymbol{r}_1(t) \approx \boldsymbol{r}_1 + \boldsymbol{v}_1 t - \frac{t^2}{2m_1}\frac{\partial V}{\partial \boldsymbol{r}_1} = \boldsymbol{r}_1 + \boldsymbol{v}_1 t + \frac{\boldsymbol{f}_1 t^2}{2m_1} = \boldsymbol{r}_1 + \boldsymbol{v}_1 t + \frac{\boldsymbol{a}_1 t^2}{2},$$

where we denote

$$\boldsymbol{f}_1(\boldsymbol{r}_1,\boldsymbol{p}_1,\boldsymbol{r}_2,\boldsymbol{p}_2) \equiv -\frac{\partial V(\boldsymbol{r}_1,\boldsymbol{p}_1,\boldsymbol{r}_2,\boldsymbol{p}_2)}{\partial \boldsymbol{r}_1}$$

the *force* with which particle 2 acts on particle 1. The vector $\boldsymbol{a}_1 \equiv \boldsymbol{f}_1/m_1$ has the meaning of *acceleration* of particle 1, in agreement with the second law of Newtonian mechanics. Similar formulas hold for particle 2. The trajectories $\boldsymbol{r}_1(t)$ and $\boldsymbol{r}_2(t)$ of the two particles are curved and depend nontrivially on each other. Such curved trajectories are not difficult to observe in macroscopic experiments with long-range electromagnetic forces. Unfortunately, these interacting trajectories, in and of itself, cannot tell us what form of relativistic dynamics the interaction belongs to. To answer this question, it is necessary to investigate other types of inertial transformations.

As an important example, we will explain what experimental observations can distinguish two popular forms of dynamics: the instant form (IF)

$$H = h_1 + h_2 + V, \tag{8.34}$$
$$\boldsymbol{P} = \boldsymbol{p}_1 + \boldsymbol{p}_2, \tag{8.35}$$
$$\boldsymbol{J} = \boldsymbol{j}_1 + \boldsymbol{j}_2, \tag{8.36}$$
$$\boldsymbol{K} = \boldsymbol{k}_1 + \boldsymbol{k}_2 + \boldsymbol{Z} \tag{8.37}$$

and the point form (PF)

$$H = h_1 + h_2 + V, \tag{8.38}$$
$$\boldsymbol{P} = \boldsymbol{p}_1 + \boldsymbol{p}_2 + \boldsymbol{U}, \tag{8.39}$$
$$\boldsymbol{J} = \boldsymbol{j}_1 + \boldsymbol{j}_2, \tag{8.40}$$
$$\boldsymbol{K} = \boldsymbol{k}_1 + \boldsymbol{k}_2. \tag{8.41}$$

8.3.2 Boosts

Similar to the above analysis of time translations, we can study boost transformations. For point form interactions (8.38)–(8.41), the potential boost $\boldsymbol{Z}$ is zero, so that the boosts of positions and velocities are the same as in the noninteracting case. We have[7]

$$r_{1x}^{PF}(\theta) = e^{-\frac{ic}{\hbar}K_{0x}\theta} r_{1x} e^{\frac{ic}{\hbar}K_{0x}\theta} = e^{-\frac{ic}{\hbar}k_{1x}\theta} r_{1x} e^{\frac{ic}{\hbar}k_{1x}\theta}$$
$$\approx r_{1x} + c\theta[k_{1x}, r_{1x}]_P + O(\theta^2) = \frac{r_{1x}}{\cosh\theta(1 - v_{1x}vc^{-2})}, \tag{8.42}$$
$$v_{1x}^{PF}(\theta) = e^{-\frac{ic}{\hbar}K_{0x}\theta} v_{1x} e^{\frac{ic}{\hbar}K_{0x}\theta} = e^{-\frac{ic}{\hbar}k_{1x}\theta} v_{1x} e^{\frac{ic}{\hbar}k_{1x}\theta}$$
$$\approx v_{1x} + c\theta[k_{1x}, v_{1x}]_P + O(\theta^2) = \frac{v_{1x} - v}{1 - v_{1x}vc^{-2}}. \tag{8.43}$$

On the other hand, in the instant form, the boosts (8.37) are dynamical, and transformations depend on the strength of interaction. For example, we get the following boosted position:

$$r_{1x}^{IF}(\theta) = e^{-\frac{ic}{\hbar}K_x\theta} r_{1x} e^{\frac{ic}{\hbar}K_x\theta}$$
$$= e^{-\frac{ic}{\hbar}(K_{0x}+Z_x)\theta} r_{1x} e^{\frac{ic}{\hbar}(K_{0x}+Z_x)\theta}$$
$$\approx r_{1x} + c\theta[k_{1x}, r_{1x}]_P + c\theta[Z_x, r_{1x}]_P + O(\theta^2)$$
$$= \frac{r_{1x}}{\cosh\theta(1 - v_{1x}vc^{-2})} + c\theta[Z_x, r_{1x}]_P + O(\theta^2). \tag{8.44}$$

The first term on the right-hand side is the same as in the interaction-free point-form formula (8.42). This term is responsible for the known relativistic length contraction effect (A.6). The second term in (8.44) is the correction due to the presence of particle 2. This correction depends on observables of both particles, and it leads to a nontrivial dependence of the boosted position on the state of the system and on interactions therein.

Similarly, we get the following effect of the boost on the instant-form velocity:

$$v_{1x}^{IF}(\theta) = e^{-\frac{ic}{\hbar}K_x\theta} v_{1x} e^{\frac{ic}{\hbar}K_x\theta} = e^{-\frac{ic}{\hbar}(K_{0x}+Z_x)\theta} v_{1x} e^{\frac{ic}{\hbar}(K_{0x}+Z_x)\theta}$$
$$= v_{1x} + c\theta[k_{1x}, v_{1x}]_P + c\theta[Z_x, v_{1x}]_P + O(\theta^2)$$
$$= \frac{v_{1x} - v}{1 - v_{1x}vc^{-2}} + c\theta[Z_x, v_{1x}]_P + O(\theta^2)$$
$$= (v_{1x} - v) + \frac{v_{1x}v(v_{1x} - v)}{c^2} + c\theta[Z_x, v_{1x}]_P + O(\theta^2). \tag{8.45}$$

All terms on the right-hand side have a clear physical meaning. The first term $v_{1x} - v$ is the usual nonrelativistic shift of velocity. This is the most obvious effect of boosts, familiar from our daily experience. The second term is a relativistic correction, applicable to both noninteracting and interacting particles. This correction is nothing

7 For simplicity, we consider only the x-components. For the general case, see (**1**-4.7)–(**1**-4.9) and (8.23)–(8.25). As usual, $v \equiv c\tanh\theta$ is the speed of the moving reference frame.

but the $(v/c)^2$-order contribution to the relativistic law of addition of velocities (**1**-4.7). Presently, there is a lot of experimental evidence about the validity of this law.[8] The third term is a correction due to the interaction between particles 1 and 2. This prediction of RQD still awaits experimental confirmation.

So, in the instant form of dynamics, there is a close analogy between time translations and boosts. Both these transformations depend on interaction, i. e., they are dynamical. However, there is a huge difference in the possibility of experimental observation of these dependencies. To see the effect of time translation, you do not have to do anything – just wait. However, in order to see the dynamical effect of boost, it is necessary to have measuring devices moving at very high speeds. Of course, this is associated with enormous technical difficulties. Therefore, boost transformations of particle positions and velocities have not yet been measured to an accuracy sufficient to detect the dynamical effects $c\theta[Z_x, r_{1x}]_P$ and $c\theta[Z_x, v_{1x}]_P$.

So, at this stage measurements of boosts cannot help us to decide the preferred form of relativistic dynamics. Let us now turn to rotations and space translations.

8.3.3 Rotations

In both instant and point forms of dynamics, rotations do not depend on interaction, so the term $\boldsymbol{Y}$ in (8.31) is equal to zero, and corresponding transformations of particle positions (and other observables) are exactly the same as in the noninteracting case, for example,[9]

$$\begin{aligned}
\boldsymbol{r}_1(\boldsymbol{\varphi}) &= e^{-\frac{i}{\hbar}\boldsymbol{J}\cdot\boldsymbol{\varphi}}\boldsymbol{r}_1 e^{\frac{i}{\hbar}\boldsymbol{J}\cdot\boldsymbol{\varphi}} = e^{-\frac{i}{\hbar}\boldsymbol{j}_1\cdot\boldsymbol{\varphi}}\boldsymbol{r}_1 e^{\frac{i}{\hbar}\boldsymbol{j}_1\cdot\boldsymbol{\varphi}} = \boldsymbol{\varphi}\boldsymbol{r}_1, \\
\boldsymbol{v}_1(\boldsymbol{\varphi}) &= \boldsymbol{\varphi}\boldsymbol{v}_1.
\end{aligned}$$

This conclusion is in full accordance with experimental observations.

8.3.4 Space translations

In the instant form of dynamics, spatial shifts are interaction-independent, i. e.,

$$\begin{aligned}
\boldsymbol{r}_1^{\mathrm{IF}}(\boldsymbol{a}) &= e^{-\frac{i}{\hbar}\boldsymbol{P}\cdot\boldsymbol{a}}\boldsymbol{r}_1 e^{\frac{i}{\hbar}\boldsymbol{P}\cdot\boldsymbol{a}} = e^{-\frac{i}{\hbar}(\boldsymbol{p}_1+\boldsymbol{p}_2)\cdot\boldsymbol{a}}\boldsymbol{r}_1 e^{\frac{i}{\hbar}(\boldsymbol{p}_1+\boldsymbol{p}_2)\cdot\boldsymbol{a}} \\
&= e^{-\frac{i}{\hbar}\boldsymbol{p}_1\cdot\boldsymbol{a}}\boldsymbol{r}_1 e^{\frac{i}{\hbar}\boldsymbol{p}_1\cdot\boldsymbol{a}} = \boldsymbol{r}_1 - \boldsymbol{a}.
\end{aligned}$$

---

8 See, e. g., Subsection 8.5.6.

9 Recall that the notation $\boldsymbol{\varphi}\boldsymbol{r}_1$ denotes a passive action of the rotation $\boldsymbol{\varphi}$ on the vector $\boldsymbol{r}_1$; see Appendix **1**-D.5.

Again, this result is confirmed by experiments and our everyday experience, when we observe interacting physical systems from various positions, i. e., in a wide range of values of the transformation parameter $\boldsymbol{a}$.

However, the situation is radically different in the point form of dynamics. The generator of spatial shifts (8.39) depends on interaction. Therefore, translations of the observer must have a nontrivial effect on the measured values of particle observables. For example, a translation along the x-axis changes the x-component of the position as follows:

$$\begin{aligned} r_{1x}^{\mathrm{PF}}(a) &= e^{-\frac{i}{\hbar}P_x a} r_{1x} e^{\frac{i}{\hbar}P_x a} = e^{-\frac{i}{\hbar}(p_{1x}+p_{2x}+U_x)a} r_{1x} e^{\frac{i}{\hbar}(p_{1x}+p_{2x}+U_x)a} \\ &\approx r_{1x} + a[(p_{1x}+U_x), r_{1x}]_P + O(a^2) \\ &= r_{1x} - a + a[U_x, r_{1x}]_P + \cdots, \end{aligned} \tag{8.46}$$

where the last term on the right-hand side is the interaction correction. Spatial translations also affect velocity, i. e.,

$$\begin{aligned} v_{1x}^{\mathrm{PF}}(a) &= e^{-\frac{i}{\hbar}P_x a} v_{1x} e^{\frac{i}{\hbar}P_x a} = e^{-\frac{i}{\hbar}(p_{1x}+p_{2x}+U_x)a} v_{1x} e^{\frac{i}{\hbar}(p_{1x}+p_{2x}+U_x)a} \\ &\approx v_{1x} + a[U_x, v_{1x}]_P + O(a^2). \end{aligned} \tag{8.47}$$

Such corrections have not been seen in experiments, despite the fact that it is not so difficult to arrange observations of the same object from very different vantage points. Therefore, there are good reasons to believe that the dynamical nature of spatial translations (8.46)–(8.47) has not been observed, simply because it does not exist.

8.3.5 Support of instant form dynamics

So, we conclude that the experimentally observed effects of spatial translations and rotations do not depend on the forces acting in physical systems. Indeed, it is very easy to connect world views of two observers related to each other by the translation vector $\boldsymbol{a}$. We should simply add $\boldsymbol{a}$ to positions of all atoms in the universe. The same is true for rotations: all atoms in the universe experience the same rotation, independent of their states and involved interactions. Thus, translations and rotations have a kinematical character, as in the instant form of dynamics

$$\begin{aligned} \boldsymbol{P} &= \boldsymbol{P}_0, \\ \boldsymbol{J} &= \boldsymbol{J}_0. \end{aligned}$$

The situation with time translations is completely different. We cannot predict their effect without precise knowledge of the system's state and its interactions. So, time translations are definitely dynamical. Then the Poincaré group structure leaves us no other choice than to accept that boosts are interaction-dependent and dynamical too.

In Postulate **1**-6.3, we assumed that interactions in nature belong to the instant form. Now we see that this was the correct choice.

Our arguments in this section were based on the assumption that trajectories of interacting particles are accessible to experimental observation. This is true for long-range interactions, such as electromagnetic forces between electric charges, whose nontrivial interacting dynamics can be observed directly.[10] Therefore, such systems should be definitely described within the instant form of relativistic dynamics.

In Chapter 4, from the analysis of particle decays, we showed that the instant form of dynamics must be also applied to short-range weak nuclear forces. The decays controlled by such interactions are rather slow, so that one can observe their time dependencies. The slowing down of the decays of moving particles is a characteristic feature of the instant form.

However, the above analysis is not applicable to systems controlled by strong nuclear forces. In this case, neither interacting trajectories nor time-dependent decay laws can be measured. The presence of strong nuclear interactions manifests itself only through scattering effects or through energies of bound states. Both these classes of phenomena are insensitive to the form of dynamics, as shown in Subsection **1**-7.2.4. This means that the form of dynamics responsible for strong nuclear interactions is yet to be determined.

8.3.6 Physical inequivalence of forms of dynamics

Postulate **1**-6.3 (instant form of dynamics) contradicts the widely held view that various forms of dynamics are physically equivalent. In the literature, one can find calculations performed in the instant, point and front forms. It is often said that one can arbitrarily choose a form of dynamics that is most convenient for calculations. Where does this idea come from? There are two main sources. The first one is the undeniable fact that various forms of dynamics are really equivalent from the point of view of scattering.[11] The second source is the questionable assumption that all physically relevant information can be obtained from the S-matrix:

> If one adopts the point of view, first expressed by Heisenberg, that all experimental information about the physical world is ultimately deduced from scattering experiments and reduces to knowledge of certain elements of the scattering matrix (or the analogous classical quantity), then different dynamical theories which lead to the same S-matrix must be regarded as physically equivalent – S. N. Sokolov and A. N. Shatnii [229].

---

10 To make such observations, it is sufficient to have a plastic comb and pieces of paper or a pair of household magnets.

11 This was explained in Subsection **1**-7.2.4.

In our opinion, it is wrong to think that the S-matrix provides a complete description of everything that can be observed. For example, the time evolution and other inertial transformations of observables, considered earlier in this section, cannot be described within the framework of the S-matrix formalism. Theoretical description of these phenomena requires knowledge of generators of the Poincaré group, in particular the total momentum $\boldsymbol{P}$ and the boost operator $\boldsymbol{K}$. Two forms of dynamics, equivalent in scattering, can have different operators $\boldsymbol{P}$ and $\boldsymbol{K}$. Consequently, these forms can have completely different transformations of states with respect to spatial translations and/or boosts. In principle, these differences can be measured in experiments. For example, as we saw in Subsections 4.4.4 and 8.3.4, in the point form of dynamics, the decays of moving particles accelerate and the results of spatial translations depend on interaction, which makes this form physically unacceptable.

8.3.7 Currie–Jordan–Sudarshan theorem

In Section 8.2, we proved Theorem 8.1, which states that in the moving reference frame, space–time coordinates of events associated with noninteracting particles are given by Lorentz formulas (A.2)–(A.5). The absence of interaction played an important role in this proof. Does this mean that Lorentz transformations are no longer applicable in interacting systems? Our answer to this question is "*yes*."

As we shall see below, trajectories of interacting particles are transformed by formulas more complicated than equations (A.2)–(A.5). In other words, such particles do not have "invariant world lines." Apparently, it was Thomas [249] who first noticed this contradiction in special relativity. Currie, Jordan and Sudarshan analyzed this problem more deeply [44] and proved the following theorem,[12] which can be regarded as the inverse of Theorem 8.1.

Theorem 8.2 (Currie, Jordan and Sudarshan). *In a two-particle system with direct interaction, trajectories (= world lines) of particles obey Lorentz transformation formulas* (A.2)–(A.5) *only if there is no interaction.*

Proof. Our plan is similar to the proof of Theorem 8.1. We will compare formulas for the positions $\boldsymbol{r}_{1,2}(\theta, t')$ and momenta $\boldsymbol{p}_{1,2}(\theta, t')$ of the two particles obtained by two methods. In the first method, we will use Lorentz–Einstein transformations of special relativity. In the second method, following the Wigner–Dirac theory, we will apply unitary operators of time translations and boosts to the observables $\boldsymbol{r}_{1,2}$ and $\boldsymbol{p}_{1,2}$, as in Subsections 8.3.1 and 8.3.2. Our goal is to show that these two methods give different results when an interacting representation of the Poincaré group is used. It will be

12 We formulate a simplified version of the theorem, confining ourselves to only two particles. For more general formulations, see [44, 43].

sufficient to verify that the difference is present in terms of the first order in t' and θ. So, we will work in this approximation.

In our first method we apply the traditional Lorentz-Einstein (LE) formulas. From equations (A.2)–(A.5) and (**1**-4.3) we obtain the following transformations for the position and momentum of particle 1 (the same formulas are valid for particle 2):

$$r_{1x}^{\mathrm{LE}}(\theta,t') \approx r_{1x}(0,t) - ct\theta, \tag{8.48}$$

$$r_{1y}^{\mathrm{LE}}(\theta,t') = r_{1y}(0,t), \tag{8.49}$$

$$r_{1z}^{\mathrm{LE}}(\theta,t') = r_{1z}(0,t), \tag{8.50}$$

$$p_{1x}^{\mathrm{LE}}(\theta,t') \approx p_{1x}(0,t) - \frac{1}{c}h_1(0,t)\theta, \tag{8.51}$$

$$p_{1y}^{\mathrm{LE}}(\theta,t') = p_{1y}(0,t), \tag{8.52}$$

$$p_{1z}^{\mathrm{LE}}(\theta,t') = p_{1z}(0,t), \tag{8.53}$$

$$t' \approx t - \frac{\theta}{c}r_{1x}(0,t). \tag{8.54}$$

While remaining within the first-order approximation in t' and θ, we can rewrite equation (8.48) as follows:

$$\begin{aligned}
r_{1x}^{\mathrm{LE}}(\theta,t') &= r_{1x}\left(0, t' + \frac{r_{1x}(0,t)}{c}\theta\right) - \left(t' + \frac{r_{1x}(0,t)}{c}\theta\right)c\theta \\
&\approx r_{1x}(0,t') + \frac{dr_{1x}(0,t')}{dt'} \cdot \frac{r_{1x}(0,t)\theta}{c} - c\theta t' \\
&\approx r_{1x}(0,t') + \frac{dr_{1x}(0,t')}{dt'} \cdot \frac{r_{1x}(0,t')\theta}{c} - c\theta t'.
\end{aligned} \tag{8.55}$$

Next we use the Wigner-Dirac (WD) method and formulas (**1**-3.65), (**1**-4.4) to obtain

$$\begin{aligned}
r_{1x}^{\mathrm{WD}}(\theta,t') &= (e^{-\frac{ic}{\hbar}K_x\theta}e^{\frac{i}{\hbar}Ht'}e^{\frac{ic}{\hbar}K_x\theta})e^{-\frac{ic}{\hbar}K_x\theta}r_{1x}(0,0)e^{\frac{ic}{\hbar}K_x\theta}(e^{-\frac{ic}{\hbar}K_x\theta}e^{-\frac{i}{\hbar}Ht'}e^{\frac{ic}{\hbar}K_x\theta}) \\
&= (e^{\frac{i}{\hbar}Ht'\cosh\theta}e^{-\frac{ic}{\hbar}P_xt'\sinh\theta})e^{-\frac{ic}{\hbar}K_x\theta}r_{1x}(0,0)e^{\frac{ic}{\hbar}K_x\theta}(e^{\frac{ic}{\hbar}P_xt'\sinh\theta}e^{-\frac{i}{\hbar}Ht'\cosh\theta}) \\
&\approx e^{\frac{i}{\hbar}Ht'}e^{-\frac{ic}{\hbar}P_xt'\theta}(r_{1x}(0,0) - c\theta[r_{1x}(0,0),K_x]_P)e^{\frac{ic}{\hbar}P_xt'\theta}e^{-\frac{i}{\hbar}Ht'} \\
&\approx e^{\frac{i}{\hbar}Ht'}(r_{1x}(0,0) - c\theta t' - c\theta[r_{1x}(0,0),K_x]_P)e^{-\frac{i}{\hbar}Ht'} \\
&= r_{1x}(0,t') - c\theta[r_{1x}(0,t'),K_x(t')]_P - c\theta t',
\end{aligned} \tag{8.56}$$

$$\begin{aligned}
p_{1x}^{\mathrm{WD}}(\theta,t') &= (e^{\frac{i}{\hbar}Ht'\cosh\theta}e^{-\frac{ic}{\hbar}P_xt'\sinh\theta})e^{-\frac{ic}{\hbar}K_x\theta}p_{1x}(0,0)e^{\frac{ic}{\hbar}K_x\theta}(e^{\frac{ic}{\hbar}P_xt'\sinh\theta}e^{-\frac{i}{\hbar}Ht'\cosh\theta}) \\
&\approx e^{\frac{i}{\hbar}Ht'}e^{-\frac{ic}{\hbar}P_xt'\theta}(p_{1x}(0,0) - c\theta[p_{1x}(0,0),K_x]_P)e^{\frac{ic}{\hbar}P_xt'\theta}e^{-\frac{i}{\hbar}Ht'} \\
&\approx e^{\frac{i}{\hbar}Ht'}(p_{1x}(0,0) - c\theta[p_{1x}(0,0),K_x]_P)e^{-\frac{i}{\hbar}Ht'} \\
&= p_{1x}(0,t') - c\theta[p_{1x}(0,t'),K_x(t')]_P.
\end{aligned} \tag{8.57}$$

If the Lorentz transformations were exact, the LE and WD results would agree with each other. Let us accept this hypothesis and show that it can be true only in the non-interacting case.

So, if our hypothesis were correct, by comparing (8.55) and (8.56) we would obtain

$$\frac{dr_{1x}(0,t)}{dt}\cdot\frac{r_{1x}(0,t)\theta}{c} = -c\theta[r_{1x}(0,t),K_x(t)]_P$$

or, using $dr_{1x}/dt = [r_{1x},H]_P = \partial H/\partial p_{1x}$ and $[r_{1x},K_x]_P = \partial K_x/\partial p_{1x}$,

$$c^2\frac{\partial K_x}{\partial p_{1x}} = -r_{1x}\frac{\partial H}{\partial p_{1x}}.$$

Similar arguments lead to the general formula $(i,j = 1,2,3)$

$$c^2\frac{\partial K_j}{\partial p_{1i}} = -r_{1j}\frac{\partial H}{\partial p_{1i}}. \tag{8.58}$$

Next, comparing (8.51) with (8.57), we obtain

$$\begin{aligned}
p_{1x}(0,t) - \frac{h_1(0,t)\theta}{c} &= p_{1x}\left(0,t-\frac{r_{1x}(0,t)}{c}\theta\right) - c\theta[p_{1x}(0,t'),K_x(t')]_P \\
&\approx p_{1x}(0,t) - \frac{r_{1x}(0,t)\theta}{c}\frac{\partial p_{1x}(0,t)}{\partial t} - c\theta[p_{1x}(0,t'),K_x(t')]_P \\
&\approx p_{1x}(0,t) - \frac{r_{1x}(0,t)\theta}{c}[p_{1x}(0,t),H]_P - c\theta[p_{1x}(0,t),K_x(t)]_P
\end{aligned}$$

and

$$c^2[p_{1x},K_x]_P = -r_{1x}[p_{1x},H]_P + h_1.$$

In the general case $(i,j = 1,2,3)$, we get

$$-c^2\frac{\partial K_j}{\partial r_{1i}} = r_{1j}\frac{\partial H}{\partial r_{1i}} + \delta_{ij}h_1. \tag{8.59}$$

Joining together (8.58)–(8.59), we conclude that in the case of assumed Lorentz-invariant world lines, the interacting generators should obey the following system of equations:

$$c^2\frac{\partial K_k}{\partial \boldsymbol{p}_1} = -r_{1k}\frac{\partial H}{\partial \boldsymbol{p}_1}, \tag{8.60}$$

$$c^2\frac{\partial K_k}{\partial \boldsymbol{p}_2} = -r_{2k}\frac{\partial H}{\partial \boldsymbol{p}_2}, \tag{8.61}$$

$$c^2\frac{\partial K_k}{\partial r_{1i}} = -r_{1k}\frac{\partial H}{\partial r_{1i}} - \delta_{ik}h_1, \tag{8.62}$$

$$c^2\frac{\partial K_k}{\partial r_{2i}} = -r_{2k}\frac{\partial H}{\partial r_{2i}} - \delta_{ik}h_2. \tag{8.63}$$

Subtracting the derivative of (8.61) with respect to $\boldsymbol{p}_1$ from the derivative of (8.60) with respect to $\boldsymbol{p}_2$, we obtain

$$(r_{2k} - r_{1k})\frac{\partial^2 H}{\partial \boldsymbol{p}_2 \partial \boldsymbol{p}_1} = 0.$$

This implies

$$\frac{\partial^2 H}{\partial \boldsymbol{p}_2 \partial \boldsymbol{p}_1} = 0.$$

In a similar manner we obtain[13]

$$\frac{\partial^2 H}{\partial \boldsymbol{r}_2 \partial \boldsymbol{r}_1} = \frac{\partial^2 H}{\partial \boldsymbol{r}_2 \partial \boldsymbol{p}_1} = \frac{\partial^2 H}{\partial \boldsymbol{p}_2 \partial \boldsymbol{r}_1} = 0.$$

Thus, there are only two nonzero cross-derivatives of the Hamiltonian, i. e.,

$$\frac{\partial^2 H}{\partial \boldsymbol{p}_1 \partial \boldsymbol{r}_1} \neq 0,$$
$$\frac{\partial^2 H}{\partial \boldsymbol{p}_2 \partial \boldsymbol{r}_2} \neq 0.$$

Therefore, only pairs of arguments $(\boldsymbol{p}_1, \boldsymbol{r}_1)$ and $(\boldsymbol{p}_2, \boldsymbol{r}_2)$ can be present simultaneously in each term in H. In the most general form, this requirement can be satisfied by writing the full Hamiltonian as

$$H = H_1(\boldsymbol{p}_1, \boldsymbol{r}_1) + H_2(\boldsymbol{p}_2, \boldsymbol{r}_2).$$

This means that the force acting on particle 1 does not depend on the state (position and momentum of particle 2), i. e.,

$$\boldsymbol{f}_1 = \frac{\partial \boldsymbol{p}_1}{\partial t} = [\boldsymbol{p}_1, H]_P = [\boldsymbol{p}_1, H_1(\boldsymbol{p}_1, \boldsymbol{r}_1)]_P,$$

and conversely, movements of particle 2 are independent of particle 1. This means that there is no interaction. □

So, our assumption about Lorentz-like (covariant) transformations of trajectories has led us to the absurd conclusion about the lack of interaction $V = 0, \boldsymbol{Z} = \boldsymbol{0}$.[14] Several alternative explanations for this paradox were tried in the literature. One idea was that the Hamiltonian dynamics is not suitable for describing relativistic interactions. Instead, various non-Hamiltonian approaches were offered [255, 256, 244, 188, 129,

---

13 Here we use $\partial h_2/\partial \boldsymbol{r}_1 = \partial\sqrt{m_2c^4 + p_2^2c^2}/\partial \boldsymbol{r}_1 = 0$ and $\partial h_2/\partial \boldsymbol{p}_1 = \partial h_1/\partial \boldsymbol{r}_2 = \partial h_1/\partial \boldsymbol{p}_2 = 0$.

14 This explains the name "no-interaction theorem" often used for the Currie–Jordan–Sudarshan result.

228], which deviated from the Poincaré-invariant recipes professed in this book. So far, the predictive power of these approaches remains rather limited. Moreover, they are not easily generalized to the quantum domain.

Another view is that the quantities $\boldsymbol{r}_i$ and $\boldsymbol{p}_i$ do not describe true particle observables, or that such observables are strictly defined only in the absence of interactions. Some researchers went even further and questioned the corpuscular description of nature, suggesting to replace it by the pure field-based approach [20] and claiming that "*there are no particles, there are only fields*" [111, 163, 270, 106, 243]. Other models admit the existence of observable particles, but in addition require that interactions are carried by certain agents (invisible fields or virtual particles) having their own degrees of freedom, momenta and energies.

However, we reject these explanations. Non-Hamiltonian versions of dynamics contradict fundamental postulates of the Poincaré–Wigner–Dirac relativistic quantum theory. We also stick to the opinion that the physical world can be described in terms of particles with well-defined positions, momenta and other observables. We maintain that these particles interact with each other by means of instantaneous potentials without any intermediaries. Therefore, for us, the only way to solve the "no-interaction" paradox is to accept that Lorentz transformations cannot be applied to positions of particles in interacting systems [215]. Contrary to the traditional Assertion A.2 (manifest covariance of physical laws), we introduce the following statement.

Statement 8.3 (interaction-dependence of boost transformations). *Transformations of observables under boosts cannot be described by a universal formula. These transformations depend on the observed system, its state and the forces acting therein. In other words, boost transformations are dynamical.*

8.4 Do instantaneous interactions violate causality?

We have found out that RQD describes interactions between particles in terms of *instantaneous potentials*. However, textbooks tell us that instantaneous forces violate the principle of causality. In this section, we will challenge this conclusion. Our claim is that if the dynamical nature of boosts is properly taken into account, then instantaneous potentials conform to causality in all frames of reference.

8.4.1 Interaction in different frames of reference

Consider two interacting particles in RQD. The dressed Hamiltonian in the two-particle Hilbert space is a function of the positions and momenta of the two particles,[15] $H^d =$

15 Here we confine ourselves to low energies and neglect the processes of radiation and pair production.

$H^d(\boldsymbol{r}_1, \boldsymbol{p}_1, \boldsymbol{r}_2, \boldsymbol{p}_2)$. Trajectories of particles can be obtained by standard formulas, i. e.,

$$
\begin{aligned}
\boldsymbol{r}_1(t) &= e^{\frac{i}{\hbar}H^d t}\boldsymbol{r}_1 e^{-\frac{i}{\hbar}H^d t} \\
\boldsymbol{p}_1(t) &= e^{\frac{i}{\hbar}H^d t}\boldsymbol{p}_1 e^{-\frac{i}{\hbar}H^d t}, \\
\boldsymbol{r}_2(t) &= e^{\frac{i}{\hbar}H^d t}\boldsymbol{r}_2 e^{-\frac{i}{\hbar}H^d t}, \\
\boldsymbol{p}_2(t) &= e^{\frac{i}{\hbar}H^d t}\boldsymbol{p}_2 e^{-\frac{i}{\hbar}H^d t}
\end{aligned}
$$

and the force[16] acting on particle 2,

$$\boldsymbol{f}_2(t) = \frac{d}{dt}\boldsymbol{p}_2(t) = -\frac{i}{\hbar}[\boldsymbol{p}_2(t), H^d],$$

depends on the positions and momenta of both particles at the same time instant t. Thus, it is expressed by a function

$$\boldsymbol{f}_2(t) \equiv \boldsymbol{f}_2(\boldsymbol{r}_1(t), \boldsymbol{p}_1(t); \boldsymbol{r}_2(t), \boldsymbol{p}_2(t)), \tag{8.64}$$

which is characteristic for interactions propagating instantaneously in the stationary frame of reference O.

The special theory of relativity forbids propagation of signals faster than the speed of light. This prohibition is usually explained by the fact that when switching to a moving frame by Lorentz formulas, we can get a situation in which the effect occurs earlier than the cause (see Appendix A.4 and Subsection 8.1.4). However, we also know from Subsection 8.3.7 that for interacting systems the usual Lorentz transformations do not work, and therefore the justification of the superluminal ban should be reconsidered.

Let us now take a look at our two-particle system from the point of view of the moving observer O'. According to the Wigner–Dirac theory, this observer sees the following particle trajectories[17]:

$$
\begin{aligned}
\boldsymbol{r}_1(\boldsymbol{\theta}, t') &= e^{-\frac{ic}{\hbar}\boldsymbol{K}^d\cdot\boldsymbol{\theta}} e^{\frac{i}{\hbar}H^d t'}\boldsymbol{r}_1 e^{-\frac{i}{\hbar}H^d t'} e^{\frac{ic}{\hbar}\boldsymbol{K}^d\cdot\boldsymbol{\theta}}, \\
\boldsymbol{p}_1(\boldsymbol{\theta}, t') &= e^{-\frac{ic}{\hbar}\boldsymbol{K}^d\cdot\boldsymbol{\theta}} e^{\frac{i}{\hbar}H^d t'}\boldsymbol{p}_1 e^{-\frac{i}{\hbar}H^d t'} e^{\frac{ic}{\hbar}\boldsymbol{K}^d\cdot\boldsymbol{\theta}}, \\
\boldsymbol{r}_2(\boldsymbol{\theta}, t') &= e^{-\frac{ic}{\hbar}\boldsymbol{K}^d\cdot\boldsymbol{\theta}} e^{\frac{i}{\hbar}H^d t'}\boldsymbol{r}_2 e^{-\frac{i}{\hbar}H^d t'} e^{\frac{ic}{\hbar}\boldsymbol{K}^d\cdot\boldsymbol{\theta}}, \\
\boldsymbol{p}_2(\boldsymbol{\theta}, t') &= e^{-\frac{ic}{\hbar}\boldsymbol{K}^d\cdot\boldsymbol{\theta}} e^{\frac{i}{\hbar}H^d t'}\boldsymbol{p}_2 e^{-\frac{ic}{\hbar}H^d t'} e^{\frac{i}{\hbar}\boldsymbol{K}^d\cdot\boldsymbol{\theta}}.
\end{aligned}
$$

The Hamiltonian in the frame O' is (**1**-3.64)

$$H^d(\boldsymbol{\theta}) = e^{-\frac{ic}{\hbar}\boldsymbol{K}^d\cdot\boldsymbol{\theta}} H^d e^{\frac{ic}{\hbar}\boldsymbol{K}^d\cdot\boldsymbol{\theta}}, \tag{8.65}$$

---

16 In this case, we define the force as the time derivative of the momentum. Our conclusions will not change if we use the alternative definition $\boldsymbol{f} = m \times d^2\boldsymbol{r}/dt^2$, as in Subsection 6.2.3.

17 See formula (**1**-3.65). As usual, here t' is the time measured by the clock of the observer O', $\boldsymbol{\theta}$ is the rapidity of this observer and $\boldsymbol{K}^d$ is the interaction-dependent dressed boost operator introduced in Subsection 2.3.6.

so the force acting on particle 2 in this frame,

$$\begin{aligned}
\boldsymbol{f}_2(\boldsymbol{\theta},t') &= \frac{d}{dt'}\boldsymbol{p}_2(\boldsymbol{\theta},t') = -\frac{i}{\hbar}[\boldsymbol{p}_2(\boldsymbol{\theta},t'), H^d(\boldsymbol{\theta})] \\
&= -\frac{i}{\hbar}[e^{-\frac{ic}{\hbar}\boldsymbol{K}^d\cdot\boldsymbol{\theta}} e^{\frac{i}{\hbar}H^d t'} \boldsymbol{p}_2 e^{-\frac{i}{\hbar}H^d t'} e^{\frac{ic}{\hbar}\boldsymbol{K}^d\cdot\boldsymbol{\theta}}, e^{-\frac{ic}{\hbar}\boldsymbol{K}^d\cdot\boldsymbol{\theta}} H^d e^{\frac{ic}{\hbar}\boldsymbol{K}^d\cdot\boldsymbol{\theta}}] \\
&= -\frac{i}{\hbar} e^{-\frac{ic}{\hbar}\boldsymbol{K}^d\cdot\boldsymbol{\theta}} [e^{\frac{i}{\hbar}H^d t'} \boldsymbol{p}_2 e^{-\frac{i}{\hbar}H^d t'}, H^d] e^{\frac{ic}{\hbar}\boldsymbol{K}^d\cdot\boldsymbol{\theta}} \\
&= -\frac{i}{\hbar} e^{-\frac{ic}{\hbar}\boldsymbol{K}^d\cdot\boldsymbol{\theta}} [\boldsymbol{p}_2(0,t'), H^d] e^{\frac{ic}{\hbar}\boldsymbol{K}^d\cdot\boldsymbol{\theta}} \\
&= e^{-\frac{ic}{\hbar}\boldsymbol{K}^d\cdot\boldsymbol{\theta}} \boldsymbol{f}_2(0,t') e^{\frac{ic}{\hbar}\boldsymbol{K}^d\cdot\boldsymbol{\theta}} \\
&= e^{-\frac{ic}{\hbar}\boldsymbol{K}^d\cdot\boldsymbol{\theta}} \boldsymbol{f}_2(\boldsymbol{r}_1(0,t'), \boldsymbol{p}_1(0,t'); \boldsymbol{r}_2(0,t'), \boldsymbol{p}_2(0,t')) e^{\frac{ic}{\hbar}\boldsymbol{K}^d\cdot\boldsymbol{\theta}} \\
&= \boldsymbol{f}_2(\boldsymbol{r}_1(\boldsymbol{\theta},t'), \boldsymbol{p}_1(\boldsymbol{\theta},t'); \boldsymbol{r}_2(\boldsymbol{\theta},t'), \boldsymbol{p}_2(\boldsymbol{\theta},t')),
\end{aligned} \tag{8.66}$$

is a function of positions and momenta of both particles at the same time instant t'. Moreover, in agreement with the principle of relativity, this function $\boldsymbol{f}_2$ has exactly the same form as in the rest frame (8.64). Therefore, the moving observer O' also thinks that interaction propagates instantaneously.

This means that if two distant events are connected to each other by an instantaneous potential, then they occur simultaneously in all reference frames. There is no frame in which these two events change their time order. First, this tells us that instantaneous interactions do not violate the principle of causality. Second, for such events linked by interaction, the notion of "relativity of simultaneity" does not apply. Actually, for such events there is no clear distinction between "causes" and "effects", as all of them are governed by the same time evolution operator. So, one can say that all of them have a common cause.

8.4.2 Frascati experiment in moving reference frame

Let us illustrate the above findings on the example of the Frascati experiment described in Subsection 7.4.5. Recall that this experiment studied the Coulomb field of a relativistic electron beam emerging from the accelerator pipe. The rest-frame evolution of the Coulomb field predicted in RQD is shown on three consecutive panels in Figures 8.2 (a)→(b)→(c). The snapshot in Figure 8.2 (a) is taken shortly before the electron beam leaves the pipe, the panel in Figure 8.2 (b) shows the immediate moment of the beam ejection and Figure 8.2 (c) – just a little bit later.

According to RQD, the Coulomb field disk is formed instantly (panel in Figure 8.2 (b)), so that three events A (beam's exit from the pipe), B and C (sensors click) occur simultaneously, despite the fact that B and C are consequences of the event A and separated from A by considerable distances.

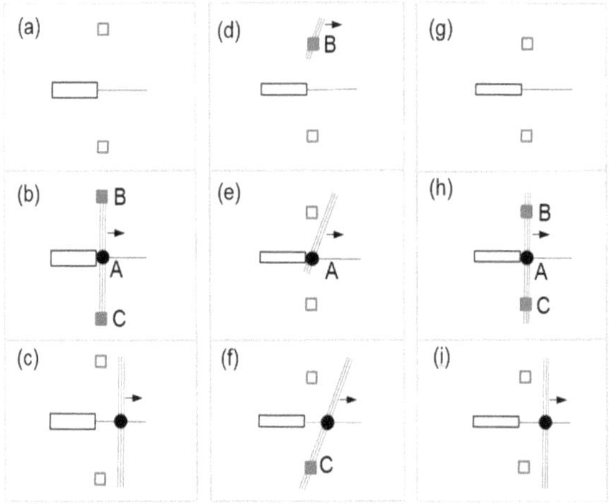

Figure 8.2: Time evolution of the instantaneous Coulomb field of the electron beam leaving the accelerator pipe: (a)→(b)→(c) in the reference frame at rest; (d)→(e)→(f) in the moving frame as predicted by Lorentz formulas (A.2)–(A.5); (g)→(h)→(i) in the moving frame as predicted by the Wigner–Dirac theory. Squares indicate locations of electric field sensors. Filled squares are sensors registering signals.

How does this situation look from the point of view of the moving observer O'? Let us first take the side of special relativity. Suppose that the observer O' moves at a high speed from the bottom of the page up.[18] Then, applying Lorentz formulas (A.2)–(A.5), we get three panels, Figures 8.2 (d)→(e)→(f).[19]

The absurdity of these drawings is clear already from the panel in Figure 8.2 (d), where a part of the Coulomb field disk emerged even before the beam left the accelerator pipe. This means that the "effect" B has occurred earlier than its "cause" A, in apparent contradiction with the principle of causality. Two other panels show the further development of events: the electron beam leaves the accelerator pipe in Figure 8.2 (e), the lower sensor clicks in Figure 8.2 (f).

In order to avoid such gross violations of causality, special relativity forbids the superluminal formation of the Coulomb field disk. However, this solution does not satisfy us, because it contradicts both the experiment and RQD.

8.4.3 Does Frascati experiment violate causality?

Our explanation of the causality in the Frascati experiment is based on the reasoning from Subsection 8.4.1. The key point is that the involved physical system is interacting.[20] Therefore, the classical Lorentz transformation formulas are not applicable. The transition to the moving frame of reference must be performed with the help of boost

18 We call this the x-direction.

19 For more realism, we squeezed these drawings along the x-axis to show the relativistic length contraction effect.

20 Signals on the sensors appear due to the direct Coulomb interaction of their charges with electron beam charges.

generators containing interaction terms. An exact mathematical analysis of such a transition is very difficult. However, even without complex calculations, it is possible to say with certainty that the result of this transition will not resemble the one shown in Figures 8.2 (d)→(e)→(f). This is clear already from the fact that, according to (8.66), in a moving frame, the forces experienced by charges are completely determined by their instantaneous positions. Most likely, a properly performed boost transformation will produce the picture shown in the sequence of Figures 8.2 (g)→(h)→(i). In this picture, the Coulomb potential propagates instantly in both reference frames O and O'.

Thus, the instantaneous Coulomb potential adequately describes the Frascati experiments with relativistic electron beams. The causality is not violated in these experiments.

8.5 Comparison with special relativity

Our conclusions about the dynamical character of boosts and the instantaneous propagation of electromagnetic interactions contradict the consensus developed for over hundred years since Einstein formulated his theory of relativity. As we consider our approach more consistent and preferable, we have to explain why the special-relativistic arguments were erroneous.

In Subsections 8.5.1–8.5.2 we shall analyze existing proofs of Lorentz transformations and find the exact places where they err. In Subsections 8.5.3–8.5.5, we will conclude that it is necessary to reject the manifest covariance of special relativity. In addition, we will claim that Minkowski's four-dimensional space–time is just an approximation, though quite a good one.

In Subsection 8.5.6 we will see that our approach does not contradict existing experiments.

8.5.1 Are Lorentz transformations universal?

Einstein founded his special theory [63] on two postulates. The first postulate was the principle of relativity. The second postulate established the independence of the speed of light from the motion of the light source or the observer. Both these statements remain valid in our theory.[21] Based on his postulates, Einstein considered a series of thought experiments with rulers, clocks, light beams and moving trains. These reasonings demonstrated the relativity of simultaneity, the contraction of moving objects and the slowing down of moving clocks. These conclusions were generalized in the Lorentz formulas (A.2)–(A.5), which linked the times and positions of localized events in different frames of reference. As we showed in Theorem 8.1, our approach leads exactly

21 See Postulate **1**-2.1 and Statement **1**-5.1.

to the same transformation laws for events occurring with noninteracting particles, such as photons. Up to this point, our RQD is in complete agreement with the special theory of relativity.

Although Einstein's postulate of relativity has universal applicability to all events and processes in nature, his postulate about the "invariance of the speed of light" relates only to freely propagating light pulses. So, strictly speaking, all the conclusions drawn from the two postulates in [63] can be applied only to events connected in some way with light propagation. Nevertheless, in his work Einstein tacitly assumed[22] that the same conclusions can be extended to all events, regardless of their physical nature and involved interactions. From our point of view, this assumption is the main mistake (or, rather, an approximation) of special relativity.

8.5.2 About "derivations" of Lorentz transformations

There are numerous publications [155, 158, 207, 211, 79, 187, 91] whose authors claim that Lorentz transformation formulas (A.2)–(A.5) can be derived even without the second Einstein postulate. However, these works do not look convincing. Usually they make a dubious assumption about the universality of the desired transformations.

For example, the very first equality (1) in [211],

$$x'_\mu = \sum_\nu a_{\mu\nu} x_\nu \quad (\mu, \nu = 0, 1, 2, 3), \; x_0 \equiv t,$$

expects that transformations of the 4-coordinates (x_ν) are given by the universal matrix $(a_{\mu\nu})$, which is independent of the nature of the event occurring at the point x_ν. In fact, similar formulas were used as starting points in other articles [155, 91]. The same logical error is also made in equation (1) of [158],

$$\begin{aligned} x' &= f(x, t; a_1, \ldots, a_N), \\ t' &= g(x, t; a_1, \ldots, a_N), \end{aligned}$$

where transformations $(x, t) \to (x', t')$ are supposed to depend only on the 4-coordinates themselves and on the parameters $a_1, \ldots, a_N$ of the inertial transformation. The physical nature of the event at (x, t) was not even contemplated.

In a somewhat more consistent approach [207], an attempt was made to find transformations of 4-coordinates of a particular system – the trajectory of a freely moving particle. It is not surprising that in the absence of interaction, the usual linear Lorentz formula was obtained as in Subsection 8.2.4. However, there is absolutely no reason to generalize this result to more complex systems, such as trajectories of interacting particles.

22 And this unjustified assumption has since been repeated in all textbooks.

So, the main problem with mentioned derivations is that the universal and kinematical nature of boost transformations was supposed to be true *a priori*. However, as we saw in Section 8.3, the structure of the Poincaré group requires the boost transformations to be dynamical. This means that boost transformations of 4-coordinates of events cannot be universal: they must depend on the state of the system and on the interactions between the system's components.

The irony is that the usual special-relativistic assumptions about kinematical boosts, universal Lorentz transformations and invariant world lines in interacting systems are in contradiction with the Poincaré group structure, which is accepted as the basis of relativistic physics by everyone.

8.5.3 Poincaré invariance vs. manifest covariance

From what was said above it should be clear that there are two rather different points of view on the formulation of relativistic theories. The traditional approach of Einstein and Minkowski takes on faith Assertion A.1 (the universality of Lorentz transformations). This implies important consequences, such as the prohibition of superluminal signals (Assertion A.3). The kinematical character of boosts finds its mathematical expression in the ideas about the 4D space–time and 4-tensor transformations of physical observables (Assertion A.2).

Special relativity demands that all interactions must conform to these severe restrictions of *manifest covariance*. No wonder that the allowed class of interacting models appears empty (see the Currie–Jordan–Sudarshan theorem).

In our book, we adhere to a different view on relativity. We call our approach *Poincaré-invariant*. It is based on two fundamental postulates: the principle of relativity (Postulate **1**-2.1) and laws of quantum mechanics from Sections **1**-1.4 and **1**-1.5. From these two premises we derived Statement **1**-3.2 (unitary representations U_g of the Poincaré group). By adding Postulate **1**-6.3 (instant form of dynamics), we arrived at our key Statement 8.3 (the dynamical character of boosts).

The representation U_g fully determines how observables change with respect to inertial transformations. Applying the interaction boost generator $\boldsymbol{K}$ to various observables $\mathfrak{g}$, we obtain formulas for their transformations into moving reference frame:

$$\mathfrak{g}(\boldsymbol{\theta}) = e^{-\frac{ic}{\hbar}\boldsymbol{K}\cdot\boldsymbol{\theta}}\mathfrak{g}e^{\frac{ic}{\hbar}\boldsymbol{K}\cdot\boldsymbol{\theta}}.$$

These formulas depend on the state of the multiparticle system and on interactions acting there. Of course, these transformations are different from universal 4-tensor formulas of special relativity. The relativistic invariance of our approach is proved by explicit verification of commutation relations of the Poincaré Lie algebra, as we did in Appendices **2**-E and E.

A deep conflict between the manifest covariance of special relativity and the Poincaré invariance of quantum physics was noticed by a number of researchers. For example, Foldy wrote:

> To begin our discussion of relativistic covariance, we would like first to make clear that we are not in the least concerned with appropriate tensor or spinor equations, or with "manifest covariance" or with any other mathematical apparatus which is intended to exploit the space-time symmetry of relativity, useful as such may be. We are instead concerned with the *group* of inhomogeneous Lorentz transformations as expressing the inter-relationship of physical phenomena as viewed by different equivalent observers in un-accelerated reference frames. That this group has its basis in the symmetry properties of an underlying space-time continuum is interesting, important, but not directly relevant to the considerations we have in mind – L. Foldy [84].

This issue was also raised by Bacry, who came to a similar conclusion:

> The *Minkowski manifest covariance* cannot be present in quantum theory but we want to preserve the *Poincaré covariance* – H. Bacry [9].

8.5.4 About time measurements

Special relativity, with its manifestly covariant approach, takes a purely "geometric" point of view on Lorentz transformations. In this theory, time and three Cartesian coordinates are treated on an equal basis as components of a single 4-vector. Such unification suggests that there must be a certain similarity between these two types of physical quantities. However, in quantum mechanics (as in our daily experience), there is a huge difference between space and time. The spatial coordinates x, y, z are attributes (observables) of physical systems (particles or groups of particles). In quantum mechanics, these coordinates are expectation values of the Newton–Wigner position operator.

On the other hand, time is simply a numerical label, assigned to measurements in accordance with the readings of the laboratory clock at the time of the experiment. The clock is separated from the physical system.[23] The readings of the clock do not depend on which system is being observed and what is its state. The clock readings can be recorded even if there is no physical system in our laboratory. Therefore, time is not an observable, and there is no quantum Hermitian operator corresponding to time.

8.5.5 Is world's geometry four-dimensional?

Our position is that there is no "symmetry" or "analogy" between space and time coordinates. Therefore, there is no need for the four-dimensional Minkowski space

23 See Figure 1 in the Introduction to Volume 1.

postulated in special relativity. Similarly, there is no basis for classifying physical quantities into 4-scalars, 4-vectors, 4-tensors, etc. Yes, in some cases, observables behave in such a specific way with respect to boosts and rotations.[24] But most often boosted observables have a complex dependence on the particular state of the system, interactions, etc.

Brown's book [22] discussed historical and philosophical roots of the idea that relativistic effects (such as length contraction and time dilation) come from the dynamical behavior of specific physical systems rather than from kinematical (geometrical) properties of the universal "space–time continuum". In general, we share this view. However, in our conclusions we go even further and argue that the difference between "dynamical" and "kinematical" approaches to boosts is not just philosophical. It has real observable consequences. As we saw in Section 8.3, boosts depend on interactions and their results cannot be reduced to universal Lorentz formulas or "pseudo-rotations" in the Minkowski space–time. Thus, the pseudo-Euclidean four-dimensional continuum and the manifest covariance of physical laws can be understood only as approximations.

Additional arguments against attributing a fundamental role to the Minkowski space–time can be found in [10].

8.5.6 Experimental checks of special relativity

Of course, it is true that predictions of special relativity have been confirmed by numerous experiments with unsurpassed accuracy. However, at a closer look it turns out that most of these experiments cannot distinguish special relativity from our RQD approach. In some cases this is due to the fact that predictions of the two theories do coincide. In other cases, the differences are so small that the required accuracy is not accessible with modern instruments.

From our discussions in this chapter it should be clear that special-relativistic Lorentz formulas can be applied to observables of noninteracting (free) particles and to *total* observables of composite physical systems, both interacting and noninteracting. It turns out that practically all experimental tests of special relativity [175, 162, 210, 196] deal with these types of measurements.

Experimenters routinely check the relativistic kinematics (the relationship between momenta, velocities and energies of free particles) and the validity of energy–momentum conservation laws in collisions, reactions and decays. Obviously, in these cases, there is no difference between predictions of RQD and Einstein's theory.

---

24 See, for example, the 4-vector transformation (**1**-4.3)–(**1**-4.4) of the *total* energy momentum of an isolated system.

Another class of experiments measures the frequency (energy) of light and its dependence on the motion of the source or the observer [116, 117, 125, 107]. We presented our interpretations of the Doppler effect in Subsections **1**-5.4.5 and **1**-6.5.3. Again, here we do not expect to find any surprises.

Yet another group of experiments deals with measuring the speed of light and checking its independence from the movement of the source and/or observer. This class includes interference experiments of Michelson–Morley and Kennedy–Thorndike as well as direct measurements of the photon speed [4]. All these measurements are performed with free photons, so again predictions of the two theories coincide exactly.

There is an exceptional class of experiments where one *can*, at least in principle, observe differences between the two theories. These are measurements of decays of rapidly moving unstable particles. In this case we are dealing with a physical system whose behavior (decay) is controlled by interaction for a sufficiently long period of time. However, in Chapter 4 we found that modern experimental techniques are not sensitive enough to register extremely small differences between Einstein's "time dilation" formula and RQD predictions.[25]

The special theory of relativity forbids the existence of absolutely rigid bodies. But this prohibition is apparently violated in the famous *Mössbauer effect*, where the recoil energy of a photon-emitting nucleus is instantaneously distributed among all atoms in a macroscopic crystal [5].

In Sections 7.4 and 8.4, we described the Frascati experiment, which established the superluminal dynamics of the electric field of relativistic charges. This experiment violates the special-relativistic ban on superluminal propagation, but it fully agrees with RQD predictions.

8.6 Why do we need quantum fields?

The main idea of RQD is that particles are the most fundamental elements of nature, and all physics can be explained as manifestations of the quantum behavior of point particles interacting by instantaneous potentials with one another. If this idea is confirmed, then the concept of fields will become superfluous, as once the concept of the ether was found redundant. On the other hand, it is also true that (quantum) fields are at the center of all modern theories of elementary particles, and in fact we started our formulation of RQD from the quantum-field version of QED in Section **2**-3.1. Is there any contradiction here? In other words, *what is the role of quantum fields in relativistic quantum theory?*

25 Interesting ideas of experiments with moving unstable particles can be found in [253, 254].

8.6.1 Are quantum fields measurable?

The idea of traditional quantum field theory is that particles observed in experiments – photons, electrons, protons, etc. – are not the fundamental ingredients of nature. It is argued that the most fundamental role is played by fields. Each type of particle corresponds to its own field – a continuous all-pervasive substance, spreading throughout the universe. Dyson called it "a single fluid that fills all space-time" [60]. Fields are present even in situations where there is not a single particle, i. e., in the vacuum.

If the fields are such important components of physical reality, then one would like to be able to measure them directly. However, the quantities measured in physical experiments are closely related to particles rather than fields. For example, in the classical limit, we can measure (the expectation values of) positions, momenta, velocities and energies of particles as functions of time (= trajectories). In interacting particle systems, it is possible to measure energies of bound states and their probability distributions (= squares of the modulus of particle wave functions). Important information is obtained by studying the relationships between observables of particles before and after their collisions (S-matrix, scattering cross section). All these experimental data have a transparent and natural description in the language of particles, their wave functions and their (Hermitian operators of) observables.

On the other hand, the properties of fields (their values at a point, their spatial and temporal derivatives, etc.) cannot be observed directly. Fermion fields cannot even formally be attributed to observables, since they are not Hermitian operators. The electromagnetic field is usually cited as an exception, whose components $\boldsymbol{E}$ and $\boldsymbol{B}$ are definitely observable. But even this statement is questionable. When we say that we "measured an electric field" at a certain point in space, we actually put a test charge there and measured the force,[26] exerted on this charge by surrounding charges. Nobody measures electric and magnetic fields as such. Therefore, the ideas of electric and magnetic fields without loss of generality can be expressed in the equivalent language of electric charges and instantaneous forces acting between them. Of course, such a description may be consistent only if there is no specific energy and momentum associated with fields and if interactions between charges propagate without retardation. Both these conditions are fulfilled in our theory (RQD).

8.6.2 Quantum fields and space–time

In Subsection 8.1.5 we have already noted that the quantum field $\psi(0, \boldsymbol{x})$ cannot be interpreted as an operator that creates or annihilates a particle at the space point $\boldsymbol{x}$.

26 Or rather acceleration, which is proportional to the applied force (6.17).

The formal nature of quantum fields $\psi(t, \boldsymbol{x})$ is also clear from the fact that their arguments t and $\boldsymbol{x}$ are not related to the actual measured times and positions of physical events. The variable t is a parameter that we used in (**1**-7.10) and (**2**-1.60) to describe the "t-dependence" of regular operators. This t-dependence is generated by the non-interacting Hamiltonian H_0 and, as we explained in Subsection **1**-7.1.2, it has no relation to the observed time-dependent behavior of physical quantities. In fact, the t-dependence of operators was introduced into the theory only as a convenient computational technique. For example, when we compute the S-matrix (**1**-7.11), we integrate with respect to the dummy variable t' from $-\infty$ to $+\infty$, so that the t'-dependence vanishes in the final result.

Also, we should not confuse the three field arguments $\boldsymbol{x}$ with physical positions of particles or with eigenvalues of the Newton–Wigner position operator. The coordinates x, y, z are treated as dummy integration variables when interaction operators are constructed from quantum fields in (**2**-3.14)+(**2**-3.15). So, it seems appropriate to regard the parameters $(t, \boldsymbol{x})$ as abstract coordinates in the formal Minkowski space–time $\mathcal{M}$, which is unrelated to the physical space and time.

Of course, we agree with the following two quotes:

> ...it was possible to give up Minkowski space-time without rejecting the Poincaré group and its attached energy-momentum space. Every physicist would easily convince himself that all quantum calculations are made in the energy-momentum space and that the Minkowski x_μ are just dummy variables without physical meaning (although almost all textbooks insist on the fact that these variables are not related with position, they use them to express *locality* of interactions!) – H. Bacry [9].

> It is important to note that the $\boldsymbol{x}$ and t that appear in the quantized field $A(\boldsymbol{x}, t)$ are not quantum-mechanical variables but just parameters on which the field operator depends. In particular, $\boldsymbol{x}$ and t should not be regarded as the space-time coordinates of the photon – J. Sakurai [205].

Thus, we come to the conclusion that quantum fields $\psi(t, \boldsymbol{x})$ are simply formal linear combinations of creation and annihilation operators of particles. Also, we should not regard quantum fields as a kind of "generalized" or "second-quantized" versions of wave functions. The role of quantum fields is more technical than fundamental. They are mathematical objects that turned out to be very convenient for constructing Poincaré-invariant operators of the potential energy V and the potential boost $\boldsymbol{Z}$. Indeed, in Volume 2 we saw that if we construct these operators as polynomials in fields,[27] then the three main requirements to a successful physical theory formulated in the Preface to Volume 2 (relativistic invariance, cluster separability and description of processes with a variable number of particles) are automatically satisfied.

However, it would be wrong to think that these three requirements can be valid *only* within the field framework. Our dressed formalism (RQD) also satisfies all three

27 See equations (**2**-3.6)–(**2**-3.7) and (**2**-3.13)–(**2**-3.18).

requirements, despite the fact that it uses the fieldless language of creation and annihilation operators.

It seems appropriate to conclude this subsection with the following quote:

> But what is the ontological status of those quantum fields that quantum field theory describes? Does reality consist of a four-dimensional spacetime at every point of which there is a collection of operators on an infinite-dimensional Hilbert space? ... But I hope you will agree that *you* are not a continuous field of operators on an infinite-dimensional Hilbert space. Nor, for that matter, is the page you are reading or the chair you are sitting in. Quantum fields are useful mathematical tools. They enable us to calculate things – N. D. Mermin [168].

8.6.3 Renormalization and dressing in a nutshell

So, we have abandoned the formalism of quantum fields in favor of dressed particles. When we took this step, we acquired something and lost something. Before discussing our gains and losses, let us briefly summarize the way which brought us to the dressed Hamiltonian $H^d = H_0 + V^d$ in Sections 3.1 and 5.2.

We started with the naïve field Hamiltonian QED $H^n = H_0 + V^n$ in Subsection **2**-3.1.3 (the upper left box in Figure 8.3) and demonstrated some of its good properties, such as the Poincaré invariance and cluster separability. However, when we decided to calculate the S-operator beyond the lowest nonvanishing (second) perturbation order (arrow (1) in Figure 8.3), we got a meaningless infinite result. This problem was partially

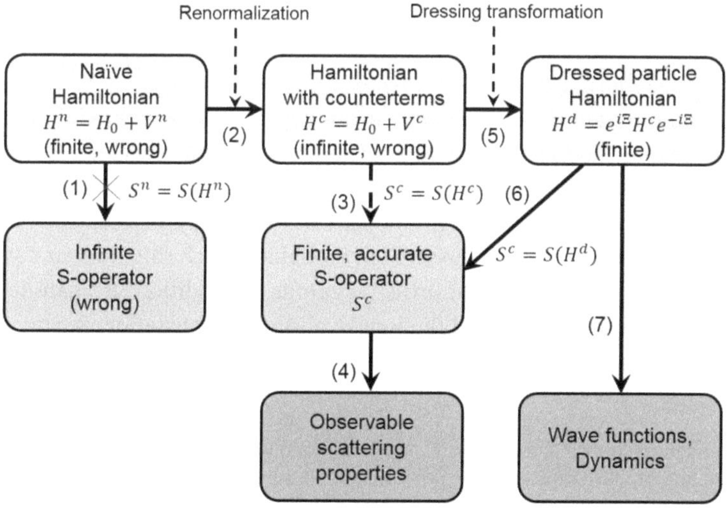

Figure 8.3: The logic of construction of the dressed Hamiltonian $H^d = H_0 + V^d$. Here $S(H)$ is the perturbation series (**1**-7.14)–(**1**-7.15), which allows us to calculate the S-operator from a generic Hamiltonian H. The same formula is applied to the Hamiltonian H^c in step (3) and to the dressed Hamiltonian H^d in step (6).

solved by renormalization theory in Chapter **2**-4 (arrow (2)): infinite counterterms were added to the Hamiltonian H^n, and thus a new Hamiltonian $H^c = H_0 + V^c$ was formulated. Although H^c was formally infinite, all divergences were successfully canceled in the S-operator (dashed arrow (3)). Moreover, after this cancellation some small residual terms (radiative corrections) remained in each perturbation order. By taking them into account, one could obtain very accurate results for observable scattering cross sections and energies of bound states (arrow (4)).

In our opinion, the above renormalization program of Tomonaga, Schwinger and Feynman, represented by arrow (2), was only the first step towards complete elimination of ultraviolet divergences from QFT. As a result of renormalization, QED divergences were "swept under the rug," and this rug turned out to be the Hamiltonian H^c with counterterms. This divergent operator was absolutely unacceptable as a generator of time translations. First, in the limit of infinite cutoff, the matrix elements of H^c on particle states were infinite. Second, the Hamiltonian H^c contained *unphys* terms like $a^\dagger b^\dagger c^\dagger$ and $a^\dagger c^\dagger a$, leading to unphysical time evolutions even in the simplest vacuum and one-particle states.[28]

In RQD, we proposed a way out of this difficulty. To remove *unphys* interactions, we applied a unitary dressing transformation to the Hamiltonian H^c (arrow (5)) and got a new "dressed" Hamiltonian,

$$H^d = e^{i\Xi} H^c e^{-i\Xi}, \tag{8.67}$$

with purely *phys* interactions. We succeeded in choosing a unitary operator $e^{i\Xi}$ in such a way that all divergences in the Hamiltonian canceled out.[29] In addition, after this transformation, the Poincaré invariance and cluster separability of the theory did not suffer, and the S-operator calculated with the help of H^d was exactly the same as the usual S-operator of the renormalized QED (arrow (6)).

The dressed Hamiltonian H^d has several advantages over the field Hamiltonian H^c. Unlike "trilinear" interactions in H^c, all terms in H^d have a clear, direct physical meaning and correspond to real observable physical processes (see Table 2.1). Both Hamiltonians H^c and H^d can be used to calculate scattering amplitudes and energies of bound states. However, only in the dressed formalism, such calculations can be carried out without regularization and artificial cutoffs. In addition, H^d can describe wave functions of bound states and the time evolution of nonstationary states in a straightforward manner (arrow (7)).

---

28 Although the divergences in the Hamiltonian H^c can be avoided by using the "similarity renormalization" approach [98, 97, 262], or within the framework of effective QED with a finite cutoff parameter Λ, the problem of the unphysical time evolutions (see Subsection 1.1.2) remains in all QED formulations that operate with *unphys* interactions, i. e., do not use dressing.

29 We can say that our approach has swept the divergence under another rug. This time the rug is the phase Ξ of the dressing transformation, but this operator has no physical meaning, so there is no harm if it turns out to be infinite.

It is also important that our "quantum theory of elementary particles" (which is based on the Hamiltonian H^d) is conceptually much simpler than the "theory of quantum fields" (based on the Hamiltonian H^c). RQD is very similar to ordinary quantum mechanics: states are described by normalized wave functions, the time evolution and scattering amplitudes are controlled by a finite well-defined Hamiltonian H^d without divergences, stationary states and their energies can be found by diagonalizing this Hamiltonian, the vacuum and individual particles are among its eigenstates, etc. The only significant difference from conventional quantum mechanics is the nonconservation of the number of particles in RQD, i. e., the processes of particle production and absorption are described adequately in this approach.

In Table 8.1 we have collected properties of the three main approaches to QED. We were interested in how the principal components of these theories (the S-operator, the Hamiltonian H as well as its eigenvectors $|\psi_i\rangle$ and eigenvalues E_i) depend on the ultraviolet cutoff Λ. Obviously, only the dressed RQD is completely cutoff-independent, as can be expected from a self-consistent theory.

Table 8.1: Comparison of the three approaches to QED.

Theory	**Naïve QED**	**Renormalized QED**	**RQD**
Hamiltonian	H^n	H^c	H^d
S depends on Λ?	yes	no	no
H depends on Λ?	no	yes	no
$|\psi_i\rangle$ depend on Λ?	yes	yes	no
E_i depend on Λ?	yes	no	no

8.6.4 What is next?

The above derivation of the dressed Hamiltonian H^d involved a series of questionable steps: "quantization of fields $\to$ renormalization $\to$ dressing." In other words, we first chose a wrong (field) Hamiltonian H^n and then, at the cost of huge efforts and complicated corrections, we formulated the new (dressed) Hamiltonian H^d, which is better suitable for describing physical reality. Perhaps, historically such a tortuous path was inevitable, but are all these steps absolutely necessary for constructing a realistic physical theory? Was nature designed to be so difficult? Maybe now, when we know the right answer (H^d), there is an opportunity to somehow shorten the procedure for constructing the final theory by omitting the problematic stages H^n and H^c?

Unfortunately, such a path has not been found yet, and we must rely on quantum fields and on clumsy renormalization and dressing procedures to arrive at a satisfactory theory of interacting elementary particles.

In fact, we are left with an unpleasant dilemma. On the one hand, we know that renormalized fields are unacceptable, because of known difficulties. But at the same time, this traditional field-based approach provides a relatively simple, convenient and accurate diagrammatic method for calculating scattering amplitudes in high orders of perturbation theory. On the other hand, the theory of dressed particles is physically transparent and mathematically flawless. But, unfortunately, in this approach, the rules for constructing realistic interactions are not known. In our book, we proposed to fit these interactions to the scattering amplitudes known from the traditional QED. But our approach is certainly unsatisfactory, because it does not allow the dressed theory to stand on its own two feet. Some new ideas are desperately needed.

9 Summary

Don't worry about people stealing your ideas. If your ideas are any good, you'll have to ram them down people's throats.
Howard Aiken

In this book we presented a new relativistic quantum theory of interactions. Our approach was based on two statements that contradict textbook wisdom:
(1) *Elementary particles are the primary constituents of matter.* These particles (electrons, protons, photons, etc.) obey the laws of quantum mechanics and interact with each other through instantaneously propagating potentials that depend on the distances between the particles and their velocities. These potentials can also change the number of particles.
(2) *The dynamical nature of boosts.* The perception of an interacting system by moving observers differs from the predictions of the special theory of relativity. Relativistic effects, such as length contraction and time dilation, are not universal. We claim the dependence of these effects on the state of the observed system and on interactions acting therein.

Our first statement about the primary role of particles and their direct interactions contradicts the fundamental assumptions of QFT and Maxwell's electrodynamics. We agree that quantum fields are a useful mathematical tool for the construction of invariant interactions and calculation of scattering amplitudes. However, in order to solve more general problems involving the time evolution of observables and properties of bound states, we recommend switching to the dressed Hamiltonian of particles with action-at-a-distance. This approach, among other things, completely solves the problem of ultraviolet divergences in QFT. In the classical limit, the Hamiltonian theory of directly interacting charges and magnetic moments is a viable alternative to the Maxwell field electrodynamics.

The most common argument against instantaneous interactions is their conflict with the special-relativistic prohibition of superluminal signaling. We explain this imaginary paradox, using our second statement about the dynamical (= interaction-dependent) character of boost transformations. This implies, in particular, that the textbook universal Lorentz transformations can be applied only to noninteracting systems. In interacting systems, boost transformations must contain small – but fundamentally important – corrections that depend on the particular state of the system and on interactions. These corrections are sufficient to make invalid the traditional "proof" of the incompatibility between instantaneous potentials and the principle of causality. In fact, there is no such incompatibility, and the instantaneous action-at-a-distance does not contradict any fundamental principle.

In most experimental situations, predictions of RQD do not differ from traditional approaches, or these differences are too small to be observed by modern means. There-

https://doi.org/10.1515/9783110493221-009

fore, experimental verification of RQD is a nontrivial task. The most convincing evidence in favor of RQD is the superluminal propagation of electromagnetic forces discovered in experiments performed by professor Pizzella's team. For further progress in understanding the nature of electromagnetic phenomena, a careful analysis of these results and their independent verification are urgently needed.

A Special theory of relativity

In this appendix, we briefly recall the main statements of Einstein's special theory of relativity [63, 64]. This theory is criticized in the main text of this volume.

A.1 Lorentz transformations for time and position

The fundamental idea of special relativity is that Minkowski space–time[1] $\mathcal{M}$ is the true representative of the physical space and time. In particular, it is assumed that the coordinates (ct, x, y, z) introduced in Appendix **1**-J.1 can be interpreted as space–time coordinates of real localized physical events. Minkowski himself described this union with the following words:

> From henceforth, space by itself and time by itself, have vanished into the merest shadows and only a kind of blend of the two exists in its own right.

Special relativity claims that 4×4 matrices (**1**-J.10) describe exactly boost transformations of 4-coordinates (ct, x, y, z) of all physical events. Suppose that observer O' moves with respect to O with rapidity $\boldsymbol{\theta}$. Suppose also that $(ct, \boldsymbol{x})$ are space–time coordinates of an event observed in the reference frame O. Then, according to special relativity, the 4-coordinates $(ct', \boldsymbol{x}')$ of the same event from the point of view of O' are given by equation (**1**-J.9), which is called the *Lorentz transformation for the time and position of the event.* In particular, if the rapidity vector is directed along the x-axis ($\boldsymbol{\theta} = (\theta, 0, 0)$), then the boost matrix becomes (**1**-J.11)

$$\tilde{\boldsymbol{\theta}}_x^{-1} = \begin{bmatrix} \cosh\theta & -\sinh\theta & 0 & 0 \\ -\sinh\theta & \cosh\theta & 0 & 0 \\ 0 & 0 & 1 & 0 \\ 0 & 0 & 0 & 1 \end{bmatrix} \tag{A.1}$$

and Lorentz transformations (**1**-J.9) can be written in a more familiar form,

$$ct' = ct\cosh\theta - x\sinh\theta, \tag{A.2}$$

$$x' = x\cosh\theta - ct\sinh\theta, \tag{A.3}$$

$$y' = y, \tag{A.4}$$

$$z' = z. \tag{A.5}$$

It is important to note that special relativity makes the following claim.

1 Formally introduced in Appendix **1**-J as a space of the pseudo-orthogonal representation of the Lorentz group.

https://doi.org/10.1515/9783110493221-010

Assertion A.1 (universality of Lorentz transformations). *Lorentz transformations* (A.2)–(A.5) *are exact and universal; they are applicable to all types of events in all physical systems; they do not depend on the physical nature of the events, on the composition of the physical system and on forces acting in the system.*

A.2 Manifest covariance

According to Assertion A.1, Lorentz transformations (A.2)–(A.5) are universal and kinematical. Therefore, in analogy with familiar three-dimensional scalars, vectors and tensors,[2] special relativity demands that all basic physical quantities must transform in a linear "manifestly covariant" way, like 4-scalars, 4-vectors or 4-tensors.

Assertion A.2 (manifest covariance of physical laws [64]). *Every general law of nature must be so constituted that it is transformed into a law of exactly the same form when, instead of the space–time variables t, x, y, z of the original coordinate system K, we introduce new space–time variables t', x', y', z' of a coordinate system K'. In this connection the relation between the ordinary and the accented magnitudes is given by the Lorentz transformation, or in brief, general laws of nature are covariant with respect to Lorentz transformations.*

As explained in many textbooks, Assertions A.1 and A.2 imply various secondary predictions of special relativity. One important consequence is the length contraction of all objects by a universal factor,

$$l' = l/\cosh\theta, \tag{A.6}$$

in the moving frame. So far, we do not have direct experimental confirmation of this prediction.

The other consequence is the well-established law of addition of velocities (**1**-4.7)–(**1**-4.9).

A.3 Decay of moving particle in special relativity

Another well-known result is that the time intervals measured by a moving clock increase $\cosh\theta$ times, i. e.,

$$\Delta t' = \Delta t \cosh\theta. \tag{A.7}$$

Experimentally, this *time dilation* effect was confirmed in observations of decays of fast-moving particles.

Suppose that from the point of view of O, an unstable particle is prepared at rest in the origin $x = y = z = 0$ of the coordinate system. Also suppose that the particle prepa-

2 See Appendices **1**-D.2–**1**-D.4.

ration happened at the time $t = 0$ ($\Upsilon(0,0) = 1$).[3] Then the observer O can associate the space–time point

$$(ct, x, y, z)_{\text{prep}} = (0, 0, 0, 0) \tag{A.8}$$

with the event "preparation of the particle." We know that the nondecay probability decreases with time according to the (almost) exponential law

$$\Upsilon(0, t) \approx \exp\left(-\frac{t}{\tau_0}\right). \tag{A.9}$$

At time $t = \tau_0$, the nondecay probability is $\Upsilon(0, \tau_0) = e^{-1}$. This "one lifetime" event has 4-coordinates

$$(ct, x, y, z)_{\text{life}} = (c\tau_0, 0, 0, 0) \tag{A.10}$$

in the frame O.

Let us now take the point of view of the moving observer O'. In accordance with the special theory of relativity, this observer also sees the two events "particle preparation" and "one lifetime," when the decay probabilities are equal to 1 and e^{-1}, respectively. However, O' attributes other 4-coordinates to these events. Substituting (A.8) and (A.10) into Lorentz formulas (A.2)–(A.5), we see that, from the point of view of O', the "particle preparation" event has the 4-coordinate $(0, 0, 0, 0)$, and the "one lifetime" event has the 4-coordinate $(c\tau_0 \cosh\theta, -c\tau_0 \sinh\theta, 0, 0)$. Hence, the time elapsed between these two events is $\cosh\theta$ times as long as in the reference frame O. In other words, the decay looks exactly $\cosh\theta$ times as slow as from the point of view of O'. This result is expressed by the famous Einstein "time dilation" formula,

$$\Upsilon(\theta, t) = \Upsilon(0, t/\cosh\theta), \tag{A.11}$$

which was confirmed in numerous experiments [200, 7, 199]. The most detailed verification was performed in experiments with relativistic muons [11, 75].[4]

A.4 Ban on superluminal signals

Perhaps, the most famous claim of special relativity is the following assertion.

Assertion A.3 (ban on superluminal signals). *No signal can propagate faster than the speed of light c.*

---

3 Here we denote $\Upsilon(\theta, t)$ the nondecay probability observed at time t from the frame O' moving with respect to O with the speed θ; see Chapter 4.

4 RQD calculations in Chapter 4 demonstrate the approximate character of equation (A.11).

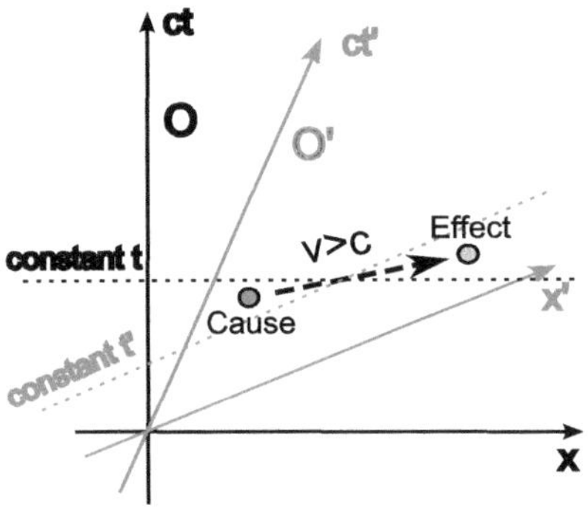

Figure A.1: Illustration to the "proof" that superluminal signals violate causality. In the rest frame O, the "cause" is below the line of constant time t, while the "effect" is above this line, as expected. In the moving frame O', the positions of these two events with respect to the line of constant t' are interchanged. This is a violation of causality.

The "proof" of this assertion [204] uses the *causality principle*, stating that *the cause precedes the effect in all reference frames*. Suppose that two events "cause" and "effect" are connected by a cause–effect relationship, but are separated by a space-like interval[5] in the reference frame O with the coordinate axes (ct, x), as shown in Figure A.1. In special relativity, 4-coordinates of the two events in the moving frame O' are obtained by Lorentz formulas (A.2)–(A.5). Graphically, this transformation can be represented by a *pseudo-rotation* of the coordinate axes $(ct, x) \to (ct', x')$, as shown in the figure. If the speed of the observer O' is high enough, then she can see the "effect" before the "cause," which clearly violates the principle of causality. Thus, to remain causal, Einstein's theory forbids superluminal signaling.[6]

5 This means that $|\Delta x| > c\Delta t$, i. e., the signal propagates from the "cause" to the "effect" at a speed exceeding the speed of light.

6 In Sections 7.3 and 7.4 we discuss a number of experiments that violate this prohibition and thus challenge special relativity.

B Unitary dressing transformation

In Section 2.3 we will need the following result.

Theorem B.1 (transformations preserving the S-operator). *A unitary transformation of the Hamiltonian*

$$H' = e^{i\Xi} H e^{-i\Xi}$$

preserves the S-operator if the Hermitian operator Ξ has the standard form (**2**-1.57)–(**2**-1.58)

$$\Xi = \sum_{N=0}^{\infty} \sum_{M=0}^{\infty} \Xi_{NM}, \tag{B.1}$$

$$\Xi_{NM} = \sum_{\{\eta,\eta'\}} \int d\boldsymbol{q}'_1 \dots d\boldsymbol{q}'_N d\boldsymbol{q}_1 \dots d\boldsymbol{q}_M D_{NM}(\boldsymbol{q}'_1\eta'_1; \dots; \boldsymbol{q}'_N\eta'_N; \boldsymbol{q}_1\eta_1; \dots; \boldsymbol{q}_M\eta_M)$$
$$\times \delta\left(\sum_{i=1}^{N} \boldsymbol{q}'_i - \sum_{j=1}^{M} \boldsymbol{q}_j\right) \alpha^\dagger_{\boldsymbol{q}'_1\eta'_1} \dots \alpha^\dagger_{\boldsymbol{q}'_N\eta'_N} \alpha_{\boldsymbol{q}_1\eta_1} \dots \alpha_{\boldsymbol{q}_M\eta_M},$$

where all terms Ξ_{NM} have smooth coefficient functions and are either (i) unphys *or* (ii) phys *and vanishing on their energy shells. An admixture of* renorm *terms in Ξ_{NM} is not acceptable, because it would destroy the scattering equivalence of the Hamiltonians H and H'.*

Idea of the proof. In our proof, we will use condition (**1**-7.32) of the scattering equivalence of two Hamiltonians. Since operator Ξ has the form (B.1), the left-hand side of (**1**-7.32) for each term Ξ_{NM} is

$$\lim_{t\to\pm\infty} e^{\frac{i}{\hbar}H_0 t} \Xi_{NM} e^{-\frac{i}{\hbar}H_0 t}$$
$$= \lim_{t\to\pm\infty} \sum_{\{\eta,\eta'\}} \int d\boldsymbol{q}'_1 \dots d\boldsymbol{q}'_N d\boldsymbol{q}_1 \dots d\boldsymbol{q}_M D_{NM}(\boldsymbol{q}'_1\eta'_1; \dots; \boldsymbol{q}'_N\eta'_N; \boldsymbol{q}_1\eta_1; \dots; \boldsymbol{q}_M\eta_M)$$
$$\times \delta\left(\sum_{i=1}^{N} \boldsymbol{q}'_i - \sum_{j=1}^{M} \boldsymbol{q}_j\right) e^{\frac{i}{\hbar}\mathcal{E}_{NM} t} \alpha^\dagger_{\boldsymbol{q}'_1\eta'_1} \dots \alpha^\dagger_{\boldsymbol{q}'_N\eta'_N} \alpha_{\boldsymbol{q}_1\eta_1} \dots \alpha_{\boldsymbol{q}_M\eta_M}, \tag{B.2}$$

where $\mathcal{E}_{NM}$ is the energy function. In the limit $t \to \pm\infty$, momentum integrals tend to zero by the Riemann–Lebesgue lemma **2**-A.1, because the coefficient functions D_{NM} are smooth, while the factor $e^{\frac{i}{\hbar}\mathcal{E}_{NM} t}$ rapidly oscillates in the momentum space.[1] So,

1 The oscillations are absent on the energy shell ($\mathcal{E}_{NM} = 0$), but this region is irrelevant, because $D_{NM} = 0$ there, according to the theorem's condition.

https://doi.org/10.1515/9783110493221-011

the limit in (B.2) is zero, and due to (**1**-7.34), the Hamiltonians H and H' are scattering-equivalent.

The operator Ξ cannot contain *renorm* terms, because their energy functions $\mathcal{E}_{NM}$ are identically zero everywhere, the products $e^{\frac{i}{\hbar}H_0 t}\Xi_{NM}e^{-\frac{i}{\hbar}H_0 t}$ are independent on t and the scattering equivalence condition (**1**-7.32) is violated. □

Theorem B.2. *If operator* Ξ *has a smooth coefficient function and is either* (i) unphys *or* (ii) phys *and vanishing on its energy shell,*[2] *then operator* (**2**-1.65)

$$\underline{\Xi} \equiv \frac{-i}{\hbar}\int_{-\infty}^{0} \Xi(t')dt' = \Xi \circ \frac{-1}{\mathcal{E}_{\Xi}}$$

is smooth.

Proof. The only possible source of singularity in $\underline{\Xi}$ is the factor $\mathcal{E}_{\Xi}^{-1}$ that diverges on the energy shell. However, operators Ξ, satisfying the theorem's condition, either do not have the energy shell at all, or vanish on the energy shell. In both cases the operator $\underline{\Xi}$ is nonsingular. □

2 That is, exactly as in the condition of Theorem B.1.

C Integral for decay law

Here we are interested in the integral (4.92), which occurs in the decay law (4.91) observed from a moving frame, i. e.,

$$I = \int_{-\infty}^{\infty} \frac{dm}{\Gamma^2/4 + (m - m_A)^2} e^{-\frac{it}{\hbar}\sqrt{m^2c^4 + p^2c^2}}$$
$$= c^2 \int_{-\infty}^{\infty} \frac{dz}{\Gamma^2 c^4/4 + (z - m_A c^2)^2} e^{-\frac{it}{\hbar}\sqrt{z^2 + p^2c^2}}. \tag{C.1}$$

To calculate this integral, let us go over to the complex plane of the variable $z = x + iy$, as shown in Figure C.1. Because of the two-valued square root in the exponent, we need to determine the sheet of the Riemann surface on which the integration will be performed. To do this, we make the cuts $[-ipc, +i\infty)$ and $[ipc, +i\infty)$ and define the desired integration contour $A \to B \to CL \to CR \to D \to E \to F \to A$. Within this contour, the argument $\arg(\sqrt{z^2 + p^2c^2})$ of the square root

$$\sqrt{z^2 + p^2c^2} \equiv \left|\sqrt{z^2 + p^2c^2}\right| e^{i \arg(\sqrt{z^2 + p^2c^2})}$$
$$\arg(\sqrt{z^2 + p^2c^2}) = \frac{1}{2}(\arg(z - ipc) + \arg(z + ipc))$$

changes continuously. For each segment of the contour, the quantities $\arg(z - ipc)$ and $\arg(z + ipc)$ are shown in Table C.1.

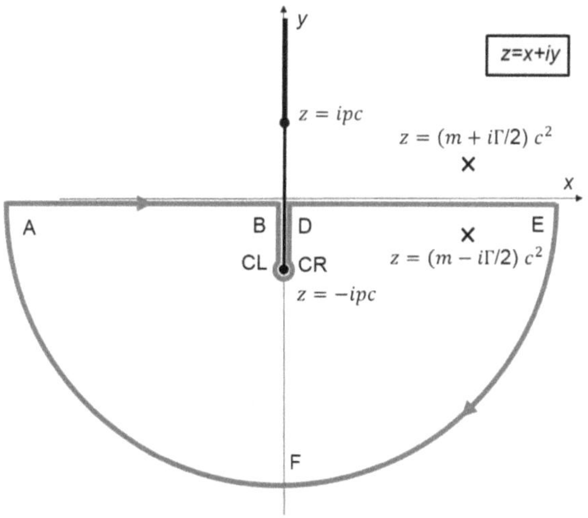

Figure C.1: The contour in the complex energy plane for calculating integral (C.1).

https://doi.org/10.1515/9783110493221-012

Table C.1: Elements of the integral (C.1) along the contour $A \to B \to CL \to CR \to D \to E \to F \to A$. We are interested in the limits $r \to 0$, $R \to \infty$.

Contour's segment	arg($z + ipc$)	arg($z - ipc$)	Change of variables
$A \to B$	$-\pi \to -3\pi/2$	$-\pi \to -\pi/2$	$z = x$
$B \to CL$	$-3\pi/2$	$-\pi/2$	$z = iy$
$CL \to CR$	$-3\pi/2 \to \pi/2$	$-\pi/2$	$z = -ipc + re^{i\phi}$
$CR \to D$	$\pi/2$	$-\pi/2$	$z = iy$
$D \to E$	$\pi/2 \to 0$	$-\pi/2 \to 0$	$z = x$
$E \to F \to A$	$0 \to -\pi/2 \to -\pi$	$0 \to -\pi/2 \to -\pi$	$z = Re^{i\phi}$

There is one pole[1] $z = mc^2 - i\Gamma c^2/2$ of the integrand inside this contour. Therefore, the whole integral is equal to the residue[2]

$$\int\limits_{A\to B} \dots dz + \int\limits_{B\to CL} \dots dz + \int\limits_{CL\to CR} \dots dz + \int\limits_{CR\to D} \dots dz + \int\limits_{D\to E} \dots dz + \int\limits_{E\to F\to A} \dots dz = -\frac{2\pi i}{-i\Gamma} e^{-\frac{it}{\hbar}\sqrt{p^2c^2+(m_A-i\Gamma/2)^2c^4}}.$$

Taking note that the sum $\int_{A\to B} + \int_{D\to E}$ is equal to our desired integral (C.1), we get

$$I = -\int\limits_{B\to CL} \dots dz - \int\limits_{CL\to CR} \dots dz - \int\limits_{CR\to D} \dots dz - \int\limits_{E\to F\to A} \dots dz + \frac{2\pi}{\Gamma} e^{\frac{-it}{\hbar}\sqrt{p^2c^2+(m_A-i\Gamma/2)^2c^4}}.$$

In this sum, the integral over the small circle $\int_{CL\to CR}$ is equal to zero, because the radius of the circle tends to zero, and the integrand is finite. The integral over the large half-circle $\int_{E\to F\to A}$ also tends to zero, because the argument of the exponent in the integrand has a large negative contribution. So we get

$$\begin{aligned} I = &-c^2 \int\limits_0^{-pc} \frac{idy}{\Gamma^2c^4/4 + (iy - m_Ac^2)^2} e^{\frac{it}{\hbar}\sqrt{p^2c^2-y^2}} \\ &- c^2 \int\limits_{-pc}^{0} \frac{idy}{\Gamma^2c^4/4 + (iy - m_Ac^2)^2} e^{-\frac{it}{\hbar}\sqrt{p^2c^2-y^2}} - \frac{2\pi i}{-i\Gamma} e^{-\frac{it}{\hbar}\sqrt{p^2c^2+(m_A-i\Gamma/2)^2c^4}} \\ = &\; ic^2 \int\limits_0^{pc} \frac{(e^{\frac{it}{\hbar}\sqrt{p^2c^2-s^2}} - e^{-\frac{it}{\hbar}\sqrt{p^2c^2-s^2}})}{\Gamma^2c^4/4 + (is + m_Ac^2)^2} ds + \frac{2\pi}{\Gamma} e^{-\frac{it}{\hbar}\sqrt{p^2c^2+(m_A-i\Gamma/2)^2c^4}} \end{aligned}$$

1 Indicated by a cross in Figure C.1.

2 An extra factor of (−1) is introduced, because the integration contour is traversed clockwise.

$$\approx -\int_0^{pc} \frac{2c^2 ds}{\Gamma^2 c^4/4 + (is + m_A c^2)^2} \sin\left(\frac{t}{\hbar}\sqrt{p^2c^2 - s^2}\right) + \frac{2\pi}{\Gamma} e^{-\frac{it}{\hbar}\sqrt{p^2c^2 + (m_A - i\Gamma/2)^2 c^4}}.$$

The integral in this expression cannot be calculated analytically. Therefore, it is necessary to introduce approximations [221]. In addition to conditions (4.90), suppose that $m_A c \ll p$, i. e., that the speed of the unstable particle is rather high. Then in the integration interval $s \in [0, pc]$ the factor $(\Gamma^2 c^4/4 + (is + m_A c^2)^2)^{-1}$ is a smooth function, whose maximum value $\approx 1/(m_A^2 c^4)$ is achieved near the left boundary of the interval and which decreases to $\approx -1/(p^2 c^2)$ at the right boundary. However, if the momentum pc is not too large, then the main contribution to the integral at large times t comes from the right boundary, because it got the slowest oscillations of the sinusoidal factor at $s \approx pc$. Therefore, this is not a big mistake if in the entire integration interval we set

$$\frac{2c^2}{\Gamma^2 c^4/4 + (is + m_A c^2)^2} \approx -\frac{2}{p^2}.$$

The resulting integral

$$\frac{2}{p^2} \int_0^{pc} ds \sin\left(\frac{t}{\hbar}\sqrt{p^2c^2 - s^2}\right)$$

can be found in tables,[3] so that finally

$$I \approx \frac{2\pi}{\Gamma} e^{-\frac{it}{\hbar}\sqrt{p^2c^2 + (m_A - i\Gamma/2)^2 c^4}} + \frac{\pi c}{p} J_1\left(\frac{pct}{\hbar}\right), \tag{C.2}$$

where J_1 is the Bessel function.

In the other extreme, we consider the case of low momentum, when $p \ll m_A c$ and in the entire integration interval $s \in [0, pc]$ the following approximation can be applied:

$$\frac{2c^2}{\Gamma^2 c^4/4 + (is + m_A c^2)^2} \approx \frac{2}{m_A^2 c^2}.$$

Then the desired integral is

$$\begin{aligned} I &\approx \frac{2\pi}{\Gamma} e^{-\frac{it}{\hbar}\sqrt{p^2c^2 + (m_A - i\Gamma/2)^2 c^4}} - \frac{2}{m_A^2 c^2} \int_0^{pc} dx \sin(t\sqrt{p^2c^2 - x^2}) \\ &= \frac{2\pi}{\Gamma} e^{-it\sqrt{p^2c^2 + (m_A - i\Gamma/2)^2 c^4}} - \frac{\pi p}{m_A^2 c} J_1\left(\frac{pct}{\hbar}\right). \end{aligned} \tag{C.3}$$

3 See formula 3.711 in [99] and equation 2.5.25.1 in [190].

D Coulomb scattering integral in fourth order

Here we would like to calculate the 3D integral

$$w(\boldsymbol{q},\boldsymbol{q}') = \int \frac{d\boldsymbol{s}}{[(\boldsymbol{q}-\boldsymbol{s})^2+\lambda^2c^2][s^2-q^2][(\boldsymbol{q}'-\boldsymbol{s})^2+\lambda^2c^2]},$$

which appeared in equation (5.16) for the commutator term in the fourth-order electron–proton potential. In this integral, we are interested in dominant terms that survive in the limit $\lambda \to 0$. The calculation method was taken from § 121 in [14].[1]

First we use (**2**-F.4)[2] and the elastic scattering condition $(q')^2 = q^2$ to write

$$\begin{aligned} w(\boldsymbol{q},\boldsymbol{q}') &= 2\int_0^1 dx \int_0^{1-x} dy \int d\boldsymbol{s} \\ &\quad \times [((\boldsymbol{q}-\boldsymbol{s})^2+\lambda^2c^2)x + ((\boldsymbol{q}'-\boldsymbol{s})^2+\lambda^2c^2)y + (s^2-q^2)(1-x-y)]^{-3} \\ &= 2\int_0^1 dx \int_0^{1-x} dy \int d\boldsymbol{s} \\ &\quad \times [s^2 - 2(\boldsymbol{q}\cdot\boldsymbol{s})x - 2(\boldsymbol{q}'\cdot\boldsymbol{s})y + \lambda^2c^2(x+y) + q^2(2x+2y-1)]^{-3}. \end{aligned}$$

Next we shift integration variables $\boldsymbol{s} \to \boldsymbol{h} \equiv \boldsymbol{s} - x\boldsymbol{q} - y\boldsymbol{q}'$ and take into account that the square of the transferred momentum is $k^2 \equiv (\boldsymbol{q}'-\boldsymbol{q})^2 = 2q^2 - 2(\boldsymbol{q}\cdot\boldsymbol{q}')$, so that

$$\begin{aligned} w(\boldsymbol{q},\boldsymbol{q}') &= 2\int_0^1 dx \int_0^{1-x} dy \int d\boldsymbol{h} \\ &\quad \times [h^2 + q^2(-x^2-y^2+2x+2y-1) - 2(\boldsymbol{q}\cdot\boldsymbol{q}')xy + \lambda^2c^2(x+y)]^{-3} \\ &= 2\int_0^1 dx \int_0^{1-x} dy \int \frac{d\boldsymbol{h}}{[h^2 - q^2(x+y-1)^2 + k^2xy + \lambda^2c^2(x+y)]^3}. \end{aligned}$$

For the integral with respect to $\boldsymbol{h}$ we use the table formula

$$\int \frac{d\boldsymbol{h}}{(h^2-a)^3} = 4\pi \int_0^\infty \frac{h^2 dh}{(h^2-a)^3} = -\frac{i\pi^2}{4a^{3/2}},$$

---

1 See also [124].

2 Where we set $a = (\boldsymbol{q}-\boldsymbol{s})^2 + \lambda^2c^2$, $b = (\boldsymbol{q}'-\boldsymbol{s})^2 + \lambda^2c^2$, $c = s^2 - q^2$.

https://doi.org/10.1515/9783110493221-013

so that

$$w(\boldsymbol{q},\boldsymbol{q}') = -\frac{i\pi^2}{2}\int_0^1 dx \int_0^{1-x} dy \frac{1}{[q^2(x+y-1)^2 - k^2xy - \lambda^2c^2(x+y)]^{3/2}}. \tag{D.1}$$

Our next change of variables $\xi = x + y$, $\zeta = x - y$ has the Jacobian

$$J = \det\left[\frac{\partial(\xi,\zeta)}{(x,y)}\right] = -2,$$

so that

$$\int \ldots dxdy = \int \ldots \frac{1}{|J|} d\xi d\zeta = \frac{1}{2}\int \ldots d\xi d\zeta.$$

The integration area is shown by the triangle $0AC$ in Figure D.1. This area does not have a simple description in terms of the variables ξ and ζ. However, the integrand is an even function of ζ, so it is sufficient to integrate only inside the (hatched) half $0BC$ of the triangle and multiply the result by 2. Taking into account all these considerations, we obtain

$$\begin{aligned} w(\boldsymbol{q},\boldsymbol{q}') &= -\frac{i\pi^2}{2}\int_0^1 d\xi \int_0^{\xi} \frac{d\zeta}{(q^2(\xi-1)^2 - k^2\xi^2/4 + k^2\zeta^2/4 - \lambda^2c^2\xi)^{3/2}} \\ &= -\frac{i\pi^2}{2}\int_0^1 \xi d\xi \frac{1}{[q^2(\xi-1)^2 - k^2\xi^2/4 - \lambda^2c^2\xi^2]\sqrt{q^2(\xi-1)^2 - \lambda^2c^2\xi}}. \end{aligned}$$

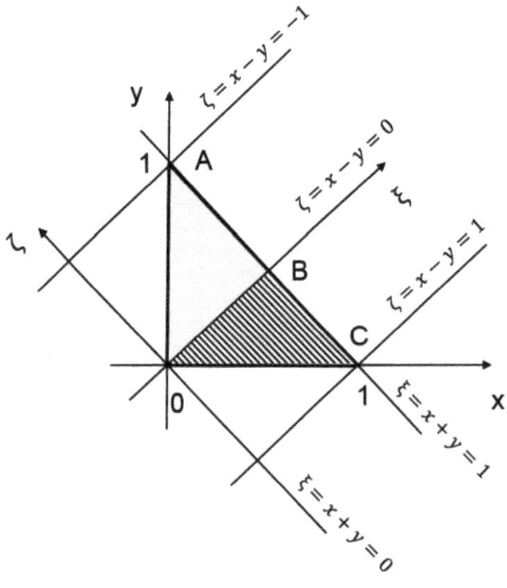

Figure D.1: To the calculation of the integral (D.1).

Next, introduce a parameter δ such that $1 \gg \delta \gg \lambda^2 c^2/q^2$ and divide the integration interval into two parts, i. e.,

$$w(\boldsymbol{q},\boldsymbol{q}') = w^{(1)}(\boldsymbol{q},\boldsymbol{q}') + w^{(2)}(\boldsymbol{q},\boldsymbol{q}') = -\frac{i\pi^2}{2}\int_0^{1-\delta}\dots d\xi - \frac{i\pi^2}{2}\int_{1-\delta}^{1}\dots d\xi.$$

In the first integral, we neglect the λ-term, i. e.,

$$\begin{aligned} w^{(1)}(\boldsymbol{q},\boldsymbol{q}') &\approx -\frac{i\pi^2}{2q^3}\int_0^{1-\delta}\frac{\xi d\xi}{[(\xi-1)^2 - k^2\xi^2/(4q^2)](\xi-1)} \\ &= \frac{i\pi^2}{2q^3}\cdot\frac{2q^2}{k^2}\ln\left(\frac{-k^2\xi^2/(4q^2) + \xi^2 - 2\xi + 1}{(1-\xi)^2}\right)\Bigg|_{\xi=0}^{\xi=1-\delta} \\ &\approx \frac{i\pi^2}{qk^2}\ln\left(\frac{-k^2}{4q^2\delta^2}\right). \end{aligned}$$

In the second integral, change the variable $y = \xi - 1$, i. e.,

$$\begin{aligned} w^{(2)}(\boldsymbol{q},\boldsymbol{q}') &\approx -\frac{i\pi^2}{2}\int_{-\delta}^{0}\frac{dy(y+1)}{(q^2y^2 - k^2/4)\sqrt{q^2y^2 - \lambda^2c^2}} \approx \frac{2i\pi^2}{k^2}\int_0^{\delta}\frac{dy}{\sqrt{q^2y^2-\lambda^2c^2}} \\ &= \frac{2i\pi^2}{qk^2}\ln(q\sqrt{q^2y^2-\lambda^2c^2} + q^2y)\Bigg|_{y=0}^{y=\delta} \\ &= \frac{2i\pi^2}{qk^2}\ln\left(\frac{q\sqrt{q^2\delta^2-\lambda^2c^2} + q^2\delta}{iq\lambda c}\right) \approx \frac{i\pi^2}{qk^2}\ln\left(\frac{-4q^2\delta^2}{\lambda^2c^2}\right). \end{aligned}$$

Putting together both parts, we finally obtain

$$w(\boldsymbol{q},\boldsymbol{q}') \approx \frac{i\pi^2}{qk^2}\ln\left(\frac{-k^2}{4q^2\delta^2}\cdot\frac{-4q^2\delta^2}{\lambda^2c^2}\right) = \frac{i\pi^2}{qk^2}\ln\left(\frac{k^2}{\lambda^2c^2}\right), \tag{D.2}$$

which coincides with equation (121.16) in [14].

E Relativistic invariance of Coulomb–Darwin–Breit electrodynamics

In this appendix we are going to prove the relativistic invariance of the classical limit of RQD discussed in Subsection 6.2.1.

From the derivation presented there, it follows that the Coulomb–Darwin–Breit Hamiltonian (6.6) + (6.7) is a part of a relativistically invariant theory in the instant form of dynamics. This means that there must exist an interacting boost operator $\boldsymbol{K}^d$ that satisfies all commutation relations of the Poincaré Lie algebra together with the Hamiltonian H^d. In principle, we could obtain an explicit form of the operator $\boldsymbol{K}^d$ by applying the unitary dressing transformation to the boost operator of QED (see Subsection 2.3.6). However, here we decided to choose an easier way. Together with [36, 34, 152], we simply postulate the form of the operator $\boldsymbol{K}^d$ and check the validity of Poincaré commutators in the $(v/c)^2$ approximation.

Let us first write two-particle noninteracting generators of the Poincaré group as sums of single-particle generators, i. e.,[1]

$$\boldsymbol{P}_0 = \boldsymbol{p}_1 + \boldsymbol{p}_2, \tag{E.1}$$

$$\boldsymbol{J}_0 = [\boldsymbol{r}_1 \times \boldsymbol{p}_1] + \boldsymbol{s}_1 + [\boldsymbol{r}_2 \times \boldsymbol{p}_2] + \boldsymbol{s}_2, \tag{E.2}$$

$$H_0 = h_1 + h_2 \approx m_1c^2 + m_2c^2 + \frac{p_1^2}{2m_1} + \frac{p_2^2}{2m_2} - \frac{p_1^4}{8m_1^3c^2} - \frac{p_2^4}{8m_2^3c^2}, \tag{E.3}$$

$$\begin{aligned}\boldsymbol{K}_0 &= -\frac{h_1\boldsymbol{r}_1}{c^2} - \frac{[\boldsymbol{p}_1 \times \boldsymbol{s}_1]}{m_1c^2 + h_1} - \frac{h_2\boldsymbol{r}_2}{c^2} - \frac{[\boldsymbol{p}_2 \times \boldsymbol{s}_2]}{m_2c^2 + h_2} \\ &\approx -m_1\boldsymbol{r}_1 - m_2\boldsymbol{r}_2 - \frac{p_1^2\boldsymbol{r}_1}{2m_1c^2} - \frac{p_2^2\boldsymbol{r}_2}{2m_2c^2} + \frac{1}{2c^2}\left(\frac{[\boldsymbol{s}_1 \times \boldsymbol{p}_1]}{m_1} + \frac{[\boldsymbol{s}_2 \times \boldsymbol{p}_2]}{m_2}\right).\end{aligned} \tag{E.4}$$

The full interacting generators are

$$\begin{aligned}H^d &= H_0 + V^d, \\ \boldsymbol{K}^d &= \boldsymbol{K}_0 + \boldsymbol{Z}^d,\end{aligned} \tag{E.5}$$

where the potential energy V^d is given by equation (6.7) and the potential boost was postulated in [36, 34, 152] as follows:

$$\boldsymbol{Z}^d \approx -\frac{q_1q_2(\boldsymbol{r}_1 + \boldsymbol{r}_2)}{8\pi c^2 r}. \tag{E.6}$$

We only need to verify nontrivial Poisson brackets of the Poincaré Lie algebra (**1**-3.49)–(**1**-3.55) that contain interacting generators H^d and $\boldsymbol{K}^d$, i. e.,

$$[J_{0i}, K_j^d]_P = \sum_{k=1}^{3} \epsilon_{ijk}K_k^d, \tag{E.7}$$

1 See equations (8.18)–(8.21) and (**1**-4.46)–(**1**-4.48).

https://doi.org/10.1515/9783110493221-014

$$[\boldsymbol{J}_0, H^d]_P = [\boldsymbol{P}_0, H^d]_P = 0, \tag{E.8}$$

$$[K_i^d, K_j^d]_P = -\frac{1}{c^2}\sum_{k=1}^{3} \epsilon_{ijk} J_{0k}, \tag{E.9}$$

$$[K_i^d, P_{0j}]_P = -\frac{1}{c^2} H^d \delta_{ij}, \tag{E.10}$$

$$[\boldsymbol{K}^d, H^d]_P = -\boldsymbol{P}_0, \tag{E.11}$$

where $i, j, k = (x, y, z)$.

Equations (E.7)–(E.8) mean that $\boldsymbol{K}^d$ is a 3-vector and H^d is a 3-scalar invariant with respect to translations. These results follow easily from Poisson brackets (8.13)–(8.17) of single-particle observables. This proof is left as an exercise for the reader.

For more complex brackets (E.9)–(E.11), it will be convenient to write H^d and $\boldsymbol{K}^d$ as series in powers of $(v/c)^2$ (upper indices in parentheses indicate the power of $(v/c)^2$), i. e.,

$$H^d \approx H^{(-1)} + H^{(0)} + H^{(1)}_{\text{orb}} + H^{(1)}_{\text{spin-orb}} + H^{(1)}_{\text{spin-spin}},$$
$$\boldsymbol{K}^d \approx \boldsymbol{K}^{(0)} + \boldsymbol{K}^{(1)}_{\text{orb}} + \boldsymbol{K}^{(1)}_{\text{spin-orb}},$$

where

$$\begin{aligned}
H^{(-1)} &= m_1c^2 + m_2c^2, \\
H^{(0)} &= \frac{p_1^2}{2m_1} + \frac{p_2^2}{2m_2} + \frac{q_1q_2}{4\pi r}, \\
H^{(1)}_{\text{orb}} &= -\frac{p_1^4}{8m_1^3c^2} - \frac{p_2^4}{8m_2^3c^2} - \frac{q_1q_2}{8\pi m_1m_2c^2r}\left((\boldsymbol{p}_1\cdot\boldsymbol{p}_2) + \frac{(\boldsymbol{p}_1\cdot\boldsymbol{r})(\boldsymbol{p}_2\cdot\boldsymbol{r})}{r^2}\right), \\
H^{(1)}_{\text{spin-orb}} &= \frac{q_1q_2[\boldsymbol{r}\times\boldsymbol{p}_1]\cdot\boldsymbol{s}_1}{8\pi m_1^2c^2r^3} - \frac{q_1q_2[\boldsymbol{r}\times\boldsymbol{p}_2]\cdot\boldsymbol{s}_2}{8\pi m_2^2c^2r^3} \\
&\quad - \frac{q_1q_2[\boldsymbol{r}\times\boldsymbol{p}_2]\cdot\boldsymbol{s}_1}{4\pi m_1m_2c^2r^3} + \frac{q_1q_2[\boldsymbol{r}\times\boldsymbol{p}_1]\cdot\boldsymbol{s}_2}{4\pi m_1m_2c^2r^3}, \\
H^{(1)}_{\text{spin-spin}} &= \frac{q_1q_2(\boldsymbol{s}_1\cdot\boldsymbol{s}_2)}{4\pi m_1m_2c^2r^3} - \frac{3q_1q_2(\boldsymbol{s}_1\cdot\boldsymbol{r})(\boldsymbol{s}_2\cdot\boldsymbol{r})}{4\pi m_1m_2c^2r^5}, \\
\boldsymbol{K}^{(0)} &= -m_1\boldsymbol{r}_1 - m_2\boldsymbol{r}_2, \\
\boldsymbol{K}^{(1)}_{\text{orb}} &= -\frac{p_1^2\boldsymbol{r}_1}{2m_1c^2} - \frac{p_2^2\boldsymbol{r}_2}{2m_2c^2} - \frac{q_1q_2(\boldsymbol{r}_1+\boldsymbol{r}_2)}{8\pi c^2 r}, \\
\boldsymbol{K}^{(1)}_{\text{spin-orb}} &= \frac{1}{2c^2}\left(\frac{[\boldsymbol{s}_1\times\boldsymbol{p}_1]}{m_1} + \frac{[\boldsymbol{s}_2\times\boldsymbol{p}_2]}{m_2}\right).
\end{aligned}$$

So, we have to prove the following relations:

$$-\frac{1}{c^2}H^{(-1)}\delta_{ij} = [K_i^{(0)}, P_{0j}]_P, \tag{E.12}$$

$$0 = [K_i^{(0)}, H^{(-1)}]_P, \tag{E.13}$$

$$-P_{0i} = [K_i^{(0)}, H^{(0)}]_P, \tag{E.14}$$

$$0 = [K_i^{(0)}, K_j^{(0)}]_P, \tag{E.15}$$

$$-\frac{1}{c^2} H^{(0)} \delta_{ij} = [K_{i\text{-orb}}^{(1)}, P_{0j}]_P + [K_{i\text{-spin-orb}}^{(1)}, P_{0j}]_P, \tag{E.16}$$

$$\begin{aligned} 0 = {} & [K_{i\text{-orb}}^{(1)}, H^{(0)}]_P + [K_{i\text{-spin-orb}}^{(1)}, H^{(0)}]_P \\ & + [K_i^{(0)}, H_{\text{orb}}^{(1)}]_P + [K_i^{(0)}, H_{\text{spin-orb}}^{(1)}]_P \\ & + [K_i^{(0)}, H_{\text{spin-spin}}^{(1)}]_P, \end{aligned} \tag{E.17}$$

$$\begin{aligned} -\frac{1}{c^2} \sum_{k=1}^{3} \epsilon_{ijk} J_{0k} = {} & [K_{i\text{-orb}}^{(1)}, K_j^{(0)}]_P + [K_{i\text{-spin-orb}}^{(1)}, K_j^{(0)}]_P \\ & + [K_i^{(0)}, K_{j\text{-orb}}^{(1)}]_P + [K_i^{(0)}, K_{j\text{-spin-orb}}^{(1)}]_P. \end{aligned} \tag{E.18}$$

Again, we omit the rather simple proofs of formulas (E.12)–(E.15). For equation (E.16) we obtain

$$\begin{aligned} & [(K_{x\text{-orb}}^{(1)} + K_{x\text{-spin-orb}}^{(1)}), P_{0x}]_P \\ & = -\left[\frac{p_1^2 r_{1x}}{2m_1 c^2}, p_{1x}\right]_P - \left[\frac{p_2^2 r_{2x}}{2m_2 c^2}, p_{2x}\right]_P - \left[\frac{q_1 q_2}{8\pi c^2} \frac{r_{1x} + r_{2x}}{r}, p_{1x}\right]_P \\ & \quad - \left[\frac{q_1 q_2}{8\pi c^2} \frac{r_{1x} + r_{2x}}{r}, p_{2x}\right]_P \\ & = -\frac{p_1^2}{2m_1 c^2} - \frac{p_2^2}{2m_2 c^2} - \frac{q_1 q_2}{4\pi c^2 r} \\ & = -\frac{1}{c^2} H^{(0)}. \end{aligned}$$

Individual terms on the right-hand side of (E.17) are

$$\begin{aligned} [K_{x\text{-orb}}^{(1)}, H^{(0)}]_P = {} & -\frac{p_1^2}{4m_1^2 c^2}[r_{1x}, p_1^2]_P - \frac{q_1 q_2 r_{1x}}{8\pi m_1 c^2}\left[p_1^2, \frac{1}{r}\right]_P - \frac{p_2^2}{4m_2^2 c^2}[r_{2x}, p_2^2]_P \\ & - \frac{q_1 q_2 r_{2x}}{8\pi m_2 c^2}\left[p_2^2, \frac{1}{r}\right]_P - \frac{q_1 q_2}{16\pi m_1 c^2 r}[r_{1x}, p_1^2]_P - \frac{q_1 q_2 r_{1x}}{16\pi m_1 c^2}\left[\frac{1}{r}, p_1^2\right]_P \\ & - \frac{q_1 q_2 r_{2x}}{16\pi m_1 c^2}\left[\frac{1}{r}, p_1^2\right]_P - \frac{q_1 q_2 r_{1x}}{16\pi m_2 c^2}\left[\frac{1}{r}, p_2^2\right]_P \\ & - \frac{q_1 q_2}{16\pi m_2 c^2 r}[r_{2x}, p_2^2]_P - \frac{q_1 q_2 r_{2x}}{16\pi m_2 c^2}\left[\frac{1}{r}, p_2^2\right]_P \\ = {} & -\frac{p_1^2 p_{1x}}{2m_1^2 c^2} - \frac{q_1 q_2 (r_{1x} - r_{2x})}{8\pi m_1 c^2} \frac{(\boldsymbol{p}_1 \cdot \boldsymbol{r})}{r^3} - \frac{p_2^2 p_{2x}}{2m_2^2 c^2} \\ & - \frac{q_1 q_2 (r_{1x} - r_{2x})}{8\pi m_2 c^2} \frac{(\boldsymbol{p}_2 \cdot \boldsymbol{r})}{r^3} - \frac{q_1 q_2 p_{1x}}{8\pi m_1 c^2 r} - \frac{q_1 q_2 p_{2x}}{8\pi m_2 c^2 r}, \end{aligned} \tag{E.19}$$

$$[K^{(1)}_{x\text{-spin-orb}}, H^{(0)}]_P = \frac{1}{2c^2}\left[\left(\frac{1}{m_1}[\boldsymbol{s}_1 \times \boldsymbol{p}_1]_x + \frac{1}{m_2}[\boldsymbol{s}_2 \times \boldsymbol{p}_2]_x\right), \frac{q_1 q_2}{4\pi r}\right]_P$$
$$= -\frac{q_1 q_2[\boldsymbol{s}_1 \times \boldsymbol{r}]_x}{8\pi m_1 c^2 r^3} + \frac{q_1 q_2[\boldsymbol{s}_2 \times \boldsymbol{r}]_x}{8\pi m_2 c^2 r^3}, \tag{E.20}$$

$$[K^{(0)}_x, H^{(1)}_{\text{orb}}]_P = \frac{1}{8m_1^2 c^2}[r_{1x}, p_1^4]_P + \frac{q_1 q_2}{8\pi m_2 c^2 r}\left[r_{1x}, \left((\boldsymbol{p}_1 \cdot \boldsymbol{p}_2) + \frac{(\boldsymbol{p}_1 \cdot \boldsymbol{r})(\boldsymbol{p}_2 \cdot \boldsymbol{r})}{r^2}\right)\right]_P$$
$$+ \frac{1}{8m_2^2 c^2}[r_{2x}, p_2^4]_P + \frac{q_1 q_2}{8\pi m_1 c^2 r}\left[r_{2x}, \left((\boldsymbol{p}_1 \cdot \boldsymbol{p}_2) + \frac{(\boldsymbol{p}_1 \cdot \boldsymbol{r})(\boldsymbol{p}_2 \cdot \boldsymbol{r})}{r^2}\right)\right]_P$$
$$= \frac{p_1^2 p_{1x}}{2m_1^2 c^2} + \frac{q_1 q_2}{8\pi m_2 c^2 r}\left(p_{2x} + \frac{(r_{1x} - r_{2x})(\boldsymbol{p}_2 \cdot \boldsymbol{r})}{r^2}\right)$$
$$+ \frac{p_2^2 p_{2x}}{2m_2^2 c^2} + \frac{q_1 q_2}{8\pi m_1 c^2 r}\left(p_{1x} + \frac{(\boldsymbol{p}_1 \cdot \boldsymbol{r})(r_{1x} - r_{2x})}{r^2}\right), \tag{E.21}$$

$$[K^{(0)}_x, H^{(1)}_{\text{spin-orb}}]_P = \left[(-m_1 r_{1x} - m_2 r_{2x}), \left(\frac{q_1 q_2[\boldsymbol{r} \times \boldsymbol{p}_1] \cdot \boldsymbol{s}_1}{8\pi m_1^2 c^2 r^3}\right.\right.$$
$$\left.\left. - \frac{q_1 q_2[\boldsymbol{r} \times \boldsymbol{p}_2] \cdot \boldsymbol{s}_2}{8\pi m_2^2 c^2 r^3} - \frac{q_1 q_2[\boldsymbol{r} \times \boldsymbol{p}_2] \cdot \boldsymbol{s}_1}{4\pi m_1 m_2 c^2 r^3} + \frac{q_1 q_2[\boldsymbol{r} \times \boldsymbol{p}_1] \cdot \boldsymbol{s}_2}{4\pi m_1 m_2 c^2 r^3}\right)\right]_P$$
$$= \frac{q_1 q_2[\boldsymbol{s}_2 \times \boldsymbol{r}]_x}{8\pi m_2 c^2 r^3} - \frac{q_1 q_2[\boldsymbol{s}_1 \times \boldsymbol{r}]_x}{8\pi m_1 c^2 r^3} - \frac{q_1 q_2[\boldsymbol{s}_2 \times \boldsymbol{r}]_x}{4\pi m_2 c^2 r^3} + \frac{q_1 q_2[\boldsymbol{s}_1 \times \boldsymbol{r}]_x}{4\pi m_1 c^2 r^3}$$
$$= \frac{q_1 q_2[\boldsymbol{s}_1 \times \boldsymbol{r}]_x}{8\pi m_1 c r^3} - \frac{q_1 q_2[\boldsymbol{s}_2 \times \boldsymbol{r}]_x}{8\pi m_2 c r^3}, \tag{E.22}$$

$$[K^{(0)}_x, H^{(1)}_{\text{spin-spin}}]_P = 0. \tag{E.23}$$

Adding together right-hand sides of equations (E.19)–(E.23), we see that equation (E.17) is, indeed, correct. Equation (E.18) is derived as follows:

$$[K^{(1)}_{x\text{-orb}}, K^{(0)}_y]_P + [K^{(0)}_x, K^{(1)}_{y\text{-orb}}]_P + [K^{(1)}_{x\text{-spin-orb}}, K^{(0)}_y]_P + [K^{(0)}_x, K^{(1)}_{y\text{-spin-orb}}]_P$$
$$= \frac{r_{1x}}{2c^2}[p_1^2, r_{1y}]_P + \frac{r_{2x}}{2c^2}[p_2^2, r_{2y}]_P + \frac{r_{1y}}{2c^2}[r_{1x}, p_1^2]_P + \frac{r_{2y}}{2c^2}[r_{2x}, p_2^2]_P$$
$$- \frac{1}{2c^2}\left[\left(-\frac{1}{m_1}s_{1z}p_{1y} - \frac{1}{m_2}s_{2z}p_{2y}\right), (m_1 r_{1y} + m_2 r_{2y})\right]_P$$
$$- \frac{1}{2c^2}\left[(m_1 r_{1x} + m_2 r_{2x}), \left(\frac{1}{m_1}s_{1z}p_{1x} + \frac{1}{m_2}s_{2z}p_{2x}\right)\right]_P$$
$$= -\frac{1}{c^2}[\boldsymbol{r}_1 \times \boldsymbol{p}_1]_z - \frac{1}{c^2}[\boldsymbol{r}_2 \times \boldsymbol{p}_2]_z - \frac{1}{c^2}(s_{1z} + s_{2z}) = -\frac{J_{0z}}{c^2}.$$

Bibliography

[1] Aguirregabiria JM, Hernández A, Rivas M. Linear momentum density in quasistatic electromagnetic systems. Eur J Phys. 2004;25:555. arXiv:physics/0404139v1.

[2] Aharonov Y, Bohm D. Significance of electromagnetic potentials in quantum mechanics. Phys Rev. 1959;115:485.

[3] Aharonov Y, Pearle P, Vaidman L. Comment on "Proposed Aharonov–Casher effect: another example of an Aharonov–Bohm effect arising from a classical lag". Phys Rev A. 1988;37:4052.

[4] Alväger T, Farley FJM, Kjellman J, Wallin I. Test of the second postulate of special relativity in the GeV region. Phys Lett. 1964;12:260.

[5] Aničin I. Heuristic arguments in favor of the view that the forces which form the bound systems of elementary particles propagate instantaneously. 2016. arXiv:1603.02907v1.

[6] Aref'eva IYa. Renormalized scattering theory for the Lee model. Theor Math Phys. 1972;12:859.

[7] Ayres DS, Cormack AM, Greenberg AJ, Kenney RW, Cladwell DO, Elings VB, Hesse WP, Morrison RJ. Measurements of the lifetime of positive and negative pions. Phys Rev D. 1971;3:1051.

[8] Babson D, Reynolds SP, Bjorkquist R, Griffiths DJ. Hidden momentum, field momentum and electromagnetic impulse. Am J Phys. 2009;77:826.

[9] Bacry H. The notions of localizability and space: from Eugene Wigner to Alain Connes. Nucl Phys B, Proc Suppl. 1989;6:222.

[10] Bacry H. The foundations of the Poincaré group and the validity of general relativity. Rep Math Phys. 2004;53:443.

[11] Bailey J, Borer K, Combley F, Drumm H, Kreinen F, Lange F, Picasso E, von Ruden W, Farley FJM, Field JH, Flegel W, Hattersley PM. Measurements of relativistic time dilatation for positive and negative muons in a circular orbit. Nature. 1977;268:301.

[12] Bakamjian B, Thomas LH. Relativistic particle dynamics. II. Phys Rev. 1953;92:1300.

[13] Ballentine LE. Quantum mechanics: a modern development. Singapore: World Scientific; 1998.

[14] Berestetskii VB, Livshitz EM, Pitaevskii LP. Quantum electrodynamics. Oxford: Elsevier; 1982.

[15] Bjorken JD, Drell SD. Relativistic quantum mechanics. New York: McGraw-Hill; 1964.

[16] Boyer TH. Darwin–Lagrangian analysis for the interaction of a point charge and a magnet: Considerations related to the controversy regarding the Aharonov–Bohm and Aharonov–Casher phase shifts. J Phys A, Math Gen. 2006;39:3455. arXiv:physics/0506181v1.

[17] Boyer TH. The paradoxical forces for the classical electromagnetic lag associated with the Aharonov–Bohm phase shift. Found Phys. 2006;19:491. arXiv:physics/0506180v1.

[18] Boyer TH. Unresolved classical electromagnetic aspects of the Aharonov–Bohm phase shift. AIP Conf Proc. 2007;962:39. arXiv:0709.0661v1.

[19] Boyer TH. Comment on experiments related to the Aharonov–Bohm phase shift. Found Phys. 2008;38:498. arXiv:0708.3194v1.

[20] Boyer TH. Illustrating some implications of the conservation laws in relativistic mechanics. Am J Phys. 2009;77:562. arXiv:0812.1017v1.

[21] Breitenberger E. Magnetic interactions between charged particles. Am J Phys. 1968;36:505.

[22] Brown HR. Physical relativity: Space-time structure from a dynamical perspective. Oxford: Oxford University Press; 2005.

[23] Buchholz D, Yngvason J. There are no causality problems with Fermi's two atom system. Phys Rev Lett. 1994;73:613. arXiv:hep-th/9403027v1.

https://doi.org/10.1515/9783110493221-015

[24] Butler JW. A proposed electromagnetic momentum-energy 4-vector for charged bodies. Am J Phys. 1969;37:1258.
[25] Caprez A, Batelaan H. Feynman's relativistic electrodynamics paradox and the Aharonov–Bohm effect. Found Phys. 2009;39:295.
[26] Caprez A, Barwick B, Batelaan H. A macroscopic test of the Aharonov–Bohm effect. Phys Rev Lett. 2007;99:210401. arXiv:0708.2428v1.
[27] Carlip S. Aberration and the speed of gravity. Phys Lett A. 2000;267:81. arXiv:gr-qc/9909087v2.
[28] Carroll R. Remarks on photons and the aether. 2005. arXiv:physics/0507027v1.
[29] Carron NJ. Fields of particles and beams exiting a conductor. Prog Electromagn Res. 2000;28:147.
[30] Cavalleri G, Cesaroni E, Tonni E, Spavieri G. Interpretation of the longitudinal forces detected in a recent experiment of electrodynamics. Eur Phys J D. 2013;26:221.
[31] Chambers RG. Shift of an electron interference pattern by enclosed magnetic flux. Phys Rev Lett. 1960;5:3.
[32] Cheng T, Su Q, Grobe R. Dynamics of virtual bosons created by classical and quantum sources. Opt Commun. 2010;283:1008.
[33] Chubykalo AE, Smirnov-Rueda R. Action at a distance as a full-value solution of Maxwell equations: basis and application of separated potential's method. Phys Rev E. 1996;53:5373.
[34] Close FE, Osborn H. Relativistic center-of-mass motion and the electromagnetic interaction of systems of charged particles. Phys Rev D. 1970;2:2127.
[35] Coester F, Polyzou WN. Relativistic quantum mechanics of particles with direct interactions. Phys Rev D. 1982;26:1348.
[36] Coleman S, Van Vleck JH. Origin of "hidden momentum forces" on magnets. Phys Rev. 1968;171:1370.
[37] Comay E. Exposing "hidden momentum". Am J Phys. 1996;64:1028.
[38] Comay E. Decomposition of electromagnetic fields into radiation and bound components. Am J Phys. 1997;65:862.
[39] Comay E. Lorentz transformation of a system carrying "hidden momentum". Am J Phys. 2000;68:1007.
[40] Cornish FHJ. An electric dipole in self-accelerated transverse motion. Am J Phys. 1986;54:166.
[41] Cremaschini C, Tessarotto M. Hamiltonian formulation for the classical EM radiation-reaction problem: Application to the kinetic theory for relativistic collisionless plasmas. Eur Phys J Plus. 2011;126:63. arXiv:1201.1816v1.
[42] Cullwick EG. Electromagnetic momentum and Newton's third law. Nature. 1952;170:425.
[43] Currie DG. Interaction contra classical relativistic Hamiltonian particle mechanics. J Math Phys. 1963;4:1470.
[44] Currie DG, Jordan TF, Sudarshan ECG. Relativistic invariance and Hamiltonian theories of interacting particles. Rev Mod Phys. 1963;35:350.
[45] Dahlström JM, L'Huillier A, Maquet A. Introduction to attosecond delays in photoionization. J Phys B, At Mol Opt Phys. 2012;45:183001.
[46] Dancoff SM. Non-adiabatic meson theory of nuclear forces. Phys Rev. 1950;78:382.
[47] Das Gupta AK. Unipolar machines, association of the magnetic field with the field-producing magnet. Am J Phys. 1963;31:428.
[48] de la Torre AC. Understanding light quanta: The photon. 2004. arXiv:quant-ph/0410179v2.
[49] de la Torre AC. Understanding light quanta: Construction of the free electromagnetic field. 2005. arXiv:quant-ph/0503023v2.
[50] de la Torre AC. Understanding light quanta: First quantization of the free electromagnetic field. Eur J Phys. 2005;26:457. arXiv:quant-ph/0410171v1.

[51] de Sangro R, Finocchiaro G, Patteri P, Piccolo M, Pizzella G. Measuring propagation speed of Coulomb fields. Eur Phys J C. 2015;75:137. arXiv:1211.2913v2.
[52] de Sangro R, Finocchiaro G, Patteri P, Piccolo M, Pizzella G. Experimental result on the propagation of Coulomb fields. J Phys Conf Ser. 2017;845:012015.
[53] de Sangro R, Finocchiaro G, Patteri P, Piccolo M, Pizzella G. Why the interpretation of "Measuring propagation speed of Coulomb fields" stands. Eur Phys J C. 2017;77:75. arXiv:1611.06935v1.
[54] Dillon G, Giannini MM. On the clothing transformation in the Lee model. Nuovo Cimento. 1973;18A:31.
[55] Dillon G, Giannini MM. On the potential description of the $VN - NN\theta$ sector of the Lee model. Nuovo Cimento. 1975;27A:106.
[56] Dirac PAM. Forms of relativistic dynamics. Rev Mod Phys. 1949;21:392.
[57] Drescher M, Hentschel M, Kienberger R, Uiberacker M, Yakovlev V, Scrinzi A, Westerwalbesloh Th, Kleineberg U, Heinzmann U, Krausz F. Time-resolved atomic inner-shell spectroscopy. Nature. 2002;419:803.
[58] Dubovyk I, Shebeko O. The method of unitary clothing transformations in the theory of nucleon–nucleon scattering. Few-Body Syst. 2010;48:109.
[59] Duncan A. The conceptual framework of quantum field theory. Oxford: Oxford University; 2012.
[60] Dyson FJ. The renormalization method in quantum electrodynamics. Proc R Soc Lond Ser A. 1951;207:395.
[61] Earman J, Fraser D. Haag's theorem and its implications for the foundations of quantum field theory. Erkenntnis. 2006;64:305. http://philsci-archive.pitt.edu/2673/1/earmanfraserfinalrevd.pdf.
[62] Edwards WF, Kenyon CS, Lemon DK. Continuing investigation into possible electric fields arising from steady conductor currents. Phys Rev D. 1976;14:922.
[63] Einstein A. Zur Electrodynamik bewegter Körper. Ann Phys. 1905;17:891.
[64] Einstein A. Relativity: the special and general theory. London: Methuen and Co; 1920.
[65] Engelhardt W. Instantaneous interaction between charged particles. 2005. arXiv:physics/0511172v1.
[66] Engelhardt W. Relativity of time and instantaneous interaction of charged particles. Am J Mod Phys. 2015;4:15.
[67] Epelbaum E. Few-nucleon forces and systems in chiral effective field theory. Prog Part Nucl Phys. 2006;57:654. arXiv:nucl-th/0509032v1.
[68] Essén H. A study of lattice and magnetic interactions of conduction electrons. Phys Scr. 1995;52:388.
[69] Essén H. Darwin magnetic interaction energy and its macroscopic consequences. Phys Rev E. 1996;53:5228.
[70] Essén H. Magnetism of matter and phase space energy of charged particle systems. J Phys A, Math Gen. 1999;32:2297.
[71] Essén H. Circulating electrons, superconductivity and the Darwin–Breit interaction. 2013. arXiv:cond-mat/0002096v2.
[72] Essén H, Fiolhais MCN. The Darwin–Breit magnetic interaction and superconductivity. In: PIERS Proceedings, Stockholm, Sweden, Aug 12–15. 2013. arXiv:1312.1607v1.
[73] Ezawa H, Kikkawa K, Umezawa H. Potential representation in quantum field theory. Nuovo Cimento. 1962;23:751.
[74] Faddeev LD. On the separation of self-interaction and scattering effects in perturbation theory. Dokl Akad Nauk SSSR. 1963;152:573.
[75] Farley FJM. The CERN $(g - 2)$ measurements. Z Phys C. 1992;56:S88.

[76] Fateev VA, Shvarts AS. Dressing operators in quantum field theory. Sov Phys Dokl. 1973;18:165.
[77] Feinberg G, Sucher J. Two-photon-exchange force between charged systems: Spinless particles. Phys Rev D. 1988;38:3763.
[78] Feynman R, Leighton R, Sands M. The Feynman lectures on physics, volume II. California: California Institute of Technology; 2013. http://www.feynmanlectures.caltech.edu/II_toc.html.
[79] Field JH. A new kinematical derivation of the Lorentz transformation and the particle description of light. Helv Phys Acta. 1997;70:542. arXiv:physics/0410062v1.
[80] Field JH. On the relationship of quantum mechanics to classical electromagnetism and classical relativistic mechanics. Eur J Phys. 2004;25:385. arXiv:physics/0403076v1.
[81] Field JH. Classical electromagnetism as a consequence of Coulomb's law, special relativity and Hamilton's principle and its relationship to quantum electrodynamics. Phys Scr. 2006;74:702. arXiv:physics/0501130v5.
[82] Field JH. Space-time transformation properties of inter-charge forces and dipole radiation: Breakdown of the classical field concept in relativistic electrodynamics. 2008. arXiv:physics/0604089v3.
[83] Fivel DI. Solutions of the Lee model in all sectors by dynamical algebra. J Math Phys. 1970;11:699.
[84] Foldy LL. Relativistic particle systems with interaction. Phys Rev. 1961;122:275.
[85] Fonda L, Ghirardi GC, Rimini A. Decay theory of unstable quantum systems. Rep Prog Phys. 1978;41:587.
[86] Franklin J. The nature of electromagnetic energy. 2007. arXiv:0707.3421v2.
[87] Friedman A. Nonstandard extension of quantum logic and Dirac's bra-ket formalism of quantum mechanics. Int J Theor Phys. 1994;33:307.
[88] Fröhlich H. Interaction of electrons with lattice vibrations. Proc R Soc Lond Ser A 1952;215:291.
[89] Fröhlich H. The theory of the superconductive state. Rep Prog Phys. 1961;24:1.
[90] Furry WH. Examples of momentum distributions in the electromagnetic field and in matter. Am J Phys. 1969;37:621.
[91] Galiautdinov A. Derivation of the Lorentz transformation without the use of Einstein's second postulate. 2017. arXiv:1701.00270v1.
[92] Giacosa F. Decay law and time dilatation. Acta Phys Pol B. 2016;47:2135. arXiv:1512.00232v3.
[93] Giakos GC, Ishii TK. Anomalous microwave propagation in open space. Microw Opt Technol Lett. 1991;4:79.
[94] Giakos GC, Ishii TK. Rapid pulsed microwave propagation. IEEE Microw Guided Wave Lett. 1991;1:374.
[95] Gill TP, Zachary WW, Lindesay J. The classical electron problem. Found Phys. 2001;31:1299. arXiv:physics/0405131v1.
[96] Girard R, Marchildon L. Are tachyon causal paradoxes solved? Found Phys. 1984;14:535.
[97] Głazek StD. Similarity renormalization group approach to boost invariant Hamiltonian dynamics. Acta Phys Pol B. 1998;29:1979. arXiv:hep-th/9712188v1.
[98] Głazek StD, Wilson KG. Renormalization of Hamiltonians. Phys Rev D. 1993;48:5863.
[99] Gradshtein IS, Ryzhik IM. Tables of integrals, sums, series and products (in Russian). Moscow: Fizmatgiz; 1963.
[100] Graneau N, Phipps T Jr., Roscoe D. An experimental confirmation of longitudinal electrodynamic forces. Eur Phys J D. 2001;15:87.
[101] Greenberg OW, Schweber SS. Clothed particle operators in simple models of quantum field theory. Nuovo Cimento. 1958;8:378.

[102] Griffiths DJ. Electrostatic levitation of a dipole. Am J Phys. 1986;54:744.
[103] Gupta SN, Radford SF. Quantum field-theoretical electromagnetic and gravitational two-particle potentials. Phys Rev D. 1980;21:2213.
[104] Gupta SN, Repko WW. Suchyta III CJ. Muonium and positronium potentials. Phys Rev D. 1989;40:4100.
[105] Haag R. On quantum field theories. Dan Mat Fys Medd. 1955;29:1.
[106] Halvorson H, Clifton R. No place for particles in relativistic quantum theories? Philos Sci. 2002;69:1. arXiv:quant-ph/0103041v1.
[107] Hasselkamp D, Mondry E, Scharmann A. Direct observation of the transversal Doppler-shift. Z Phys A. 1979;289:151.
[108] Hegerfeldt GC. Instantaneous spreading and Einstein causality in quantum theory. Ann Phys (Leipz). 1998;7:716. arXiv:quant-ph/9809030v1.
[109] Hegerfeldt GC, Neumann JT. The Aharonov–Bohm effect: the role of tunneling and associated forces. J Phys A, Math Theor. 2008;41:155305. arXiv:0801.0799v2.
[110] Hnizdo V. On linear momentum in quasistatic electromagnetic systems. 2004. arXiv:physics/0407027v1.
[111] Hobson A. There are no particles, there are only fields. Am J Phys. 2013;81:211. arXiv:1204.4616v2.
[112] Holstein BR. Effective interactions and the hydrogen atom. Am J Phys. 2004;72:333.
[113] Howe GWO. A problem of two electrons and Newton's third law. Wirel Eng. 1944;21:105.
[114] Ibison M, Puthoff HE, Little SR. The speed of gravity revisited. 1999. arXiv:physics/9910050v1.
[115] Itoh T. Derivation of nonrelativistic Hamiltonian for electrons from quantum electrodynamics. Rev Mod Phys. 1965;37:159.
[116] Ives HE, Stilwell GR. An experimental study of the rate of a moving clock. J Opt Soc Am. 1938;28:215.
[117] Ives HE, Stilwell GR. An experimental study of the rate of a moving clock. II. J Opt Soc Am. 1941;31:369.
[118] Jackson JD. Classical electrodynamics. 3rd ed. New York: J. Wiley and Sons; 1999.
[119] Jackson JD. Torque or no torque? Simple charged particle motion observed in different inertial frames. Am J Phys. 2004;72:1484.
[120] Jaffe R. Casimir effect and the quantum vacuum. Phys Rev D. 2005;72:021301. arXiv:hep-th/0503158.
[121] Jefimenko OD. A relativistic paradox seemingly violating conservation of momentum law in electromagnetic systems. Eur J Phys. 1999;20:39.
[122] Jefimenko OD. The Trouton–Noble paradox. J Phys A, Math Gen. 1999;32:3755.
[123] Johansson L. Longitudinal electrodynamic forces – and their possible technological applications. MSc Thesis. Sweden: Lund Institute of Technology; 1996. http://www.df.lth.se/~snorkelf/LongitudinalMSc.pdf.
[124] Kacser C. Higher Born approximations in non-relativistic Coulomb scattering. Nuovo Cimento. 1959;13:303.
[125] Kaivola M, Poulsen O, Riis E. Measurement of the relativistic Doppler shift in neon. Phys Rev Lett. 1985;54:255.
[126] Kaup DJ, Rupasov VI. Exactly solvable 3D model of resonance energy transfer. J Phys A, Math Gen. 1996;29:6911.
[127] Kaup DJ, Rupasov VI. Exactly solvable one-dimensional model of resonance energy transfer. J Phys A, Math Gen. 1996;29:2149.
[128] Kazes E. Analytic theory of relativistic interactions. Phys Rev D. 1971;4:999.
[129] Keister BD. Forms of relativistic dynamics: What are the possibilities? AIP Conf Proc. 1995;334:164. arXiv:nucl-th/9406032v1.

[130] Keller JM. Newton's third law and electrodynamics. Am J Phys. 1942;10:302.
[131] Khalfin LA. Quantum theory of unstable particles and relativity. 1997. Preprint of Steklov Mathematical Institute, St. Petersburg Department, PDMI-6/1997. http://www.pdmi.ras.ru/preprint/1997/97-06.html.
[132] Kholmetskii AL. Remarks on momentum and energy flux of a non-radiating electromagnetic field. Ann Fond Louis Broglie. 2004;29:549.
[133] Kholmetskii AL. On momentum and energy of a non-radiating electromagnetic field. 2005. arXiv:physics/0501148v2.
[134] Kholmetskii AL. Momentum-energy of the non-radiating electromagnetic field: open problems? Phys Scr. 2006;73:620.
[135] Kholmetskii AL. The author's collection of relativistic paradoxes in classical electrodynamics. http://www.space-lab.ru/files/pages/PIRT_VII-XII/pages/text/PIRT_IX/Kholmetskii_1.pdf.
[136] Kholmetskii AL, Yarman T. Apparent paradoxes in classical electrodynamics: relativistic transformation of force. Eur J Phys. 2007;28:537.
[137] Kholmetskii AL, Yarman T. Relativistic transformation of force: resolution of apparent paradoxes. Eur J Phys. 2007;28:1081.
[138] Kholmetskii AL, Yarman T. Energy flow in a bound electromagnetic field: resolution of apparent paradoxes. Eur J Phys. 2008;29:1135.
[139] Kholmetskii AL, Missevitch OV, Smirnov-Rueda R. Measurement of propagation velocity of bound electromagnetic fields in near zone. J Appl Phys. 2007;102:013529.
[140] Kholmetskii AL, Missevitch OV, Smirnov-Rueda R, Tzonchev RI, Chubykalo AE, Moreno I. Experimental test on the applicability of the standard retardation condition to bound magnetic fields. J Appl Phys. 2007;101:023532. arXiv:physics/0601084v1.
[141] Kijowski J, Chruściński D. Variational principle for electrodynamics of moving particles. Gen Relativ Gravit. 1995;27:267.
[142] Kislev A, Vaidman L. Relativistic causality and conservation of energy in classical electromagnetic theory. Am J Phys. 2002;70:1216. arXiv:physics/0201042v1.
[143] Kita H. A non-trivial example of a relativistic quantum theory of particles without divergence difficulties. Prog Theor Phys. 1966;35:934.
[144] Kita H. Another convergent relativistic model theory of interacting particles. Prog Theor Phys. 1968;39:1333.
[145] Kita H. Structure of the state space in a convergent relativistic model theory of interacting particles. Prog Theor Phys. 1970;43:1364.
[146] Kita H. A model of relativistic quantum mechanics of interacting particles. Prog Theor Phys. 1972;48:2422.
[147] Kita H. Vertex functions in convergent relativistic model theories. Prog Theor Phys. 1972;47:2140.
[148] Kita H. A realistic model of convergent relativistic quantum mechanics of interacting particles. Prog Theor Phys. 1973;49:1704.
[149] Kobayashi M, Sato T, Ohtsubo H. Effective interactions for mesons and baryons in nuclei. Prog Theor Phys. 1997;98:927.
[150] Korda VYu, Shebeko AV. Clothed particle representation in quantum field theory: Mass renormalization. Phys Rev D. 2004;70:085011.
[151] Korda VYu, Canton L, Shebeko AV. Relativistic interactions for the meson-two-nucleon system in the clothed-particle unitary representation. Ann Phys. 2007;322:736. arXiv:nucl-th/0603025v1.
[152] Krajcik RA, Foldy LL. Relativistic center-of-mass variables for composite systems with arbitrary internal interactions. Phys Rev D. 1974;10:1777.
[153] Krausz F, Ivanov M. Attosecond physics. Rev Mod Phys. 2009;81:163.

[154] Landau L, Lifshitz E. Course of theoretical physics, Volume 3, Quantum mechanics, Non-relativistic theory. Elmsford: Pergamon; 1977.

[155] Lee AR, Kalotas TM. Lorentz transformations from the first postulate. Am J Phys. 1975;43:434.

[156] Lemon DK, Edwards WF, Kenyon CS. Electric potentials associated with steady conduction currents in superconducting coils. Phys Lett A. 1992;162:105.

[157] Lepage GP. What is renormalization? In: DeGrand T, Toussaint D, editors. Proceedings of TASI'89: From actions to answers. Singapore: World Scientific; 1989. arXiv:hep-ph/0506330v1.

[158] Lévy-Leblond J-M. One more derivation of the Lorentz transformation. Am J Phys. 1976;44:271.

[159] Lopuszański J. The Ruijgrok-van Hove model of field theory in terms of "dressed" operators. Physica. 1959;25:745.

[160] Lv QZ, Norris S, Brennan R, Stefanovich E, Su Q, Grobe R. Computational approach for calculating bound states in quantum field theory. Phys Rev A. 2016;94:032110.

[161] Lv QZ, Stefanovich E, Su Q, Grobe R. Mass renormalization and binding energies in quantum field theory. Laser Phys. 2017;27:105301.

[162] MacArthur DW. Special relativity: Understanding experimental tests and formulations. Phys Rev A. 1986;33:1.

[163] Malament DB. In defence of dogma: Why there cannot be a relativistic quantum mechanics of (localizable) particles. In: Clifton R, editor. Perspectives on quantum reality. Dordrecht: Springer; 1996.

[164] Mansuripur M. Trouble with the Lorentz law of force: Incompatibility with special relativity and momentum conservation. Phys Rev Lett. 2012;108:193901. arXiv:1205.0096v1.

[165] McDonald KT. Limits on the applicability of classical electromagnetic fields as inferred from the radiation reaction. 2000. arXiv:physics/0003062v1.

[166] McDonald KT. Cullwick's paradox: Charged particle on the axis of a toroidal magnet. http://puhep1.princeton.edu/~mcdonald/examples/cullwick.pdf.

[167] McDonald KT. Onoochin's paradox. 2006. http://puhep1.princeton.edu/~mcdonald/examples/onoochin.pdf.

[168] Mermin ND. What's bad about this habit. Phys Today. 2009;May:8.

[169] Missevitch OV, Kholmetskii AL, Smirnov-Rueda R. Anomalously small retardation of bound (force) electromagnetic fields in antenna near zone. Europhys Lett. 2011;93:64004.

[170] Monahan AH, McMillan M. Lorentz boost of the Newton–Wigner–Pryce position operator. Phys Rev D. 1997;56:2563.

[171] Mugnai D, Ranfagni A, Ruggeri R. Observation of superluminal behaviors in wave propagation. Phys Rev Lett. 2000;84:4830.

[172] Mutze U. Relativistic quantum mechanics of *n*-particle systems with cluster-separable interactions. Phys Rev D. 1984;29:2255.

[173] Naumenko G, Popov Yu, Shevelev M. Direct observation of a semi-bare electron Coulomb field recover. J Phys Conf Ser. 2012;357:01205. arXiv:1112.1649v1.

[174] Naumenko GA, Potylitsyn AP, Sukhikh LG, Popov YuA. The investigation of relativistic electron electromagnetic field features during interaction with matter. 2010. arXiv:1006.2477v1.

[175] Newman D, Ford GW, Rich A, Sweetman E. Precision experimental verification of special relativity. Phys Rev Lett. 1978;21:1355.

[176] Okubo S. Diagonalization of Hamiltonian and Tamm–Dancoff equation. Prog Theor Phys. 1954;12:603.

[177] Oloumi D, Rambabu K. Studying the superluminal behavior of UWB antennas and its effect on near field imaging. IEEE Trans Antennas Propag. 2016;64:5084.

[178] Osakabe N, Matsuda T, Kawasaki T, Endo J, Tonomura A, Yano S, Yamada H. Experimental confirmation of Aharonov–Bohm effect using a toroidal magnetic field confined by a superconductor. Phys Rev A, Math Gen. 1986;34:815.

[179] Page L, Adams NI Jr. Action and reaction between moving charges. Am J Phys. 1945;13:141.

[180] Parrott S. Radiation from a uniformly accelerated charge and the equivalence principle. Found Phys. 2002;32:407. arXiv:gr-qc/9303025v8.

[181] Parrott S. Variant forms of Eliezer's theorem. 2005. arXiv:math-ph/0505042v1.

[182] Pavšič M. Localized states in quantum field theories. 2017. arXiv:1705.02774v1.

[183] Peskin ME, Schroeder DV. An introduction to quantum field theory. Boulder: Westview Press; 1995.

[184] Pineda A, Soto J. The Lamb shift in dimensional regularization. Phys Lett B. 1998;420:391.

[185] Pineda A, Soto J. Potential NRQED: The positronium case. Phys Rev D. 1999;59:016005. arXiv:hep-ph/9805424v2.

[186] Pinheiro MJ. Do Maxwell's equations need a revision? – A methodological note. Phys Essays. 2007;20:267. arXiv:physics/0511103v3.

[187] Polishchuk R. Derivation of the Lorentz transformations. 2001. arXiv:physics/0110076v1.

[188] Polyzou WN. Manifestly covariant, Poincaré invariant quantum theories of directly interacting particles. Phys Rev D. 1985;32:995.

[189] Polyzou WN. Relativistic quantum mechanics – particle production and cluster properties. Phys Rev C. 2003;68:015202. arXiv:nucl-th/0302023v1.

[190] Prudnikov AP, Brychkov YuA, Marichev OI. Integrals and series (in Russian). Moscow: Nauka; 1981.

[191] Ranfagni A, Mugnai D. Anomalous pulse delay in microwave propagation: A case of superluminal behavior. Phys Rev E. 1996;54:5692.

[192] Ranfagni A, Fabeni P, Pazzi GP, Mugnai D. Anomalous pulse delay in microwave propagation: A plausible connection to the tunneling time. Phys Rev E. 1993;48:1453.

[193] Ranfagni A, Fabeni P, Pazzi GP, Ricci AM, Trinci R, Mignani R, Ruggeria R, Cardone F, Agresti A. Observation of Zenneck-type waves in microwave propagation experiments. J Appl Phys. 2006;100:024910.

[194] Recami E. Superluminal waves and objects: an overview of the relevant experiments. J Phys Conf Ser. 2009;196:012020.

[195] Reck RA, Fry DL. Orbital and spin magnetization in *Fe-Co*, *Fe-Ni* and *Ni-Co*. Phys Rev. 1969;184:492.

[196] Roberts T. What is the experimental basis of special relativity? 2000. http://math.ucr.edu/home/baez/physics/Relativity/SR/experiments.html.

[197] Rohrlich F. Self-energy and stability of the classical electron. Am J Phys. 1960;28:639.

[198] Rohrlich F. The theory of the electron. http://www.philsoc.org/1962Spring/1526transcript.html.

[199] Roos CE, Marraffino J, Reucroft S, Waters J, Webster MS, Williams EGH, Manz A, Settles R, Wolf G. $\Sigma^{\pm}$ lifetimes and longitudinal acceleration. Nature. 1980;286:244.

[200] Rossi B, Hall DB. Variation of the rate of decay of mesotron with momentum. Phys Rev. 1941;59:223.

[201] Rubio H, Getino JM, Rojo O. The Aharonov–Bohm effect as a classical electromagnetic effect using electromagnetic potentials. Nuovo Cimento B. 1991;106:407.

[202] Ruijgrok ThW. Exactly renormalizable model in quantum field theory. III. Renormalization in the case of two *V*-particles. Physica. 1959;25:357.

[203] Ruijgrok ThW. General requirements for a relativistic quantum theory. Few-Body Syst. 1998;25:5.

[204] Russo G. Conditions for the generation of casual paradoxes from superluminal signals. Electron J Theor Phys. 2005;8:36.
[205] Sakurai J. Advanced quantum mechanics. Reading, Mass: Addison-Wesley; 1967.
[206] Sard RD. The forces between moving charges. Electr Eng. 1947;January:61.
[207] Sardelis DA. Unified derivation of the Galileo and the Lorentz transformations. Eur J Phys. 1982;3:96.
[208] Sato S. Some remarks on the formulation of the theory of elementary particles. Prog Theor Phys. 1966;35:540.
[209] Scharf G. Finite quantum electrodynamics. The causal approach. 2nd ed. Berlin: Springer; 1995.
[210] Schleif S. What is the experimental basis of the special relativity theory? http://www.weburbia.demon.co.uk/physics/experiments.html.
[211] Schwartz HM. Deduction of the general Lorentz transformations from a set of necessary assumptions. Am J Phys. 1984;52:346.
[212] Shebeko AV, Frolov PA. A possible way for constructing generators of the Poincaré group in quantum field theory. Few-Body Syst. 2012;52:125. arXiv:1107.5877v1.
[213] Shebeko AV, Shirokov MI. Relativistic quantum field theory (RQFT) treatment of few-body systems. Nucl Phys A. 1998;631:564.
[214] Shebeko AV, Shirokov MI. Unitary transformations in quantum field theory and bound states. Phys Part Nucl. 2001;32:15. arXiv:nucl-th/0102037v1.
[215] Shields BT, Morris MC, Ware MR, Su Q, Stefanovich EV, Grobe R. Time dilation in relativistic two-particle interactions. Phys Rev A. 2010;82:052116.
[216] Shirokov M. Quantum field theory: “dressing” contra divergencies. 1972. Preprint JINR, P2-6454, Dubna.
[217] Shirokov MI. Causality in quantum electrodynamics. Sov J Nucl Phys. 1976;23:1131.
[218] Shirokov MI. Signal velocity in quantum electrodynamics. Sov Phys Usp. 1978;21:345.
[219] Shirokov MI. Dressing and bound states in quantum field theory. 1993. Preprint JINR E4-93-55, Dubna.
[220] Shirokov MI. Bound states of “dressed” particles. 1994. Preprint JINR E2-94-82, Dubna.
[221] Shirokov MI. Decay law of moving unstable particle. Int J Theor Phys. 2004;43:1541.
[222] Shirokov MI. Evolution in time of moving unstable systems. Concepts Phys. 2006;3:193. arXiv:quant-ph/0508087v2.
[223] Shirokov MI. “Dressing” and Haag’s theorem. Report JINR-P2-7210. 2007. arXiv:math-ph/0703021v1.
[224] Shishkin GG, Shishkin AG, Smirnov AG, Dudarev AV, Barkov AV, Zagnetov PP, Rybin YuM. Investigation of possible electric potential arising from a constant current through a superconductor coil. J Phys D, Appl Phys. 2002;35:497.
[225] Shockley W, James RP. “Try simplest cases” discovery of “hidden momentum” forces on “magnetic currents”. Phys Rev Lett. 1967;18:876.
[226] Shvarts AS. Mathematical foundations of quantum field theory (in Russian). Moscow: Atomizdat; 1975.
[227] Sokolov SN. Physical equivalence of the point and instantaneous forms of relativistic dynamics. Theor Math Phys. 1975;24:799.
[228] Sokolov SN. Mechanics with retarded interactions. 1997. Preprint IHEP Protvino, IHEP 97-84. http://web.ihep.su/library/pubs/prep1997/ps/97-84.pdf.
[229] Sokolov SN, Shatnii AN. Physical equivalence of the three forms of relativistic dynamics and addition of interactions in the front and instant forms. Theor Math Phys. 1978;37:1029.
[230] Spavieri G, Cavalleri G. Interpretation of the Aharonov–Bohm and the Aharonov–Casher effects in terms of classical electromagnetic fields. Europhys Lett. 1992;18:301.

[231] Spavieri G, Gillies GT. Fundamental tests of electrodynamic theories: Conceptual investigations of the Trouton–Noble and hidden momentum effects. Nuovo Cimento. 2003;118B:205.
[232] Steane AM. The non-existence of the self-accelerating dipole, and related questions. Phys Rev D. 2014;89:125006. arXiv:1311.5798v3.
[233] Stefanovich EV. Quantum effects in relativistic decays. Int J Theor Phys. 1996;35:2539.
[234] Stefanovich EV. Quantum field theory without infinities. Ann Phys. 2001;292:139.
[235] Stefanovich EV. Is Minkowski space-time compatible with quantum mechanics? Found Phys. 2002;32:673.
[236] Stefanovich EV. Renormalization and dressing in quantum field theory. 2005. arXiv:hep-th/0503076v4.
[237] Stefanovich EV. Violations of Einstein's time dilation formula in particle decays. 2006. arXiv:physics/0603043v2.
[238] Stefanovich EV. Classical electrodynamics without fields and the Aharonov–Bohm effect. 2008. arXiv:0803.1326v2.
[239] Stefanovich EV. Lorentz boosts in interacting systems. In: Proceedings of Science, PoS, (Baldin ISHEPP XXII) 014, 2014. http://pos.sissa.it/archive/conferences/225/014/BaldinISHEPPXXII_014.pdf.
[240] Stefanovich EV. Does Pizzella's experiment violate causality? J Phys Conf Ser. 2017;845:012016.
[241] Stefanovich E. Elementary Particle Theory. Vol. 1: Quantum Mechanics. De Gruyter Stud Math Phys. Vol. 45. Berlin: De Gruyter; 2018.
[242] Stefanovich E. Elementary particle theory. Vol. 2: Quantum electrodynamics. De Gruyter Stud Math Phys. Vol. 46. Berlin: De Gruyter; 2018.
[243] Strocchi F. Relativistic quantum mechanics and field theory. Found Phys. 2004;34:501. arXiv:hep-th/0401143v1.
[244] Sudarshan ECG, Mukunda N. Forms of relativistic dynamics and world line condition and separability. Found Phys. 1983;13:385.
[245] Tamm Ig. Relativistic interaction of elementary particles. J Phys USSR. 1945;9:449. Reprinted in: Bolotovskii BM, Frenkel VYa, editors. I. E. Tamm. Selected papers. Berlin: Springer; 1991.
[246] Tani S. Formal theory of scattering in the quantum field theory. Phys Rev. 1959;115:711.
[247] Teukolsky SA. The explanation of the Trouton–Noble experiment revisited. Am J Phys. 1996;64:1104.
[248] Then JW. Experimental study of the motional electromotive force. Am J Phys. 1962;30:411.
[249] Thomas LH. The relativistic dynamics of a system of particles interacting at a distance. Phys Rev. 1952;85:868.
[250] Tonomura A, Osakabe N, Matsuda T, Kawasaki T, Endo J, Yano S, Yamada H. Evidence for Aharonov–Bohm effect with magnetic field completely shielded from electron wave. Phys Rev Lett. 1986;56:792.
[251] Trouton FT, Noble HR. The mechanical forces acting on a charged electric condenser moving through space. Philos Trans R Soc Lond A. 1904;202:165.
[252] Tuval M, Yahalom A. Newton's third law in the framework of special relativity. Eur Phys J Plus. 2014;129:240. arXiv:1302.2537v1.
[253] Urbanowski K. Decay law of relativistic particles: Quantum theory meets special relativity. Phys Lett B. 2014;737:346. arXiv:1408.6564v1.
[254] Urbanowski K. Non-classical behavior of moving relativistic unstable particles. Acta Phys Pol B. 2017;48:1411. arXiv:1506.05076v1.
[255] Van Dam H, Wigner EP. Classical relativistic mechanics of interacting point particles. Phys Rev B. 1965;138:1576.

[256] Van Dam H, Wigner EP. Instantaneous and asymptotic conservation laws for classical relativistic mechanics of interacting point particles. Phys Rev. 1966;142:838.
[257] Van Hove L. Energy corrections and persistent perturbation effects in continuous spectra. Physica. 1955;21:901.
[258] Van Hove L. Energy corrections and persistent perturbation effects in continuous spectra. II. The perturbed stationary states. Physica. 1956;22:343.
[259] Vişinesku MM, Shirokov MI. Perturbation approach to the field theory, “dressing” and divergences. Rev Roum Phys. 1974;19:461. Preprint JINR P2-8148, Dubna.
[260] Von Zuben FSG. Quantum time and spatial localization. In: Position location and navigation symposium. San Diego: IEEE; 2000.
[261] Wagner RE, Ware MR, Stefanovich EV, Su Q, Grobe R. A study of local and non-local spatial densities in quantum field theory. Phys Rev A. 2012;85:022121.
[262] Walhout TS. Similarity renormalization, Hamiltonian flow equations, and Dyson’s intermediate representation. Phys Rev D. 1999;59:065009. arXiv:hep-th/9806097v2.
[263] Walker WD. Experimental evidence of near-field superluminally propagating electromagnetic fields. 2000. arXiv:physics/0009023v1.
[264] Wallace D. Emergence of particles from bosonic quantum field theory. 2001. arXiv:quant-ph/0112149v1.
[265] Walter R. Recoil effects in scalar-field model. Nuovo Cimento. 1970;68A:426.
[266] Weinberg S. The quantum theory of fields. Vol. 1. Cambridge: University Press; 1995.
[267] Weitz M, Huber A, Schmidt-Kaler F, Leibfried D, Vassen W, Zimmermann C, Pachucki K, Hänsch TW, Julien L, Biraben F. Precision measurement of the 1S ground-state Lamb shift in atomic hydrogen and deuterium by frequency comparison. Phys Rev A. 1995;52:2664.
[268] Wesley JP. Induction produces Aharonov–Bohm effect. Apeiron. 1998;5:73.
[269] Wigner EP. On unitary representations of the inhomogeneous Lorentz group. Ann Math. 1939;40:149.
[270] Wilczek F. Quantum field theory. Rev Mod Phys. 1999;71:S58. arXiv:hep-th/9803075v2.
[271] Withayachumnankul W, Fischer BM, Ferguson B, Davis BR, Abbott D. A systemized view of superluminal wave propagation. Proc IEEE. 2010;98:1.

Index

De Gruyter Studies in Mathematical Physics

Volume 46
Eugene Stefanovich
Elementary Particle Theory: Volume 2: Quantum Electrodynamics, 2018
ISBN 978-3-11-049089-3, e-ISBN (PDF) 978-3-11-049320-7,
e-ISBN (EPUB) 978-3-11-049143-2

Volume 45
Eugene Stefanovich
Elementary Particle Theory: Volume 1: Quantum Mechanics, 2018
ISBN 978-3-11-049088-6, e-ISBN (PDF) 978-3-11-049213-2,
e-ISBN (EPUB) 978-3-11-049103-6

Volume 44
Vladimir V. Kiselev
Collective Effects in Condensed Matter Physics, 2018
ISBN 978-3-11-058509-4, e-ISBN (PDF) 978-3-11-058618-3,
e-ISBN (EPUB) 978-3-11-058513-1

Volume 43
Robert F. Snider
Irreducible Cartesian Tensors, 2017
ISBN 978-3-11-056363-4, e-ISBN (PDF) 978-3-11-056486-0,
e-ISBN (EPUB) 978-3-11-056373-3

Volume 42
Javier Roa
Regularization in Orbital Mechanics: Theory and Practice, 2017
ISBN 978-3-11-055855-5, e-ISBN (PDF) 978-3-11-055912-5,
e-ISBN (EPUB) 978-3-11-055862-3

Volume 41
Esra Russell, Oktay K. Pashaev
Oscillatory Models in General Relativity, 2017
ISBN 978-3-11-051495-7, e-ISBN (PDF) 978-3-11-051536-7,
e-ISBN (EPUB) 978-3-11-051522-0

Volume 40
Joachim Schröter
Minkowski Space: The Spacetime of the Special Relativity Theory Groups, 2017
ISBN 978-3-11-048457-1, e-ISBN (PDF) 978-3-11-048573-8,
e-ISBN (EPUB) 978-3-11-048461-8

Bei Fragen zur Produktsicherheit wenden Sie sich bitte an:
If you have any questions regarding product safety,
please contact:

Walter de Gruyter GmbH
Genthiner Straße 13
10785 Berlin
productsafety@degruyterbrill.com